First Latin-American Conference on Bioimpedance

(CLABIO 2012)

Journal of Physics: Conference Series Volume 407

Joinville, Brazil
6-9 November 2012

ISBN: 978-1-62276-873-8
ISSN: 1742-6588

Printed from e-media with permission by:

Curran Associates, Inc.
57 Morehouse Lane
Red Hook, NY 12571

Some format issues inherent in the e-media version may also appear in this print version.

Printed by Curran Associates, Inc. (2013)

For permission requests, please contact the Institute of Physics
at the address below.

Institute of Physics
Dirac House, Temple Back
Bristol BS1 6BE UK

Phone: 44 1 17 929 7481
Fax: 44 1 17 920 0979

techtracking@iop.org

Additional copies of this publication are available from:

Curran Associates, Inc.
57 Morehouse Lane
Red Hook, NY 12571 USA
Phone: 845-758-0400
Fax: 845-758-2634
Email: curran@proceedings.com
Web: www.proceedings.com

TABLE OF CONTENTS

PREFACE

The past decade has witnessed an unprecedented growth in medical technologies and a new generation of diagnostics, characterized by mobility, virtualization, homecare and costs. The ever growing demand and the rapid need for low cost tools for characterizing human tissue, and supporting intelligence and technologies for non-invasive tissue cancer investigation raise unique and evolving opportunities for research in Electrical Bioimpedance.

The CLABIO2012 – First Latin American Conference on Bioimpedance is a premier Latin-American conference on Bioimpedance for research groups working on Electrical Bioimpedance. It allows Latin American researchers to share their experiences with other groups from all over the world by presenting scientific work and potential innovations in this research area and also in the social events promoting informal get togethers in the Brazilian style. The work covers a broad range including Biomedical Engineering and Computing, Medical Physics and Medical Sciences, Environment, Biology and Chemistry. Also, the Conference is intended to give students and research groups the opportunity to learn more about Bioimpedance as an important tool in biological material characterization and also in diagnosis.

The conference is designed to showcase cutting edge research and accomplishments, and to enrich the educational and industrial experience in this field. It also represents a unique opportunity to meet colleagues and friends, exchanging ideas, and learning about new developments and best practice, while working to advance the understanding of the knowledge base that we will collectively draw upon in the years ahead to meet future challenges. Participants will attend presentations by scholars representing both institutes and academia.

The CLABIO2012 proceedings include over 25 papers selected via a peer review process. The conference program features tutorial talks by world-leading scholars and five sessions for regular paper oral presentations. In making this conference possible, we want to acknowledge our deep appreciation for the financial support of FAPESC (Foundation for Research and Innovation of Santa Catarina), CAPES (Coordination for the Improvement of Higher Level -or Education- Personnel) and also express our sincere thanks to the many individuals who contributed their time and diligence to making this conference possible. Our special thanks go to the Santa Catarina State University for contributing time and effort to organizing the conference. We also wish to express our thanks to Professors Ana Watanabe and Fabricio Noveletto for helping with the registration process, the conference desk and the diverse and sundry details that are the part of any event of this kind.

We would like also to thank all of the invited speakers and the members of the Program Committee, and wopuld like to express our gratitude to the Technological Institute of Joinville (FITEJ) for their technical co-sponsorship. We very much appreciate Orjan Martinsen, Uwe Pliquett, Fernando Martinez Seoane, Raul Gonzalez Lima, Marcio Nogueira de Souza and Carlos Augusto Gonzalez Correa for delivering keynote talks at the conference. And we would like to extend an enthusiastic round of thanks to all of our conference authors for their excellent contributions; to all the session chairs for their effort and enthusiasm; and to all the International Program Committee members and referees for their time and expertise in the paper review. Particular thanks go to Emiliano Amarante Veiga and other members of the CLABIO2012 Secretariat and organizing team for their time and outstanding work.

List of committees

General Chair
Professor Pedro Bertemes Filho (Santa Catarina State University)

Technical Program Chairs
Dr Marcio Nogueira de Souza (Rio de Janeiro Federal University)

Local Arrangement Chair
Professor Aleksander Paterno (Santa Catarina State University)
Professor Fabrício Noveletto (Santa Catarina State University)

Finance Chairman
Professor Nilson Campos (Santa Catarina State University)

Marketing & Corporate Relations
Professor Ana Watanabe (Santa Catarina State University)

Pre-Conference Coordinator
Dr Ørjan Grøttem Martinsen (University of Oslo)

Chair IEEE EMB
Dr Fernando Seoane Martinez (University of Boras)

Conference logo:

1st CLABIO
LATIN AMERICAN BIOIMPEDANCE
BRAZIL - 2012

Sponsor logos:

State University of Santa Catarina

UDESC
Joinville

Support for Scientific and Technological Research Foundation of Santa Catarina State

FAPESC

Coordination for the Improvement of Higher Level -or Education- Personnel

CAPES

IEEE
Advancing Technology for Humanity

EMB

Peer review statement

All papers published in this volume of *Journal of Physics: Conference Series* have been peer reviewed through processes administered by the proceedings Editors. Reviews were conducted by expert referees to the professional and scientific standards expected of a proceedings journal published by IOP Publishing.

First Latin-American Conference on Bioimpedance (CLABIO 2012) IOP Publishing
Journal of Physics: Conference Series **407** (2012) 012001 doi:10.1088/1742-6596/407/1/012001

ELECTROMAGNETIC INTERFERENCE REDUCTION BY DYNAMIC IMPEDANCE BALANCING APPLIED IN BIOSENSORS

João F. R. Negrão
Prof. MSc., Universidade Federal do Pará, Centro Tecnológico, Belém, BR
E-mail: jfrn@ufpa.br

Guilherme A. L. Araujo
Prof. Dr., Universidade Federal do Pará, ITEC, Engenharia Elétrica, Belém, BR
E-mail: guilh@ufpa.br

Carlos T. C. Júnior
Prof. Dr., Universidade Federal do Pará, ITEC, Engenharia Elétrica, Belém, BR
E-mail: cartav@ufpa.br

Daniel C. Souza
Prof. Dr., Universidade Federal do Pará, ITEC, Engenharia Elétrica, Belém, BR
E-mail: danielcs@ufpa.br

Marcelo O. Silva
Prof. Dr., Universidade Federal do Pará, Curso de Engenharia Industrial, Abaetetuba, BR
E-mail: mos@ufpa.br

Abstract. Electromagnetic interference, caused by the electric power line, affects the signals of electronic instruments, specifically those with low levels of amplitude. This type of signal is known as Common Mode Interference. There are many methods and architectures used to minimize or eliminate the influence of these interferences in electronic instrumentation, the most common is the use of band reject filters to eliminate them. With this objective, we present the analysis, prototyping, developing and testing of a new architecture reconfigurable filter with application in biomedical instrumentation, applied to the reduction of interference in common mode and conservation of components of the useful signal in the same range of noise, by the principle of balancing impedance dynamic using the hardware description software and simulation electronics software. The methodology was tested using a sinusoidal signal in the same standard frequency of an electrocardiogram signal in the same frequency interference (50/60 Hz). Excellent results were obtained in simulation with noise reduction of approximately 97%, while the results of experimental tests showed around 50% reduction. In both cases, the useful signal was preserved, confirming the efficiency of the proposed architecture. The method can be applied to eliminate the interference, which are in the same band the of the useful signal components..

1. Introduction

Electromagnetic Interference (EMI) is a phenomenon that affects at a greater or lesser extent the operation of electro-electronic devices [1]. In the context of noise reduction, that this work is applied. Thus was adapted a new architecture with discrete components, which reduced the interference in common mode (V_{cm}), mainly that generated by the electric power line (50/60 Hz), preserving its components in differential mode (V_d). As a measure of evaluation of this circuit, have been proposed to measure the results from Biosensors.

The circuit design is based on the reconfigurability concept [2], in which a model described in VHDL_AMS [3] allowed to validate the basic principles of operation and then used the software *Proteus Isis Schematic Capture* [4] to design a test version, using discrete components, and simulate using parameters close to reality. The experimental results came validate and prove by those obtained in simulation.

2. Concept of impedance balancing

Impedance balancing is the ability of symmetry between two circuits. When this measure of symmetry does not exist, then this system is known as an unbalanced circuit, thus causing an "attenuation factor", this factor measures the degree of divergence in relation to electrical symmetry ideal.

The umbalance is one of the factors that lead to increased noise, whitch is a reality in electronic circuits, mainly when the equipment is under the direct action from EMI. Noises can be reduced using several techniques, some electronic and other mechanical: the most trivial one is to increase the Common Mode Rejection Radio (**CMRR**) of the circuit signal processing [5]; the other technique would involve the data cables by an external metal mesh. In this context, this article proposes a new architecture using the principle of electronic impedance balancing, dynamically adjusted, to noise reduction in common mode (V_{cm}).

Before describing the proposed architecture, a brief introduction about the state of the art on impedance static balancing is presented.

2.1. Measure of unbalance

There is not a specific method for measurement of unbalance. For example, in telecommunications, the loss of signal strength due to the unbalance in telephone wires or terminals, also known as *loss unbalance longitudinal* (**LUL**), are measured according to standards of quality control, such as the ANATEL's 473 standard which, according to Volpato and Magalhes [6], using this parameter, the calculated value of unbalance in the telecommunications cables is around 19.7061 dB (LUL= - 19.7061 dB). This value tends to increase with the increasing distance between the transducer and the input circuits. To control this increase in noise due to the unbalance, it has been presented new proposals, as the architecture of AI indicated by Dobrev and Daskalov [7], which had a reduced V_{cm} range of about 200 times when he used an electrocardiogram signal [8] as a demonstration (Figure 1). The upper graph represents the noise V_{cm}, (50 Hz sine wave) and the lower, the differential signal before and after balancing, figures 1a and 1b.

The prototype aimed at eliminating V_{cm} by reducing the impedance umbalance. However, this new approach compared with other methods, has the advantage of the fact that only the common mode noise (V_{cm}) undergoes attenuation, while the useful signal components (V_d), even in the same frequency noise, is weakly influenced. The balance of the impedance is dynamically controlled by a microprocessor system, and beyond that, the proposed system can be easily programmed to operate in any other band interference, without physically changing the circuit.

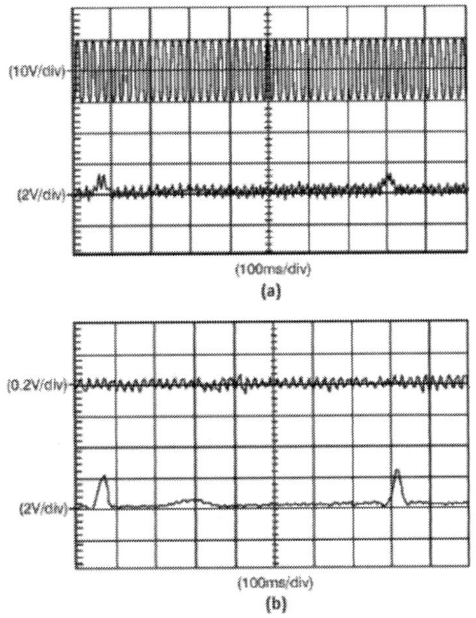

Figure 1. ECG Signal influenced by V_{cm} noise, (a) without balancing and (b) with balancing [7].

3. Theoretical fundamentation

The grounds of the prototype has their roots in Silva [9], however, a newcontrol method of noise reduction using the impedance unbalance based on a new architecture using discrete components, fully adapted and designed to reduce the effect V_{cm} interference in electronic circuits has been presented, which generated even a prototype that proved the mathematical calculations and theoretical results obtained in simulation.

The umbalance of the impedances of the electrodes is compensated dynamically using a counter circuit that controls a bank impedances, which allowed the reduction of common mode noise (V_{cm}) in an AI input impedance dynamic balancing by means of a feedback system (RECONFIGURABLE CIRCUIT), Figure 2. The composition and functioning of the prototype are described below.

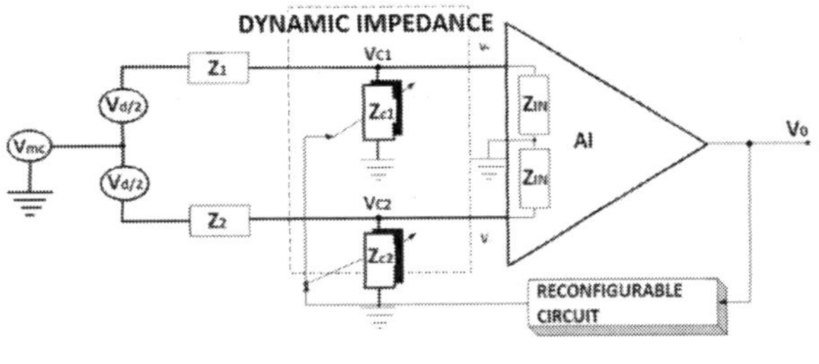

Figure 2. New architecture adapted from Silva [9].

Being:

- V_{mc} = electrical interference signal in common mode;

- $V_d/2$ = bio-potential signal at each electrode;
- V_+ e V_- = voltages at the input of AI;
- Z_1 e Z_2 = impedance of the input electrodes;
- Z_{C1} e Z_{C2} = Dynamically adjustable impedances;
- Z_{in} = input impedances of the AI amplifier;
- Reconfigurable Circuit = circuit responsible for the dynamic control of adjustable impedances;
- V_0 = adjusted output signal with minimum interference.

The reduction of the umbalance is viable through a feedback loop (Reconfigurable Circuit) that, due to the variation of maximum amplitude of the output signal (V_0), readjust the permitted values of input impedances (Z_{C1} and Z_{C2}) to obtain an output signal with minimal interference. This value is achieved when the condition of transfer function (FT), Equation 1 is satisfied, meaning that the common mode noise was canceled.

$$Z_{c1} + Z_1 = Z_{c2} + Z_2 \tag{1}$$

The general equation of a non-feedback amplifier is governed only by their internal components and displayed in Equation 2. However, as the architecture uses the principle of reconfigurability [2] by feedback, the new FT is defined by Equation (3). Still, as the main goal of analysis is the noise component, then only the second part of the equation is taken into consideration (Equation 4), it is directly influenced by the balancing of the impedances of the cables (Z_1, Z_2, and Z_{C1}, Z_{C2}), noise (V_{mc}) and the gain in differential mode (A_d) in the AI inputs. The test, the level of umbalance that can reduce this new architecture, which was the same used in longitudinal telephone cables (**LUL**), where we got the Equations (4 and 5), based on the architecture in Figure 2. And after the appropriate replacements values in Equation 6, predicting the worst case of unbalance for ($Z_{C1} = 2.5M\Omega, Z_{C2} = 10k\Omega, Z_1 = Z_2 = 1.2M\Omega$), the value found was (LUL =-0.6674 dB).

$$V_0 = V_d.A_d + V_{mc}.A_{mc} \tag{2}$$

$$V_0 = (\frac{Z_{in}}{Z_{in} + Z_1 + Z_{c1}} - \frac{Z_{in}}{Z_{in} + Z_2 + Z_{c2}})\frac{V_d}{2}A_d + \\ (\frac{Z_{in}}{Z_{in} + Z_1 + Z_{c1}} - \frac{Z_{in}}{Z_{in} + Z_2 + Z_{c2}})V_{mc}A_d + V_{mc}\frac{A_d}{CMRR} \tag{3}$$

$$V_0 = (\frac{Z_{in}}{Z_{in} + Z_1 + Z_{c1}} - \frac{Z_{in}}{Z_{in} + Z_2 + Z_{c2}}).V_{mc}.A_d \tag{4}$$

$$LUL = 20.log(\frac{V_{c2} - V_{c1}}{V_{mc}})(dB) \tag{5}$$

$$LUL = 20.log(\frac{Z_{C2}}{Z_2 + Z_{C2}} - \frac{Z_{C1}}{Z_1 + Z_{C1}})(dB) \tag{6}$$

Where: A_{mc} = is the gain of the amplifier in common mode.

4. Materials and Methods

Initially the project was described in the VHDL_AMS language and simulated in the ADVANCE platform from Mentor Graphics [3].

4.1. Project description in VHDL_AMS

In relation to the values of the sources of excitation, was only taken into consideration the common mode signal (V_{cm}). This means that when the analysis of simulation results (Figure 3), the useful signal(V_{d1} and V_{d2}) will not be present, and only the interference will be displayed, presented an frequency oscillation (f_{cm}), and $ofsset = 0$, meaning that there was no displacement of the reference point (DC level zero).

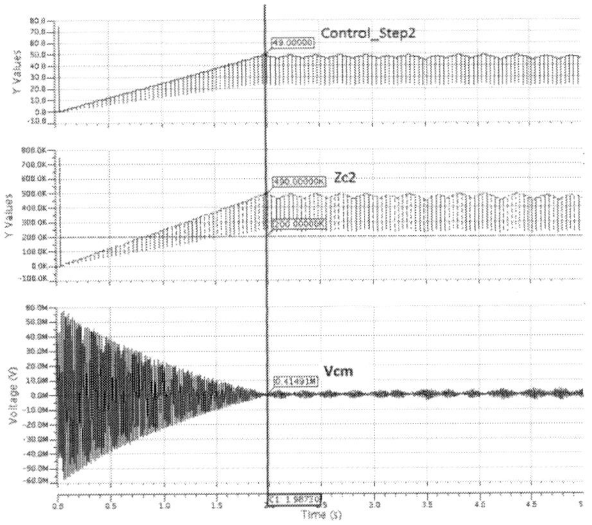

Figure 3. Simulation in VHDL_AMS with feedback control.

4.1.1. Simulation results obtained in VHDL_AMS

The results of two simulations in VHDL_AMS, given constant the input (Z_{in1}) and varying the other (Z_{in2}), are observed in the graphs in Figure 3, where three curves are displayed (listed from top to bottom): the first represents the amplitudes control ($Control - Step2$); the second, the increment or decrement the resistors bank (Z_{C2}) and the third interference signal, represented by (V_{cm}).

In the graphs of Figure 3, the interfering signal (V_{cm}), is attenuated and maintained at a level of amplitude as low as possible, through the closed loop control, which leads to the balancing of the impedances (Equation 1).

4.2. Desing and simulation using proteus software

The design and simulations were also performed using the software Proteus "ISISProfessional v7.0" [4]. The project was divided as shown in Table 1, in which are detailed the elements used in the prototype, along with a brief technical description of each electronic component.

Where:

- COMP. = trade name of the component;
- Q = numeric quantity of each component;
- VALOR COM. = informs the market values;
- DESC. = brief technical description of the element;
- IMP VAR = impedance variable block;

Table 1. List and description of components.

	$COMPONENT$	QTE	$COMMERCIAL\ VALUE$	$DESCRIPTION$
$I.VAR$	$Resistor(\Omega)$	12	$10K, 20K, 40K, 80K, 160K320K,$ $640K, 1280K, 2.55M, 2x100k, 320K$ $640K, 1280K, 2.55M, 2x100k$	$Carbon\ resistor$
	$Analog\ Switch$	8	$HEF4016$	$Analog\ switch\ circuit$
	$counter$	12	$CD4029$	$Digital\ counter\ circuit$
AI	AI	1	$INA122$	$Instrumentation\ Amplifier$
FIL	$LPF\ Filter$	1	$OP07$	$4th\ order\ active\ filter\ bandpass\ 60Hz$
	$BPF\ Filter$	2	$OP\ 07$	$Circuit\ logic\ band\ pass\ 60Hz.$
CON	$Circuit\ logic\ AND$	1	$74LS00$	$Circuit\ logic.$
	$Microcontroller\ PIC$	1	$16f877(Microchip, 2010)$	$Control\ and\ A/D\ converter\ circuit$

- AI = amplifier instrumentation block;
- FIL = filters block;
- CON = control block.

4.2.1. Functional description

The operation of the circuit will be described according to the algorithm of Figure 4.

The operating system is based on the signal compensation, comprising a component in common mode (V_{cm}) and a differential mode (V_d), which are modulated by the input impedances (Z_{VARIAB}), formed by a resistors bank. These resistors are connected (dynamically) so that will add a total of 2.55 $M\Omega$, this is possible with the use of analog switches, which are controlled by four-bit counter (4 bits) each, but together can achieve to (8 bits = 256 values). As the signal amplitude is low, it needs to be amplified, which is done by the AI. A sample of the signal after filtered by the 2nd order LPF, the output is sent (V_0) to be analyzed later. Another sample of the signal goes to the control loop input impedance because what distinguishes this prototype of which have been submitted in earlier work ([2], [5], [7]) is the analysis of noise (V_{cm}), which is separated from the useful signal (V_d) by a filter of 4th order BPF with bandwidth of 30 Hz, sufficient to pass only the components of the EMI between 40 and 70 Hz.

Then the signal passes through an A/D converter (implemented by the microcontroller (PIC 16F877)) 8-bit (which is equivalent to one resolution of 19.53 mV). If there is a sudden increase in input V_{cm}, the same microprocessor used in the A/D conversion is programmed (in C language) to perform dynamic control block CONTROL of the impedances (Z_{VARIAB}), by compensating the already discretized amplitudes of V_{cm} with two other digital reference voltages, which oscillate around the offset voltage ($\approx 2.5V$) of the microcontroller, as shown in Table 2. This type of control was more effective, using not only the clock, but also control of the count, increasing (UP) or decreasing (DW) of the counter circuits (counter), as tools (Table 1), influencing thus the increase or decreased of the variable impedance (Z_{VARIAB}) more efficiently.

4.2.2. Circuit settings

The project description of each subcircuit was distributed in the following subtopics.

Variable impedance block: The variable impedance block (Z_{variab}), consists of a group of resistances (Z_c), which are described by Equation 7. The set comes complete with analog

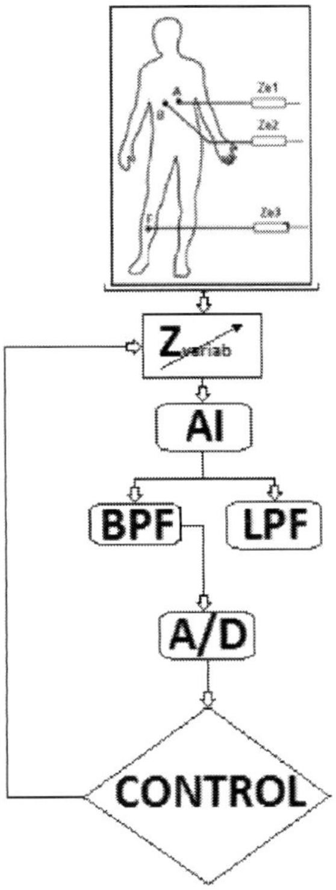

Figure 4. Circuit algorithm.

Table 2. A/D 8 bits converter.

$Voltage(mV)$	$Decimal$	$Binary$
0	0	00000000
19,53	1	00000001
...	...	...
2480,31	126	01111111
2499,84	127	10000000
2519,37	128	10000001
...	...	...
4980,15	254	11111101
4999,68	255	11111111

switches and counters, as detailed in Table 1.

$$Z_{c2} = R.\sum_{1}^{NB} 2^{(NB-1)} \tag{7}$$

Where:

- R = is the first resistance value;
- NB = the number of bits adopted in the A/D.

Instrumentation amplifier block (AI): The amplifier block consists of a high pass filter (HPF) in the input, to minimize the noise below 0.3 Hz (equation 8), beyond AI, where the gain ($G_{AI}(V/V)$) is given by equation 9.

$$H(S) = \frac{W.C.R}{\sqrt{1 + (W.C.R)^2}} \tag{8}$$

$$G_{AI} = 5 + \frac{200K}{R_G} \tag{9}$$

Where:

- RG = is the gain resistance.

Low pass/band bass filters block: The filters were dimensioned based on the RAUCH architecture, in the witch the topology of multiple feedback (MFB) is characterized by high gains and quality factor (Q), and are determined by the general transfer function (Equation 10).

$$\frac{Z_{out}}{Z_{in}} = -(\frac{Z_3}{Z_5} + \frac{Z_1}{Z_5} + \frac{Z_1}{Z_4} + \frac{Z_1.Z_3}{Z_2.Z_5} + \frac{Z_1.Z_3}{Z_4.Z_5})^{-1} \tag{10}$$

From this standard block, after replaced by their appropriate values of resistances and capacitances equivalent, have been dimensioned other filters: LPF 2nd order, and its corresponding transfer function in Equation (11,12, 13 and 14), and BPF filter, two blocks cascaded of 2nd order each regulated by a transfer function of Equation (15, 16, 17 and 18), equivalent to a 4th order. All calculated for 3 dB attenuation.

$$|H(S)|_{LPF} = \frac{a_0}{b_1 S^2 + a_1 S + (W_0)^2} \tag{11}$$

$$a_0 = -\frac{R_2}{R_1} \tag{12}$$

$$a_1 = W_0.C_1(R_2 + R_3 + \frac{R_2.R_3}{R_1}) \tag{13}$$

$$b_1 = W_0^2.C_1.C_2.R_2.R_3 \tag{14}$$

$$|H(S)|_{BPF-2?} = \frac{a_1 S}{b_1 S^2 + b_2 S + (W_0)^2} \tag{15}$$

$$a_1 = -(\frac{R_2.R_3}{R_1 + R_3}).C.W_0 \tag{16}$$

$$b_1 = (\frac{R_1.R_2.R_3}{R_1 + R_3}).C^2.W_0^2 \tag{17}$$

$$b_2 = (\frac{2.R_1.R_3}{R_1 + R_3}).C.W_0 \tag{18}$$

Control block (CONTROL): The control block is formed by the PIC microcontroller (Table 1) and a DC level shifter circuit to keep a reference of 2.5 volts.

5. Results

Figures 6 to 7 present simulation results obtained from the circuit designed in the PROTEUS software, when applying characteristic signals obtained by biosensors, in order to reduce common mode noise (V_{cm}).

5.1. Simulation results

In Figure 5, three graphs are shown, representing the result of simulation of the noise reduction system (no feedback or control), with the application of a bioelectric signal characteristic in the input. The first (top to bottom) represents the signal in common mode and differential after amplified by the AI ($V_{-AI-Out}$) constant gain, the second represents the output of the LPF biosignal ($V_{cm} + V_d$) and the third represents the noise in the output of the BPF (V_{cm}).

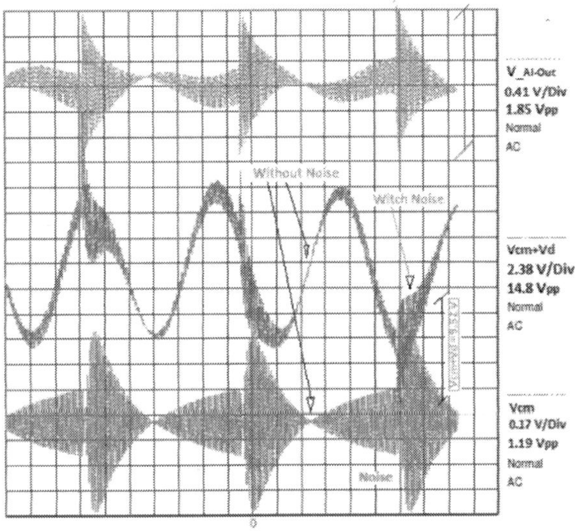

Figure 5. Biosensor: system without feedback.

In this case, the system has not been feedback and the amplitude of the interference range is approximately between a maximum value ($V_{cm\_Max} \approx 1.19Vp$) and a minimum value ($V_{cm\_Min} \approx 34.0mVp$), as previously mentioned, V_d is until influenced by V_{cm}. This interference is such that the useful signal becomes disguised (hidden) by interference ($V_{cm} \gg V_d$). This relationship begins to change as the impedance matching occurs at the input of AI, due to feedback from the common mode signal, causing a progressive loss of range of V_{cm}, without affecting the amplitude of V_d. Thus, the useful signal tends become more visible ($V_{cm} \ll V_d$) (Figure 6).

After the feedback circuit, the system generated the graphs in Figure 6. The top graph represents the output of AI (V_{AI-Out}); the second output LPF filter, and the third output BPF filter.

5.2. Experimental circuit

After the theoretical analysis and simulation results, it was made a prototype circuit (Figure 7), where the LPF and BPF filter has been implemented in PSOC microcontroller (CY8C27443).

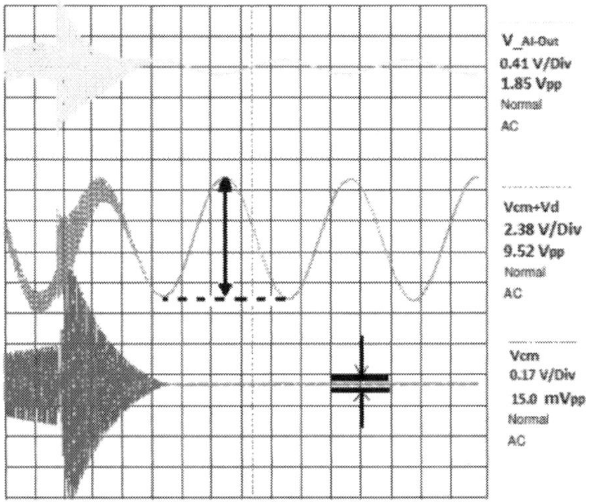

Figure 6. Biosensor: system with feedback.

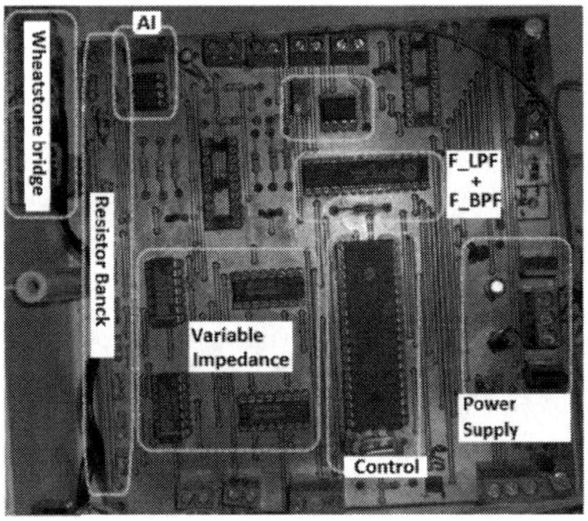

Figure 7. Printed circuit prototype.

The prototype tests were performed using as the input sine wave with a frequency of 10 Hz, in the range of an ECG signal influenced by an EMI noise of 50 Hz results were obtained by means of an oscilloscope, and are shown in Figure 8. It can be seen from this figures that the noise reduction (V_{cm}) was around 50 times, while the differential mode component (V_d), or useful signal, was minimally affected.

6. Discussion

By the analysis of the results obtained, it is concluded that the proposed circuit (described in language $VHDL\_AMS$, simulated on the platform from Mentor Graphics, designed and simulated in software PROTEUS) achieved its primary goal, which consists of reducing the common mode interference (V_{cm}) while preserving the useful components of the signals to which the system was proposed (V_{d1} and V_{d2}), as shown in Figures 6 and 8b. The gain is kept at low levels in AI, to avoid saturation in the filtering blocks. In the output of the BPF block, the signal purposely received a higher gain compared to other outlets, the need for analysis or for digital processing, more specific, if necessary. As the graph representative of the noise (V_{cm}), we find that the same was suffered attenuation by approximately 97% in biomedical signals application (Figure 6), while the useful signal is preserved(V_{AI-Out}), as shown in the graphs of Figure 6.

The same results (with reduced around 50% were confirmed when the analysis of the experimental prototype (Figure 8a, 8b).

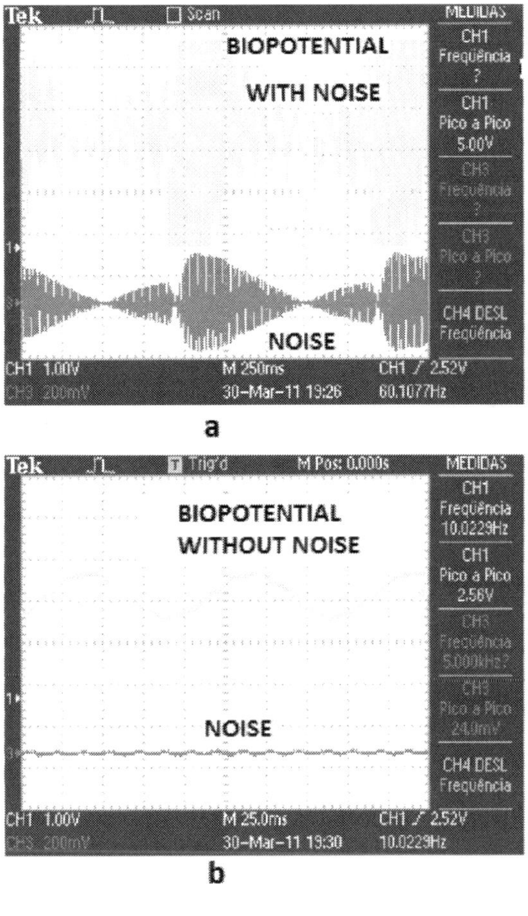

Figure 8. Results obtained with the oscilloscope, (a) circuit without feedback and (b) with feedback.

7. Conclusion

This article addressed the issue of reducing common mode noise, especially those from sources of electromagnetic energy.

Before resolution of the problem, we proceeded to introduce the concept of balancing impedance. After the initial precepts, it was proposed modification architecture of V_{cm} noise reduction by balancing dynamic input impedance (Z_{VARIAB}).

For this, a circuit was designed in hardware description language ($VHDL\_AMS$), simulated on the platform from Mentor Graphics, designed and simulated electronic software (PROTEUS), following almost real parameters of discrete circuits found in the market, and at the end it was build and tested a prototype.

Excellent results were obtained, thus confirming the idea of the proposed architecture and its application to other signs that need to be controlled without the use of complex circuits or filters that will override both the noisy signals (common mode), but also damage the components of the useful signal (or differential).

Acknowledgments
Thanks to the ALBAN program and the FADESPA by technical and financial support.

References

[1] Yamamoto A Y, T N and Kitaoka K 2000 *Instrumentation end Measurements Technology Conference. IMTC*
[2] Negrão J F R, Naviner J F and Silva I S 2006 *Journées Nationales du Réseau Doctoral de Microélectronique*
[3] 2006 *Getting Start Witch Advance MS Software* version 4.4_1 release ams tutorial ed
[4] 1989-2009 Isis labcenter electronic URL `http://www.labcenter.co.uk/`
[5] Spinelli E, Martinez M N, Mayosky M A and Pallas-Areny R A 2004 *IEEE Transactions On Biomedical Engineering* **51**
[6] Volpato R M and Magalhães G V B 2009 *VIII SEMETRO*
[7] Dobrev D and Daskalov L 2009 *Medical & Biological Engineering & Computing* **40** 122–127
[8] Berbari E J 2000 Principles of electrocardiography
[9] Silva I S 2003 *Circuits Mixtes Reconfigurables appliqués à la Mesure de Signaux Biomédicaux: Réjection de linterférence de Mode Commun* Ph.D. thesis ENST

Effects of Intense Physical Activity with Free Water Replacement on Bioimpedance Parameters and Body Fluid Estimates

E B Neves[1], L Ulbricht[1], E Krueger[2], E F R Romaneli[3], M N Souza[4]
[1] Biomedical Engineering Master Program/PPGEB, Federal Technological University of Paraná, Av. Sete de Setembro, 3165, 80230-901, Curitiba/PR, Brazil
[2] Rehabilitation Engineering Laboratory, Federal Technological University of Paraná, Av. Sete de Setembro 3165, 80230-901, Brazil
[3] Automation and Control Engineering, Academic Department of Electrotechnics, Federal Technological University of Paraná, Av. Sete de Setembro 3165, 80230-901, Brazil
[4] Biomedical Engineering Program – COPPE/UFRJ, Centro de Tecnologia, Bloco H, sala 327, 21949-900, Brazil

E-mail: borbaneves@hotmail.com

Abstract. Authors have emphasized the need for previous care in order to perform reliable bioimpedance acquisition. Despite of this need some authors have reported that intense physical training has little effect on Bioimpedance Analysis (BIA), while other ones have observed significant effects on bioimpedance parameters in the same condition, leading to body composition estimates considered incompatible with human physiology. The aim of this work was to quantify the changes in bioimpedance parameters, as well as in body fluids estimates by BIA, after four hours of intense physical activity with free water replacement in young males. Xitron Hydra 4200 equipment was used to acquire bioimpedance data before and immediately after the physical training. After data acquisition body fluids were estimates from bioimpedance parameters. Height and weight of all subjects were also acquired to the nearest 0.1 cm and 0.1 kg, respectively. Results point that among the bioimpedance parameter, extracellular resistance presented the most coherent behavior, leading to reliable estimates of the extracellular fluid and part of the total body water. Results also show decreases in height and weight of the participants, which were associated to the decrease in body hydration and in intervertebral discs.

Introduction

Bioelectrical impedance or simply bioimpedance can be defined as the frequency-dependent phenomenon associated to the electrical opposition (both real and imaginary components) a biological specimen presents to the flow of an alternating current. Bioimpedance measurement is influenced by the frequency of the electrical signal, electrochemical processes, temperature, potential of hydrogen (pH), hydration and viscosity of the fluid or biological tissue in question [1].
Among bioimpedance applications, noninvasive estimate of body fluids (extracellular fluid - ECF, intracellular fluid – ICF, and total body water - TBW) has been used in monitoring patients and athletes. Particularly, the ratio between ECF and ICF has been suggested as an effective index for

controlling health status [2]. These estimates are used independently and associated with others variables to follow-up patients with HIV [3], in the assessment of hydration status in patients undergoing hemodialysis [4], controlling the health status of patients affected by hyperthyroidism [5], dengue fever [6], in studies of hydration during cellular aging [7], and for monitoring levels of obesity and nutrition [8], among others.

Although some authors emphasize the need for previous care in order to perform reliable bioimpedance acquisition, quite different experimental protocols have been observed in the literature. Some studies control bladder emptying and type of clothing [9], while other ones advise to avoid performing intense physical exercise and to control dietary patterns, as well as alcohol intake for 24 hours before the test [10] [11]. Still others contraindicate the use of diuretic substances for 48 hours before the test [12]. One study extends for 7 days some of the recommendations listed above [13].

Regarding the influence of physical activity on the results concerning bioimpedance data, Perrella *et al.* [14] reported that two hours of rugby training had little effect on the Bioimpedance Analysis (BIA). On the other hand, Neves *et al.* [15] used several methods to assess changes in body composition of subjects after four hours of intense physical activity and observed a decrease in the mean fat percentage assessed by bioimpedance, which was considered incompatible with human physiology, indicating a strong influence of the physical activity on the results. This result reinforces the need of care protocols in assessment of body composition by BIA, as indicated by other studies [16, 17], specially under unstable conditions of hydration.

The aim of this study was to quantify the changes observed in bioimpedance parameters, as well as in bioimpedance estimates of body fluids, after four hours of intense physical activity with free water replacement in young males.

Materials and Methods

1.1. Bioimpedance data acquisition and estimates of body fluids

A Xitron Hydra 4200 equipment was used to acquire bioimpedance parameters in a tetrapolar arrangement of electrodes that considered the whole body, i.e., wrist to ankle protocol [4]. Figure 1 depicts the electrode placement, where I symbols indicate the sites where the current electrodes were placed, and V symbols indicate the sites associated to the voltage electrodes. Such electrodes placement and subject position during the measurement follow the indication established in the Xitron Hydra 4200 user's manual [4] for a whole body assessment. Stainless steel electrodes of 12 cm^2 area and electrolyte gel were used instead of the Xitron original electrodes.

The device uses 50 fixed frequencies between 5 kHz and 1 MHz, and a sinusoidal current excitation in a range from 50 to 700 μA. Based on 50 data acquisitions of impedance, the firmware of the equipment estimates values associated to a Cole model [18], composed, among other parameters, by the extracellular resistance (Re), the intracellular resistance (Ri) and the membrane capacitance (Cm).

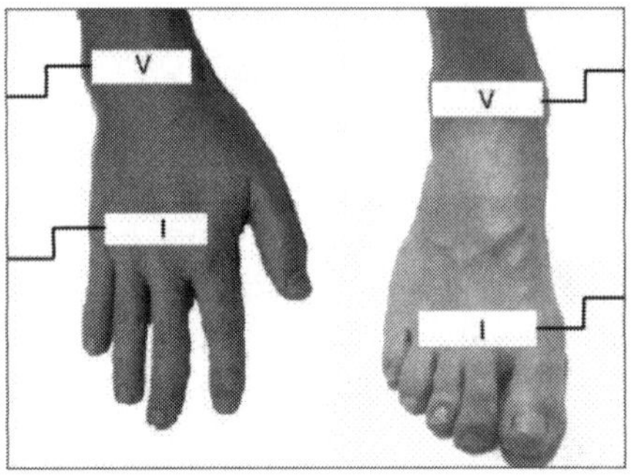

Figure 1. Electrodes placement according Xitron Hydra 4200 user's manual.

After bioimpedance parameters are determined, Re is firstly used to estimate ECF. Then, Re and Ri are used to estimate ICF and finally the total body water (TBW) is derived by adding ECF and ICF. References [19] and [20] can be seen for more details concerning the equations used to perform such estimates. Briefly, ECF and ICF can be calculated by equations (1) and (2) based on bioimpedance resistive parameters. Thus, total body water can be calculated by equation (3).

$$ECF = \frac{1}{1000} \cdot \left(\frac{K_b^2 \cdot \rho_{ECF}^2}{D_b}\right)^{\frac{1}{3}} \cdot \left(\frac{H^2 \cdot \sqrt{W}}{R_e}\right)^{\frac{2}{3}} \quad (1)$$

$$ICF = ECF \cdot \left\{\left[\frac{\rho_{TBW} \cdot (R_e + R_i)}{\rho_{ECF} \cdot R_i}\right]^{\frac{2}{3}} - 1\right\} \quad (2)$$

$$TBW = ECF + ICF \quad (3)$$

Where, K_b is a correction factor for a whole body measurement, relating the relative proportions of the leg, arm, trunk and height; ρTBW is the resistivity of the overall fluid in [Ω·m]; ρECF is the resistivity of the extracellular fluid in [Ω·m]; H is the body height in [cm]; Db is the body density in [Kg/l) and W is the body weight in [Kg].

According Xitron Operational Manual [4], the FFM can be calculated by equation 4, the hydration by equation 5, the Fat by equation 6 and the Fat % by the equation 7.

$$FFM = (d_{ECW} \cdot V_{ECW}) + (d_{ICW} \cdot V_{ICW}) \quad (4)$$

$$Hydration = \frac{TBW}{FFM} \quad (5)$$

Where, d_{ECW} (1.106) is the mean density of extracellular water and d_{ICW} (1.521) is the mean density of the intracellular water and its associated materials.

$$Fat = W_t - FFM \quad (6)$$

$$Fat\% = \frac{FAT}{W_t \cdot 100} \quad (7)$$

Where, W_t is the total weight value.

1.2. Subjects

The study sample universe was initially composed by 422 students from a military parachuting course in Brazilian Army. From such sample 92 male students were randomly selected to be evaluated during the three weeks that composed the first phase of the parachuting course, which occurred in the period from January 6th to 24th, 2009. The study was performed according to principles expressed in Declaration of Helsinki and was approved by the Ethical Committee of the Airborne Instruction Center in charge of the parachuting unit.

1.3. Assessment Protocol

During data collection each subject was evaluated twice in a single day. The first evaluation took place immediately before the four-hours training and considering that the subjects were advised to perform bladder emptying with 30 minutes early and do not drink alcohol within 72 hours before the test. The second evaluation was performed immediately after 4 hours of intense physical activity. In addition to bioimpedance data acquisition, height and weight of all subjects were acquired to the nearest 0.1 cm and 0.1 kg, respectively. The subjects were using light clothing during data collection.

1.4. Data Analysis

Statistical Package for Social Sciences for Windows, version 17.0 (SPSS Inc., Chicago, IL, EUA), was used to perform statistical analysis. Descriptive statistics were reported as mean and standard deviation (Std). Paired t-test was used to verify the difference between the studied variables before and after the period of physical activity. Inferential tests considered significant P-value < 0.05.

Results

Ninety two male military paratroopers were assessed before (index 1) and immediately after (index 2) four hours of physical training. Results concerning bioimpedance parameters are shown in Table 1.
Table 2 shows the results of ten variables that were assessed before and immediately after the physical training. Some of them were directly measured, as weight and height, while other ones were estimates by BIA. In this table one can observe that there were decreases in the mean-values of weight, height, body fat percentage (Fat %) and body mass index (BMI) after the intense physical activity period, which was expected due to a reduction in the level of hydration despite the free water replacement.
As mentioned before, despite the total body water (TBW) has remained stable; there was a reduction in the level of hydration and the ratio between the intracellular and extracellular fluid (ECF / ICF) due to a decrease in extracellular fluid (ECF) and an increase in fluid intracellular (ICF).

Table 1. Changes in bioimpedance parameters after four hours of intense physical activity. Indexes 1 and 2 stand for parameters before and immediately after the physical training, respectively.

		Mean	N	Std. Deviation	Std. Error Mean
Pair 1	Re 1[Ω]	528.98	92	44.37	4.63
	Re 2[Ω]	533.63	92	43.15	4.50
Pair 2	Ri 1[Ω]	1021.06	92	132.64	13.82
	Ri 2[Ω]	977.65	92	127.18	13.26
Pair 3	Cm 1[nF]	2.52	92	0.37	0.04
	Cm 2[nF]	2.68	92	0.39	0.04

Table 2. Results of assessment for the ten studied body characteristics after four hours of intense physical activity. Indexes 1 and 2 stand for parameters before and immediately after the physical training, respectively.

		Mean	N	Std. Deviation	Std. Error Mean
Pair 1	Weight 1 [kg]	76.35	92	7.45	0.78
	Weight 2 [kg]	74.08	92	7.13	0.74
Pair 2	Height 1 [cm]	174.61	92	5.90	0.62
	Height 2 [cm]	173.24	92	5.73	0.60
Pair 3	BMI 1 [kg/m^2]	25.04	92	2.10	0.21
	BMI 2 [kg/m^2]	24.69	92	2.07	0.22
Pair 4	ECF 1 [L]	19.50	92	1.84	0.19
	ECF 2 [L]	18.98	92	1.66	0.17
Pair 5	ICF 1 [L]	26.82	92	3.01	0.31
	ICF 2 [L]	27.38	92	3.04	0.32
Pair 6	TBW 1[L]	46.33	92	4.59	0.48
	TBW 2 [L]	46.36	92	4.40	0.46
Pair 7	FFM 1[kg]	62.37	92	6.30	0.66
	FFM 2 [kg]	62.68	92	6.14	0.64
Pair 8	Fat % 1 [%]	18.18	92	4.99	0.52
	Fat % 2 [%]	15.19	92	6.19	0.65
Pair 9	Hydration 1 [%]	74.29	92	0.41	0.04
	Hydration 2 [%]	73.99	92	0.92	0.10
Pair 10	ECF/ICF 1	0.73	92	0.05	0.01
	ECF/ICF 2	0.70	92	0.05	0.01

Table 3 shows the results of the paired t-test for all variables depicted in Tables 1 and 2. There were statistically significant differences for most of variables, but TBW and fat-free mass (%).

Table 3. Results of paired t-test for all studied variables.

		N	Pearson Correlation	P	t value	P
Pair 1	Weight 1 & Weight 2	92	0.992	0.000	21.945	0.000
Pair 2	Height 1 & Height 2	92	0.995	0.000	21.238	0.000
Pair 3	IMC 1 & IMC 2	92	0.987	0.000	10.249	0.000
Pair 4	ECF 1 & ECF 2	92	0.884	0.000	5.856	0.000
Pair 5	ICF 1 & ICF 2	92	0.735	0.000	-2.426	0.017
Pair 6	TBW 1 & TBW 2	92	0.799	0.000	-0.101	0.920
Pair 7	FFM 1 & FFM 2	92	0.787	0.000	-0.718	0.474
Pair 8	Fat % 1 & Fat % 2	92	0.472	0.000	4.904	0.000
Pair 9	Hydration 1 & Hydration 2	92	0.305	0.003	3.249	0.002
Pair 10	ECF/ICF 1 & ECF/ICF 2	92	0.647	0.000	7.182	0.000
Pair 11	Re 1 & Re 2	92	0.911	0.000	-2.411	0.018
Pair 12	Ri 1 & Ri 2	92	0.883	0.000	6.592	0.000
Pair 13	Cm 1 & Cm 2	92	0.823	0.000	-6.891	0.000

Discussion

Four hours of intense physical activity with free water replacement induced changes in body fluids in the studied population but not exactly as one could expected, because the estimates of TBW remained almost constant.

Considering the bioimpedance parameters, it was verified that Re increased continuously with the increasing of muscle activity (due to water loss from the body). However, the behavior of Ri was not clear. This fact may be related to the type of dehydration suffered by subject, where depending on the amount of water and electrolytes lost through sweat, osmotic effect within the body leads to intra and extracellular compensation, resulting in positive or negative changes in ICF that affect Ri parameter.

Change in muscle pH after exercise can also have affected BIA accuracy. There is a significant correlation between pH change and percentage change in ICF, with a significant increase in ICF during exercise [19]. At the beginning of the exercise sympathetic over activity occurs by stimuli from the motor cortex and maintained by stimuli from the central and peripheral receptors that capture the changes in pH and PCO_2. Furthermore, it is known that the increase in lactic acid levels modifies the pH and blood CO_2 [20].

Jürimäe et al. [21] investigated the influence of long-term exercise on body fluids of professional athletes (N = 12) of rowing. Each player paddled for two hours and seventeen minutes. The BIA tests were applied (I) before exercise, (II) immediately after exercise, (III) 30 min, (IV) 60 min and (V) 120 min after the end of exercise. The results showed that the weight of the participants was significantly reduced from 82.0 ± 10.8 kg to 80.6 ± 2.11 kg. However, there was no reduction in TBW immediately after exercise, but just 30 min after the end of the activity (48.7 ± 4.4 to 47.9 ± 4.1, $p < 0.05$). The amount of ECF also decreased significantly after 30 min of the end of exercise (23.3 ± 2.0 to 22.7 ± 1.91, $p < 0.05$). In the present study, the decrease in extracellular fluid (ECF) (Tables 2 and 3) was immediately observed at the end of exercise.

The decrease in height, weight, and consequently in BMI, of the participants in this study may be explained by the decrease in body hydration and in the intervertebral discs (assessed by height variation). Similar results were found by Neves et al. [15] to paratroopers and Jürimäe et al. [21] to rowing athletes.

A limitation of this study was the lack of control over temperature of the skin surface that is increased by physical activity. Beckmann [22] and Röthlingshöfer [23] reported that even after a recovery time of 15 minutes the temperature of skin remained high.

In this study, the reduction in fat percentage immediately after the exercise is a result of changes in body fluids, because, according Saunders et al. [24] a small change in balance of body fluids can be incorrectly interpreted by BIA, in form of changes in adipose tissue of athletes. Moreover, ionic losses resulting from the physical activity [23] can infer an erroneous reading of an increase in fat free mass, resulting from increased ICF.

Conclusions

We investigated changes observed in bioimpedance parameters and in estimates of body fluids by BIA after four hours of intense physical activity with free water replacement. Among the bioimpedance parameters Re exhibited the most coherent behavior, leading to reliable estimates of ECF and part of TBW. The intracellular resistance Ri seems to be more affected by changes in metabolic products of the exercise. Results also show decreases in height, weight and BMI of the participants, which can be associated to the decrease in body hydration and in intervertebral discs. Overall, our findings reinforce the need of care protocols in assessment of body composition by BIA.

References

[1] Neves EB, Pino AV, De Almeida RMVR and Souza MN 2010 Knee bioelectric impedance assessment in healthy/with osteoarthritis subjects, *Physiological Measurement,* vol. 31, pp. 207-219.

[2] Jaffrin MY, Kieffer R and Moreno MV 2005 Evaluation of a foot-to-foot impedance meter measuring extracellular fluid volume in addition to fat-free mass and fat tissue mass, *Nutrition,* vol. 21, pp. 815-824.

[3] Morrow JR, Jackson A, Disch JG and Mood D 2010 *Measurement and evaluation in human performance*: Human Kinetics Publishers.

[4] Matthie J 1997 Hydra ECF/ICF Bioimpedance Analyzer (Model 4200) Operational Manual Revision 1.01, *San Deigo, CA: Xitron Technologies Inc.*, pp. 44-8.

[5] Earthman CP, Matthie JR, Reid PM, Harper IT, Ravussin E and Howell WH 2000 A comparison of bioimpedance methods for detection of body cell mass change in HIV infection, *Journal of Applied Physiology,* vol. 88, pp. 944-956.

[6] Zhu F, Schneditz D and Levin NW 1999 Sum of segmental bioimpedance analysis during ultrafiltration and hemodialysis reduces sensitivity to changes in body position, *Kidney international,* vol. 56, pp. 692-699.

[7] Hu H and Kato Y 1995 Body composition assessed by bioelectrical impedance analysis (BIA) in patients with Graves' disease before and after treatment, *Endocrine journal,* vol. 42, p. 545.

[8] Klassen P, Mazariegos M, Deurenberg P, Solomons N and Fürst P 2000 Hydrational status assessed by bioelectrical impedance spectroscopy and dilution methods in patients with classical dengue fever, *Annals of the New York Academy of Sciences,* vol. 904, pp. 163-170.

[9] Wittchen HU and Jacobi F 2005 Size and burden of mental disorders in Europe--a critical review and appraisal of 27 studies, *European neuropsychopharmacology,* vol. 15, pp. 357-376.

[10] Ribeiro SML, Miyamoto MV, de Melo CM and Kehayias J 2011 Análise vetorial de bioimpedância e estado nutricional de idosas de acordo com o índice de massa corporal, *Rev Bras Cineantropom Desempenho Hum,* vol. 13, pp. 415-421.

[11] Moreira MM, de Souza HPC, Schwingel PA, de Sá CKC and Zoppi CC 2008 Efeitos do Exercício Aeróbico e Anaeróbico em variáveis de risco Cardíaco em Adultos com Sobrepeso, *Arq Bras Cardiol,* vol. 91, pp. 200-206.

[12] Rossi L and Tirapegui J Avaliação antropométrica segmentar comparativa de triatletas e maratonistas.

[13] Lima LD, Moraes CMB and Kirsten VR 2010 Muscle dysmorphia and the use of ergogenic supplements in athletes, *Revista Brasileira de Medicina do Esporte,* vol. 16, pp. 427-430.

[14] Perrella MM, Noriyuki PS and Rossi L 2005 Avaliação da perda hídrica durante treino intenso de rugby, *Rev Bras Med Esporte,* vol. 11, pp. 229-32.

[15] Neves EB, Oliveira AGV, Macedo ARM, Souza MN and Almeida RMVR 2008 Avaliação da hidratação dos alunos do curso de formação de Pára-quedistas Militares do Exército Brasileiro; Assessment of hydration among Brazilian army parachuting training program rookies, *Cad. saúde colet.,(Rio J.),* vol. 16, pp. 53-66.

[16] O'Brien C, Young A and Sawka M 2002 Bioelectrical impedance to estimate changes in hydration status, *International journal of sports medicine,* vol. 23, pp. 361-366.

[17] Koulmann N, *et al.* 2000 Use of bioelectrical impedance analysis to estimate body fluid compartments after acute variations of the body hydration level, *Medicine & Science in Sports & Exercise,* vol. 32, p. 857.

[18] Cole KS 1968 *Membranes, ions, and impulses: a chapter of classical biophysics* vol. 1: Univ of California Pr.

[19] Wang Z, Deurenberg P, Wang W, Pietrobelli A, Baumgartner RN and Heymsfield SB 1999 Hydration of fat-free body mass: new physiological modeling approach, *American Journal of Physiology-Endocrinology And Metabolism,* vol. 276, pp. E995-E1003.

[20] Åstrand PO 2003 *Textbook of work physiology: physiological bases of exercise*: Human Kinetics Publishers.

[21] Jürimäe J, Jürimäe T and Pihl E 2000 Changes in body fluids during endurance rowing training, presented at the Annals of the New York Academy of Sciences, 2000.

[22] Beckmann L, Hahne S, Medrano G, Kim S, Walter M and Leonhardt S, Monitoring change of body fluids during physical exercise using Bioimpedance Spectroscopy", 2009, pp. 4465-4468.

[23] Röthlingshöfer L, Ulbrich M, Hahne S and Leonhardt S 2011 Monitoring Change of Body Fluid during Physical Exercise using Bioimpedance Spectroscopy and Finite Element Simulations, *Journal of Electrical Bioimpedance,* vol. 2, pp. 79-85.

[24] Saunders MJ, Blevins JE and Broeder CE 1998 Effects of hydration changes on bioelectrical impedance in endurance trained individuals, *Medicine & Science in Sports & Exercise,* vol. 30, pp. 885-92.

Wavelet analysis of the impedance cardiogram waveforms

S. Podtaev[1], R. Stepanov[1], A. Dumler[2], S. Chugainov[2], K. Tziberkin[3]

[1] Institute of Continuous Media Mechanics, Korolyov str.1, Perm 614013, Russia
[2] Department of Introduction on Internal Diseasis, Perm State Medical Academy named after E.A.Vagner, Perm, Petropavlovskaya str. 26, Perm 614990, Russia
[3] Department of Physics, Perm State University, Bukirev str. 15, Perm 614990, Russia

E-mail: spt802@gmail.com

Abstract. Impedance cardiography has been used for diagnosing atrial and ventricular dysfunctions, valve disorders, aortic stenosis, and vascular diseases. Almost all the applications of impedance cardiography require determination of some of the characteristic points of the ICG waveform. The ICG waveform has a set of characteristic points known as A, B, E ($(dZ/dt)_{\max}$) X, Y, O and Z. These points are related to distinct physiological events in the cardiac cycle. Objective of this work is an approbation of a new method of processing and interpretation of the impedance cardiogram waveforms using wavelet analysis. A method of computer thoracic tetrapolar polyrheocardiography is used for hemodynamic registrations. Use of original wavelet differentiation algorithm allows combining filtration and calculation of the derivatives of rheocardiogram. The proposed approach can be used in clinical practice for early diagnostics of cardiovascular system remodelling in the course of different pathologies.

1. Introduction

Impedance cardiography (ICG) is a simple and cheap method for acquiring hemodynamic parameters. Unfortunately, not all physiological influences on the ICG signal have yet been identified. It has been confirmed that the main contributions to ICG signal come from ventricles, atria, aorta, and lungs. The relations between these components have been found to be dependent nonlinearly on spatial conductivity distribution. As a result, reliable and reproducible measurements of stroke volume (SV) using ICG are impossible [1]. Nevertheless, ICG can be used to monitor relative changes of SV in all cases where the spatial distribution of conductivity and geometry of the subject, during the examination, is preserved. Recent advances in waveform morphology analysis and pattern recognition software have allowed the development of reliable baseline-independent devices that can produce signal-averaged ICGs. In addition to the continuous hemodynamic monitoring of patients in intensive care, recognition of the different ICG patterns allows the rapid detection of cardiac dysfunction and the need for further cardiac evaluation. Compared with the standard ECG, the different patterns of ICG waveform are relatively easy to recognize and require considerably less time and skill to interpret than Doppler echocardiography [2].

The ICG waveform has a set of characteristic points [3] known as A, B, E ($(dZ/dt)_{\max}$) X, Y, O and Z, as shown in figure 1. These points are related to distinct physiological events in the cardiac cycle.

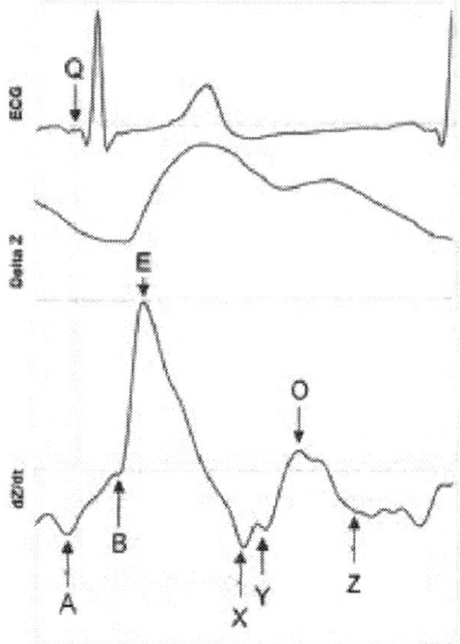

Figure 1. Characteristic ΔZ, dZ/dt and ECG signal, where A is the downward deflection due to the contraction of the atria, B is the start of ejection of blood by the left ventricle, E $((dZ/dt)_{\max})$ the major upward deflection occurring during systole, X the closure of the aortic valve, Y the closure of the pulmonic valve, O the diastolic upward deflection, Z wave describes a decreeing following the O wave, and it is associated with a third heart sound.

The A point occurred on the impedance cardiogram during the interval between the end of the P wave and the beginning of the QRS complex on the electrocardiogram, the A point occurred simultaneously with the beginning of the fourth heart sound. This wave is associated with the atrial contraction and its amplitude is correlated with the ejection fraction of the left atrium, A-wave is not observed when atrial fibrillation occurs [4]. The B point was synchronous with the maximal deflection of the first heart sound at the apex. Sometimes identifying the location of the B point is problematic because the characteristic upstroke that serves as a marker of this point is not always apparent [5]. The E point placed on the top position of the curve reflecting the maximal speed of the change in the impedance. It is associated with the maximal velocity of the ejection measured by using ultrasound methods [6]. The X point corresponds to the closure of the aortic valve (and second heart sound) and the Y point corresponds to the closure of pulmonic valve. The O wave is associated with changing of the volume during the diastolic phase of the cycle and opening snap of the mitral valve. The early diastolic O wave increases with mitral regurgitation, volume overload and increased left atrial pressure [7]. And, finally, Z wave describes a decreeing following the O wave, and it is associated with a third heart sound.

Objective of this work is an approbation of a new method of processing and interpretation of the impedance cardiogram waveforms using wavelet analysis.

2. Measuring technique and data analysis algorithm

2.1. Measurements

Ten healthy men at the age of 19 to 22 and five patients with essential hypertension participated in the study. A method of computer thoracic tetrapolar polyrheocardiography was used for hemodynamic registration [8]. For six people (three healthy and three patients), functional test with an isometric load was carried out. Static force is created by holding a raised patient's leg at an angle of 30 degrees to the horizontal plane. For twelve persons Doppler echocardiographic early E_D and late A_D diastolic waves were obtained at the mitral valve tips. Measurements were performed off-line using the average of 3 consecutive cycles.

2.2. Wavelet analysis

Wavelet transform presents a kind of "local" Fourier transform, allowing us to isolate a given structure in physical space and the Fourier space. Let us define the wavelet transform of the analyzing function $F(t)$ as

$$w_F(a,t) = \frac{1}{|a|}\int_{-\infty}^{\infty} F(t')\psi^*\left(\frac{t'-t}{a}\right)dt', \tag{1}$$

where $\psi(t)$ is the analyzing wavelet, a defines the scale and b defines the position of the wavelet. Then the coefficient w_F gives the contribution of corresponding structure into the function F.

The function F can be reconstructed using the inverse transform (see, e.g. [9])

$$F(t) = \frac{1}{C_\psi}\int_{-\infty}^{\infty}\int_{-\infty}^{\infty} \psi\left(\frac{t'-t}{a}\right) w_F\left(a,t'\right)\frac{da\,dt'}{a^2}. \tag{2}$$

The reconstruction formula (2) exists under condition that

$$C_\psi = \frac{1}{2}\int_{-\infty}^{\infty}\frac{|\hat{\psi}(k)|^2}{|k|}dk < \infty. \tag{3}$$

Here $\hat{\psi}(k) = \int \psi(t)e^{-ikt}dt$ is the Fourier transform of the analyzing wavelet $\psi(\phi)$.

Let us suppose that the function $F(t)$ is given by its derivative $G(t)$. Applying differentiation by part to equation (1) we have

$$w_G(a,t) = \frac{1}{|a|}\int_{-\infty}^{\infty} G(t')\psi^*\left(\frac{t'-t}{a}\right)dt' = \frac{1}{|a|}\int_{-\infty}^{\infty} F(t')\chi^*\left(\frac{t'-t}{a}\right)dt', \tag{4}$$

where $\chi(t) = -\psi'(t)$ is another wavelet. So the wavelet transform with wavelet family $\chi(t)$ gives the wavelet coefficients of the derivative of the analyzing function $w_G(a,t)$. Use of original wavelet differentiation algorithm [10] allows to combine filtration and calculation of the derivatives of rheocardiogram. We use as the analyzing wavelet the so-called Mexican hat $\psi(t) = (1-t^2)\exp(-t^2/2)$ for higher resolution of separated picks in time and the wavelet Morle $\psi(t) = \exp\left(-\frac{t^2}{2} + 2i\pi t\right)$ for better spectral resolution.

3. Results and discussion

The basic idea of the suggest approach is to define the characteristic points in the wavelet coefficient distribution w which are corresponds to ones are shown in figure 1. Typical distribution of wavelet coefficient is presented in figure 2. Principal difference between analysis of ordinal signal and its wavelet representation is that waves produce the local extremum in time and scale in 2D map. One can see positive defined (red) areas which correspond to E and O waves in each in the cardiac cycle. It is possible to define time, scale and amplitude for each extremum point. For the chosen case in Fig. 2 wave A has no well pronounce feature on the wavelet coefficient map but the local extremum can be found anyway for the fixed scale a. For the further analysis we take three amplitudes of wavelet coefficient A_w, E_w and O_w which correspond to A, E and O waves.

For twelve subjects Doppler echocardiographic registration of the diastolic transmitral blood flow was carried out. The first phase of the Doppler echocardiograms caused by the movement of blood through the mitral valve during the phase of rapid filling (peak E_D). The second phase (peak A_D) corresponds to the left atrium systole. Wavelet coefficient O_w correlated with Doppler peak E_D (r=0.65), and wavelet coefficient A_w did not correlate with Doppler peak A_D (r=0.23).

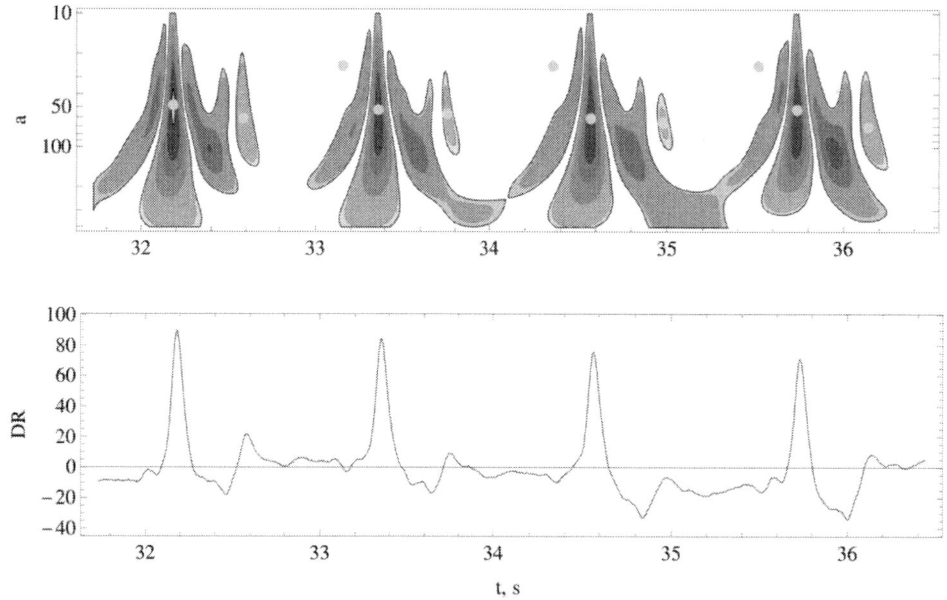

Figure 2. Wavelet coefficient map (top) of DR (bottom). Positive/negatine values correspond the red/blue area. Black means maximum absolute values and white means values close to zero. Characteristic points A, E and O are shown by green points.

Hemodynamic response to isometric functional test in healthy subjects characterized by a significant linear increase in stroke volume with increasing velocity and pressure of the left atrium filling during early diastole and increased pressure in the left atrium during atrial systole (A, E and O waves). At the end of the load, indices have decreased to normal values during the first minute of recovery period (Fig. 3a). This response to the load is explained by the action of the Frank-Starling mechanism, which allows to implement an adequate hemodynamic response to stress. For the patients an isometric stress did not cause any changes in cardiac output or

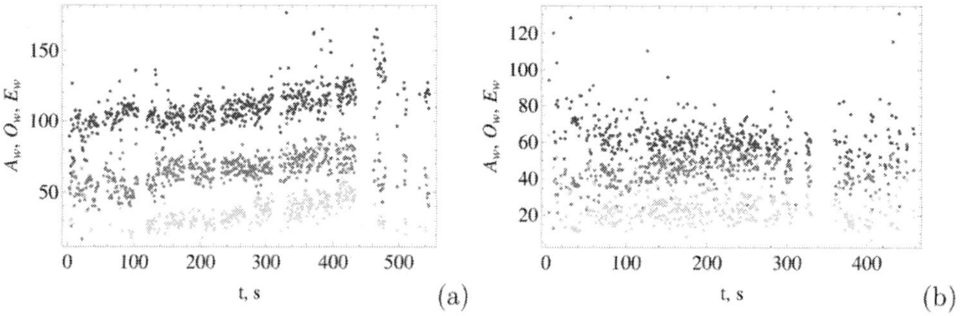

Figure 3. Typical changes of the amplitudes A_w (red points), O_w (green points) and E_w (blue points) waves during the postural test (lifting leg at an angle of 45 degrees from horizontal) (a) healthy subject, start - 100 s, the ending - 440 s (b) the patient with hypertension (55 years old, hypertension in conjunction with coronary artery disease, with fixed dilation of the left atrium and concentric left ventricular hypertrophy) start - 70 s, ending - 330 s.

diastolic left ventricular filling, which indicates the absence of compensatory mechanisms of the immediate adaptation to hemodynamic stress (Fig. 3b).

4. Conclusions

The proposed approach can be used in clinical practice for early diagnostics of cardiovascular system remodelling in the course of different pathologies. Isometric functional test with hemodynamic ICG monitoring and its subsequent wavelet analysis can be used to correct diagnosis of heart failure in hypertensive patients. The early detection system for cardiac insufficiency can be offered to practitioners.

References

[1] Wtorek J 2000 *Annals of biomedical engineering* **28** 1352–1361
[2] Bour J and Kellett J 2008 *European journal of internal medicine* **19** 399–405
[3] Lababidi Z, Ehmke D, Durnin R, Leaverton P and Lauer R 1970 *Circulation* **41** 651–658
[4] Takada K, Fujinami T, Senda K, Nakayama K and Nakano S 1977 *American Heart Journal* **94** 710–717
[5] Lozano D L, Norman G, Knox D, Wood B L, Miller B D, Emery C F and Berntson G G 2007 *Psychophysiology* **44** 113–119
[6] Kerkkamp H and Heethaar R 1999 A comparison of bioimpedance and echocardiography in measuring systolic heart function in cardiac patients *Electrical bioimpedance methods: applications to medicine and biotechnology* (*Annals of the New York Academy of Sciences* vol 873) ed Riu, PJ and Rosell, J and Bragos, R and Casas, O Int Comm Promot Res Bio Impedance (New York Acad Sciences) pp 149–154 10th International Conference on Electrical Bio-Impedance, Barcelona, Spain, apr 05-09, 1998
[7] Schieken R, Patel M, Falsetti H and Lauer R 1981 *British Heart Journal* **45** 166–172 ISSN 0007-0769
[8] Zubarev M, Dumler A, Shutov V and Popov N 1999 Assessment of left ventricular systolic function and diastolic time intervals by the bioimpedance polyrheocardiographic system *Electrical Bioimpedance Methods: Applications To Medicine And Biotechnology* (*Annals Of The New York Academy Of Sciences* vol 873) ed Riu, PJ and Rosell, J and Bragos, R and Casas, O (New York Acad Sciences) pp 191–196
[9] Daubechies I 1992 *Applied Mathematics* **61**
[10] Stepanov R, Frick P, Shukurov A and Sokoloff D 2002 *Astronomy and Astrophysics* **391** 361–368

Mesh generation and computational modeling techniques for bioimpedance measurements: an example using the VHP data

A A Danilov[1], V Yu Salamatova[2], Yu V Vassilevski[1]

[1]Institute of Numerical Mathematics, Russian Academy of Sciences
[2]Scientific-Educational Centre of the Institute of Numerical Mathematics, Russian Academy of Sciences

Email: a.a.danilov@gmail.com

Abstract. Here, a workflow for high-resolution efficient numerical modeling of bioimpedance measurements is suggested that includes 3D image segmentation, adaptive mesh generation, finite-element discretization, and the analysis of simulation results. Using the adaptive unstructured tetrahedral meshes enables to decrease significantly a number of mesh elements while keeping model accuracy. The numerical results illustrate current, potential, and sensitivity field distributions for a conventional Kubicek-like scheme of bioimpedance measurements using segmented geometric model of human torso based on Visible Human Project data. The whole body VHP man computational mesh is constructed that contains 574 thousand vertices and 3.3 million tetrahedrons.

1. Introduction

At present, computerized models of real human anatomy are widely used in various research areas such as radiography, nuclear medicine, radiation protection and other [1, 2]. The same is applied to impedance simulation studies (see, e.g., [3, 4]). Modeling of sensitivity distributions for various measurement schemes requires solving a number of well-defined computational problems using high-resolution anatomically accurate 3D models. Our aim was to describe mesh generation and computational modeling techniques for bioimpedance measurements based on high-resolution anatomical models of the human body. Using finite-element models on tetrahedral meshes, unlike conventional finite-difference schemes on cubic meshes, allows taking into account tissue anisotropy thus providing more realistic description of electric properties of the human body.

2. Segmentation and mesh generation

The original 3D image of the human torso was derived from the Visible Human Project data as a stack of 843 567×305 color images with resolution $1\times1\times1$ mm [5]. We started from a partially segmented data of a human torso from the Voxel-Man project [6]. These data were primarily aimed at visualization purposes and contained a significant amount of unclassified tissue. From a computational point of view, void spaces between tissues were not allowed in a segmented model. So, a further processing of the segmented data was undertaken.

Most of the segmentation process was performed manually using ITK-SNAP segmentation software program [7]. In some cases semi-automatic threshold-based approach was used for specific tissues like fat, bones and blood vessels. At the final stage we used semi-automatic post-processing algorithms for filling remaining gaps between tissues and final segmented data smoothing. These algorithms are based on mathematical morphology operations: erosion, dilation, opening and closing, connected-component labeling and Gaussian smoothing. Final steps are performed using Convert3D tool from ITK-SNAP package.

The geometrical model obtained from the image segmentation was used to create an unstructured tetrahedral mesh. For this, several approaches can be used, including marching cubes algorithm for surface reconstruction [8], surface triangulation smoothing and coarsening [9], 3D Delaunay triangulation, and advancing front technique for volume mesh generation [10]. Of these, we used the Delaunay triangulation algorithm from CGAL-Mesh library [11]. This algorithm permits defining a specific mesh size for each tissue and organ. So, we used a small mesh size for blood vessels and other organs and tissues with the similar geometrical features. A large mesh size was prescribed for gross tissues like fat or muscles. This approach allowed us to create computational meshes with a reasonably small number of elements still preserving the local geometric features of the segmented model.

The segmented model containing 26 labels describing major organs and tissues of the human torso and the generated mesh with 413.508 vertices and 2.315.329 tetrahedrons are shown in Fig. 1. This mesh represents most anatomical features of the human torso.

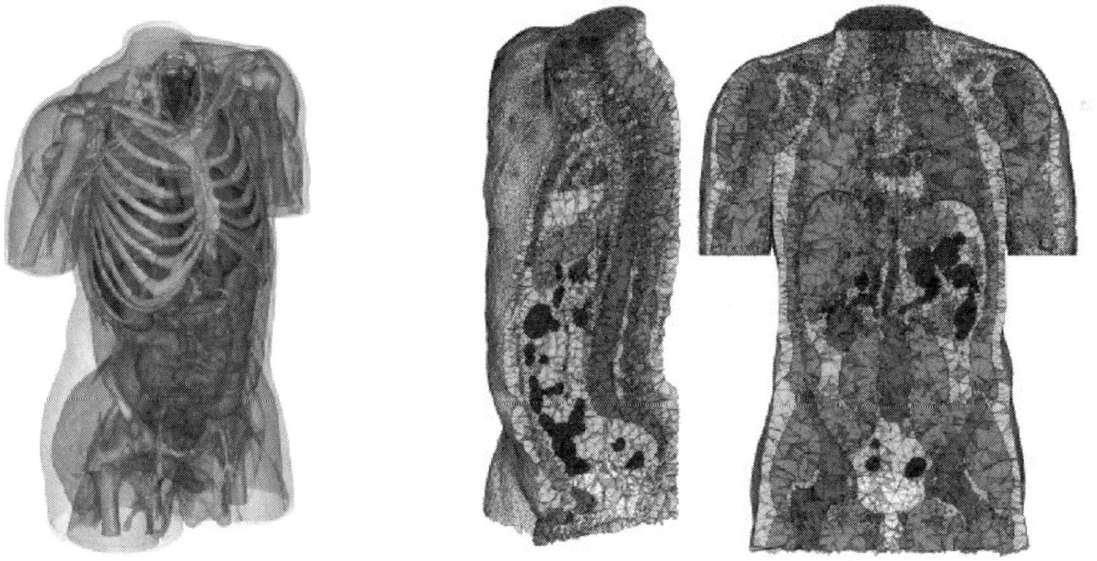

Fig. 1. Geometrical model of the segmented image (left) and unstructured tetrahedral meshes (right)

After mesh generation, we added a skin layer and multilayered electrodes to the surface of the constructed mesh. Boundary triangulation was used to create a prismatic mesh on the surface, and then each prism was split into three tetrahedrons resulting in a conformal mesh. Mesh cosmetics algorithms from the Ani3D library [12] were used to improve mesh quality. This essential step reduces discretization errors and condition number of the resulted systems of linear equations.

3. Mathematical model

The electrical fields generated during bioimpedance measurements are governed by the equation

$$\mathrm{div}(\mathbf{C}\nabla U) = 0 \quad \text{in } \Omega \qquad (1)$$

with the boundary conditions

$$(\mathbf{J},\mathbf{n}) = I_0 / S_\pm \quad \text{on } \Gamma_\pm \qquad (2)$$

$$(\mathbf{J},\mathbf{n}) = 0 \quad \text{on } \partial\Omega \backslash \Gamma_\pm \qquad (3)$$

$$U(x_0, y_0, z_0) = 0 \qquad (4)$$

$$\mathbf{J} = \mathbf{C}\nabla U \qquad (5)$$

where Ω – the computational domain, $\partial\Omega$ – its boundary, $\Gamma_{\pm}$ – electrode contact surfaces, **n** – external unit normal vector, U – electric potential, **C** – conductivity tensor, **J** – current density, I_0 – electric current, $S_{\pm}$ – area of the electrode contacts. Equation (1) determines the distribution of electric field in the domain with heterogeneous conductivity **C**. Equation (2) sets a constant current density on the electrode contact surfaces. Equation (3) defines the no-flow condition on the boundary. Uniqueness of the solution is guaranteed by equation (4), where (x_0, y_0, z_0) – some point of the domain Ω.

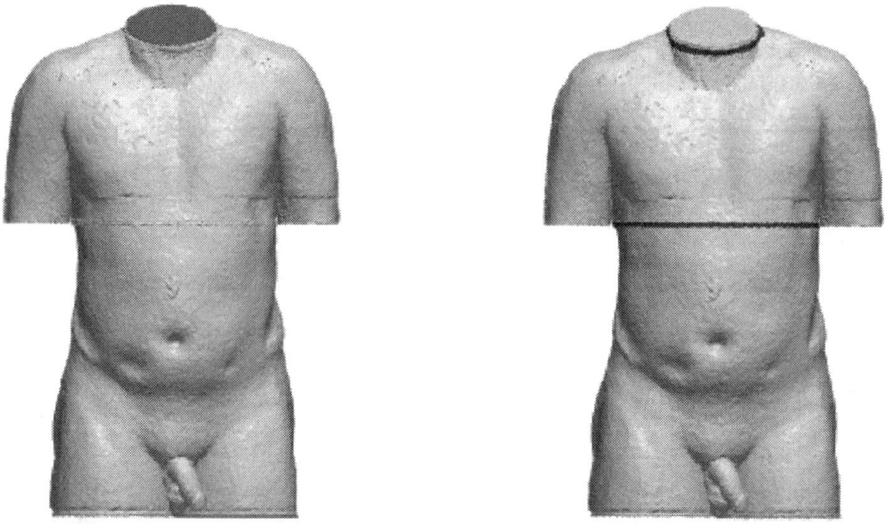

Fig. 2. Electrode configuration for Kubicek-like measurement scheme: current electrodes in red, voltage electrodes in blue

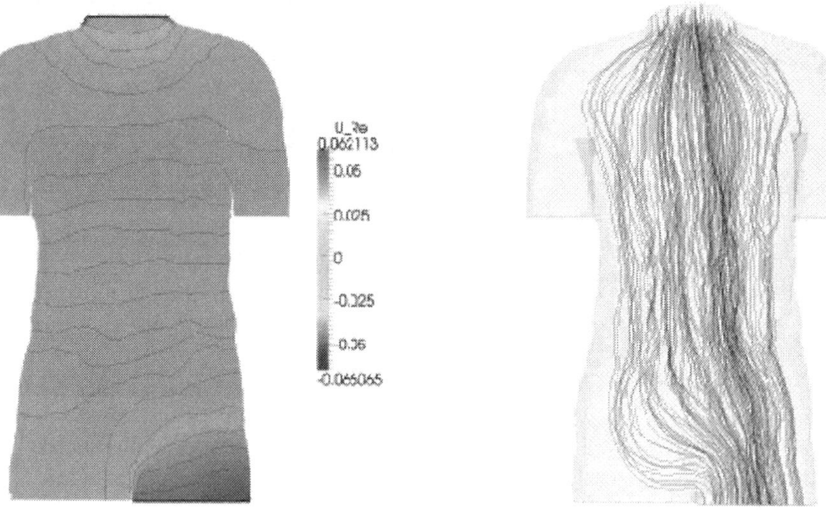

Fig. 3. Cutplane of the potential field and the respective current lines. Measurement frequency 50 kHz

Discretization of the equations (1)–(5) was obtained using finite element scheme with piecewise linear elements P1 on unstructured tetrahedral meshes. The resulting system of linear equations was solved by GMRES method with second-order ILU preconditioner [13]. Several numerical experiments were performed on simplified geometrical models for a convergence study of the proposed scheme.

These tests demonstrated first-order convergence on a series of unmatched meshes and nearly second order convergence on a series of nested meshes (data not shown).

4. Numerical results

Let us consider the Kubicek-like electrode configuration scheme [14] with a pair of current electrodes on the head and the left leg, and a pair of band voltage electrodes as depicted in Fig. 2. Since our calculations were performed on a torso segmented model, we applied the boundary conditions at the appropriate interfaces.

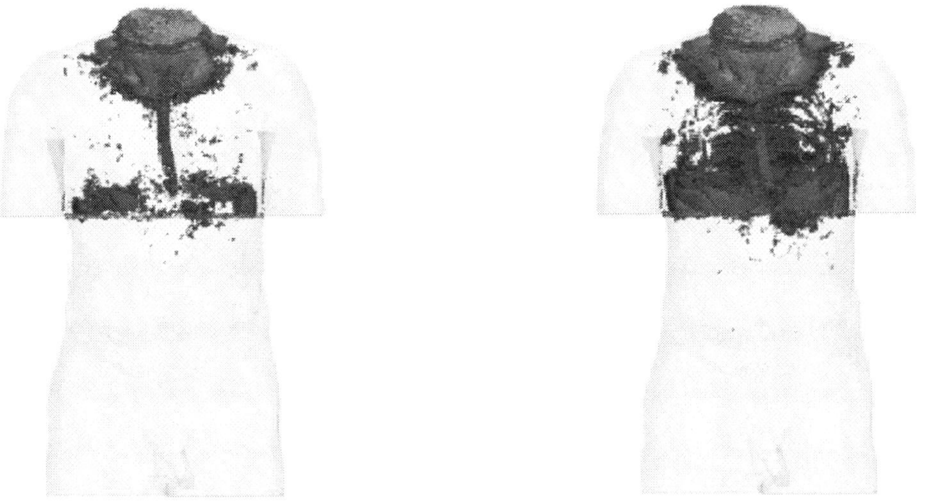

Fig. 4. Areas of high sensitivity, threshold 50% (left) and 70% (right). Measurement frequency 5 kHz

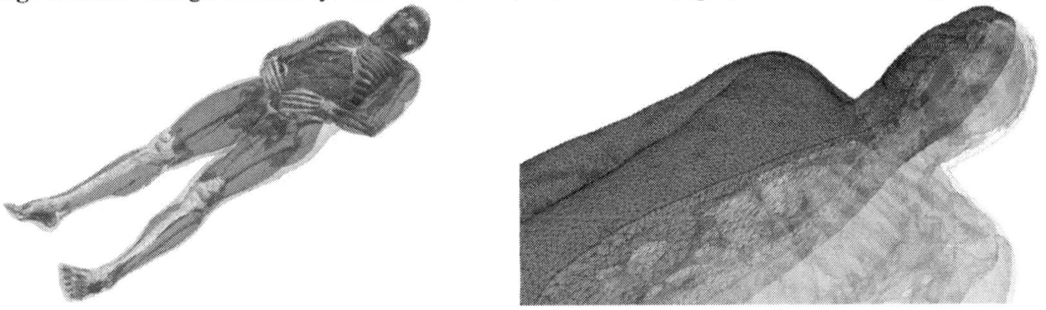

a *b*

Fig. 5. Segmented whole body model of the Visible Human Man (a) and a part of related generated mesh (b)

We defined the conductivity parameters for labeled tissues as in [15]. The simulations were performed for current frequencies 5 and 50 kHz. The computed current and potential field distributions are shown in Fig. 3. The results of sensitivity analysis are presented in Fig. 4 and Table 1 show good quantitative agreement with that of sensitivity distribution analysis made by Kauppinen et al. [3].

We used the proposed techniques to construct the computational mesh for the whole body model based on VHP data. The related segmented model containing 32 materials and generated mesh containing 574.128 vertices and 3.300.481 tetrahedrons are shown in Fig. 5.

Table 1. Relative contribution (%) of organs and tissues to the impedance signal

Organ/tissue	Measurement frequency		Organ/tissue	Measurement frequency	
	5 kHz	50 kHz		5 kHz	50 kHz
Muscles	72.56	69.44	Heart	0.64	1.01
Left lung	5.23	5.49	Bones	0.57	0.51
Right lung	4.59	5.03	Pancreas	0.47	0.45
Diaphragm	2.57	2.39	Liver	0.28	0.50
Oesophagus	2.27	2.00	Skin	0.23	0.22
Stomach	2.19	2.14	Pulmonary trunk	0.05	0.11
Thyroid gland	1.61	1.50	Pulmonary veins	0.15	0.28
Visceral fat	1.25	0.54	Pulmonary arteries	0.15	0.28
Intestine	0.91	1.12	Other arteries	0.75	1.50
Trachea	0.68	0.71	Other veins	2.21	4.18
			Other tissues	0.64	0.60

5. Conclusion

In this work an unstructured mesh generation and numerical modeling technology for bioimpedance measurements is suggested. Possible applications include an analysis of sensitivity distributions for conventional (e.g., local or polysegmental) measurement schemes, as well as the development of new ones for specific purposes, optimization of electrode properties, their placement configurations and frequency spectra. Our approach enables mesh generation for high-resolution models with a significantly less number of mesh elements as compared to rectangular meshes thus reducing the computational costs.

Acknowledgements

This work was partially supported by the Russian Foundation for Basic Research (grants no. 11-01-00971, 10-01-91055), and the Federal Program "Academic and pedagogical staff of innovative Russia".

References

[1] Caon M 2004 *Radiat. Environ. Biophys.* **42** 229-235
[2] Xu X G, Eckerman K F 2009 Handbook of anatomical models for radiation dosimetry (Boca Raton: CRC Press)
[3] Kauppinen P K, Hyttinen J A, Malmivuo J A 1998 *Ann. Biomed. Eng.* **26** 694-702
[4] Yang F, Patterson R 2010 in: 32nd Annual Int. Conf. of the IEEE EMBS. Buenos-Aires, Argentina, August 31- Sept. 4
[5] The Visible Human Project http://www.nlm.nih.gov/research/visible/visible_human.html
[6] Höhne K H, Pflesser B, Pommert A et al. 2001 *Meth. Inform. Med.* **40** 83-89
[7] Yushkevich P A, Piven J, Hazlett H C et al. 2006 *Neuroimage* **31** 1116-1128
[8] Wu Z, Sullivan J M 2003 *Int. J. Numer. Meth. Eng.* **58** 189-207
[9] Taubin G 1995 in: Proc. of the 22nd annual conference on computer graphics and interactive techniques (N.Y.: ACM) 351-358
[10] Frey P J, George P L 2000 Mesh generation: application to finite elements (Paris, Oxford: Hermes Science)
[11] Rineau L, Yvinec M 2007 *Comput. Geom. Theory Appl.* **38** 100-110
[12] 3D generator of anisotropic meshes http://sourceforge.net/projects/ani3d
[13] Kaporin I E 1998 *Numer. Linear Algebra Appl.* **5** 483-509
[14] Kubicek W G 1967 Pat. USA. №3.340.867 A61b.5/02 Sept.12
[15] Gabriel C, Peyman A, Grant E 2009 *Phys. Med. Biol.* **54** 4863-4878

Body composition in remission of childhood cancer

G Ja Tseytlin[1], A V Anisimova[2], E Z Godina[2], I A Khomyakova[2], M V Konovalova[1], D V Nikolaev[3], S G Rudnev[4], O A Starunova[3], A Yu Vashura[1]

[1]Federal Centre of Pediatric Hematology, Oncology and Immunology
[2]Institute and Museum of Anthropology, Moscow State University
[3]Scientific Research Centre 'Medas'
[4]Institute of Numerical Mathematics, Russian Academy of Sciences
all – Moscow, Russian Federation

Email: sergey.rudnev@gmail.com

Abstract. Here, we describe the results of a cross-sectional bioimpedance study of body composition in 552 Russian children and adolescents aged 7-17 years in remission of various types of cancer (remission time 0-15 years, median 4 years). A sample of 1500 apparently healthy individuals of the same age interval was used for comparison. Our data show high frequency of malnutrition in total cancer patients group depending on type of cancer. 52.7% of patients were malnourished according to phase angle and percentage fat mass z-score with the range between 42.2% in children with solid tumors located outside CNS and 76.8% in children with CNS tumors. The body mass index failed to identify the proportion of patients with malnutrition and showed diagnostic sensitivity 50.6% for obesity on the basis of high percentage body fat and even much less so for undernutrition – 13.4% as judged by low phase angle. Our results suggest an advantage of using phase angle as the most sensitive bioimpedance indicator for the assessment of metabolic alterations, associated risks, and the effectiveness of rehabilitation strategies in childhood cancer patients.

1. Introduction

An observed dramatic increase in cancer survival rates [1] have lead to the necessity of the prognosis and control of late adverse effects of treatment and quality of life in such patients during the rehabilitation period [2]. In adults, cancer development is not infrequently accompanied with a syndrome of progressive weight loss, or cachexia. Cachexia is more common in people with lung cancer (the most frequent form of malignant tumor in Russia) as well as in pancreatic and gastrointestinal cancers, and, in contrast to semistarvation, is accompanied with a pronounced depletion not only in fat mass, but also in fat-free mass [3]. Unlike this, the body mass in childhood cancer patients follows usually a normal pattern of change at the background of some growth delay, whereas in the structure of cancer morbidity in children, leukemia represents the most prevalent form [4]. Our previous study of a sample of 220 children in remission of acute lymphoblastic leukemia showed significant alterations in body composition parameters in this group relative to the same number of age- and sex-matched healthy controls [5]. Now we aim at the comparison of body composition and nutritional status in children and adolescents in remission of various cancer types, as well as at the identification and ranking of significant bioimpedance variables of patients' state.

2. Subjects and methods

552 children and adolescents (313 males and 239 females aged 7-17 years) after treatment for cancer comprised a patients' group and were assessed cross-sectionally in 2008-2011 at the Russian Field sanatorium (Chekhov area, Moscow region) by the staff members of rehabilitation department of the Federal Centre of Pediatric Hematology, Oncology and Immunology. All patients were in first remission, and none had received any hormone replacement therapy. Remission time ranged from 0 to 15 years. The patients' group was subdivided into five subgroups (denoted A to E) according to diagnosis: 64 were cured of solid (except for brain/CNS) tumors (subgroup A), 86 – of lymphomas and malignant histiocytosis (B), 320 – of acute lymphoblastic leukemia (C), 26 – of non-lymphoblastic leukemia (D), and 56 – of CNS tumors (E). In addition, 1500 apparently healthy children and adolescents (837 males and 663 females) of the same age range were measured representing a control group. The subjects were measured at schools in Moscow, Arkhangelsk, and Arkhangelsk region by the team of trained anthropologists from Moscow University.

Standing height was measured using a stadiometer in cancer patients and the GPM anthropometer in controls, respectively. Weight was measured on a digital scale in both groups. Body mass index (BMI) was calculated as body mass (BM) divided by standing height (Ht) squared.

The whole-body impedance was measured on the right hand side of the body using the bioimpedance meter ABC-01 'Medas' (SRC Medas, Russia) according to a conventional tetrapolar scheme at a frequency 50 kHz. Phase angle (PA) was calculated as $\arctan(X_C/R)\times180°/\pi$, where X_C is the reactance and R the whole-body electric resistance. Fat-free mass (FFM) was assessed using Houtkooper equation [6]: $FFM=0.61\times(Ht^2/R)+0.25\times BM+1.31$, where Ht is measured in cm. Fat mass (FM) was calculated as the difference between BM and FFM, and %FM as (FM/BM)×100. Other body composition variables, such as body cell mass (BCM) and skeletal-muscle mass (SMM), were determined by analogy using appropriate regression formulae provided by the manufacturer. Similarly to BMI, fat mass index (FMi) and fat-free mass index (FFMi) were calculated as FM/Ht^2 and FFM/Ht^2, respectively.

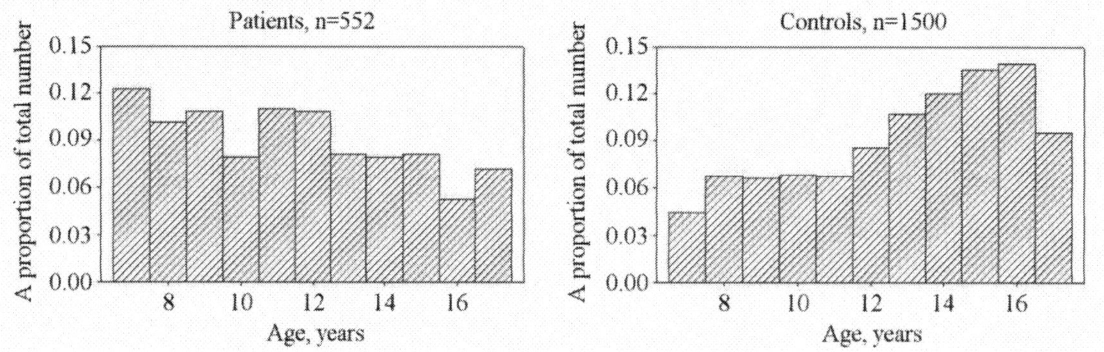

Fig. 1. Distributions of patients and controls according to age

Because of significant difference in age distributions of our patient and control groups (see Fig. 1 and Table 1), the data on healthy controls were first normalized using LMS method, and then the patients' data were expressed as z-scores according to the formula $z=[(y/\mu)^{\nu}-1]/(\sigma\nu)$ if $\nu\neq0$, and $z=\log(y/\nu)/\sigma$ if $\nu=0$, where y is the measured value of a parameter, and μ, σ and ν are, respectively, the mean, dispersion, and asymmetry of the probability distribution function for a subgroup of age-related healthy controls of the same sex. Finally, the data were drawn against the smoothing reference percentile curves representing the control group (for methodology, see [7,8]). For this, a recently developed by the one of authors (*OAS*) software program BIAStatistica [9] was used. The 85th and 95th percentiles for %FM and BMI, as well as the 5th percentiles for phase angle [10,11] and BMI in the control group, were used as a cut-offs to define malnutrition.

Two-sample Mann-Whitney rank test of the equality of two population medians was used. A *p* value of 0.05 was used to define significance.

3. Results and discussion

Basic anthropometric characteristics of cancer patients are shown in Table 1. Neither body mass, no BMI of total patients' group differ significantly from that of the control group, although in the subgroups B and D the body mass showed significant trends of opposite directions. Body height was moderately but statistically significantly decreased in patients, primarily due to pronounced stunting observed in the subgroup E (see Table 1).

Table 1. Basic anthropometric characteristics of cancer patients, mean (s.d.)

Parameter	Cancer patients (n=552)	Patient subgroups				
		A (n=64)	B (n=86)	C (n=320)	D (n=26)	E (n=56)
Age, yrs	11.4 (3.1)*	11.1 (3.2)^{*bde}	12.6 (3.3)^{ac}	11.0 (3.0)^{*bde}	12.5 (3.3)^{ac}	12.3(2.8)^{ac}
Remission, yrs	4.3 (2.5)	4.6 (2.7)^{be}	3.6 (2.4)^{ac}	4.6 (2.4)^{be}	3.9 (2.0)	3.8 (2.7)^{ac}
BM, z-score	-0.1 (1.2)	-0.2 (1.1)^{d}	-0.3 (1.3)^{*cd}	0.1 (1.2)^{b}	0.3 (1.2)^{*abe}	-0.5 (1.4)^{d}
Ht, z-score	-0.2 (1.2)*	-0.1 (1.0)	-0.3 (1.2)*	-0.1 (1.1)	-0.2 (1.2)	-0.7 (1.4)*
BMI, z-score	0.0 (1.2)	-0.2 (1.2)^{d}	-0.2 (1.3)^{d}	0.1 (1.1)	0.4 (1.3)^{ab}	-0.1 (1.2)

* significant difference compared to healthy controls
$^{a\text{-}e}$ significant difference compared to patient subgroups A-E, respectively

Table 2. Bioimpedance parameters and body composition estimates in cancer patients, mean (s.d.)

Parameter, z-score	Cancer patients (n=552)	Patient subgroups				
		A (n=64)	B (n=86)	C (n=320)	D (n=26)	E (n=56)
R	0.6 (1.2)*	0.5 (1.2)^{e}	0.7 (1.1)^{ce}	0.4 (1.2)^{be}	0.5 (1.1)^{e}	1.3 (1.3)^{**abcd}
X_C	-0.5 (1.2)*	-0.4 (1.0)^{c}	-0.4 (1.2)^{c}	-0.7 (1.1)^{abe}	-0.4 (1.2)	-0.2 (1.1)^{**c}
PA	-1.3 (1.3)*	-1.1 (1.3)^{ce}	-1.2 (1.2)	-1.3 (1.2)^{a}	-1.1 (1.3)	-1.8 (1.2)^{**a}
FM	0.3 (1.1)*	0.0 (1.2)^{**cde}	0.2 (1.1)^{d}	0.4 (1.0)^{a}	0.8 (1.0)^{**ab}	0.4 (1.0)^{a}
%FM	0.5 (1.1)*	0.1 (1.2)^{**de}	0.4 (1.1)^{de}	0.5 (1.0)^{e}	1.0 (1.0)^{ab}	0.9 (1.0)^{**abc}
FMi	0.4 (1.0)*	0.0 (1.2)^{**cde}	0.2 (1.1)^{de}	0.4 (1.0)^{a}	0.8 (1.1)^{ab}	0.6 (1.0)^{ab}
FFM	-0.3 (1.3)*	-0.3 (1.0)^{a}	-0.5 (1.2)^{ce}	-0.2 (1.2)^{be}	-0.1 (1.2)^{e}	-1.1 (1.6)^{**abcd}
FFMi	-0.4 (1.2)*	-0.4 (1.2)	-0.6 (1.1)^{c}	-0.2 (1.2)^{be}	-0.1 (1.4)^{e}	-0.9 (1.3)^{**acd}
SMM	-0.4 (1.2)*	-0.3 (0.9)^{e}	-0.6 (1.1)^{ce}	-0.3 (1.2)^{be}	-0.4 (1.1)^{e}	-1.2 (1.6)^{**abcd}
%SMM	-0.2 (1.1)*	0.0 (1.1)^{e}	-0.2 (1.1)^{e}	-0.2 (1.0)^{e}	-0.6 (1.4)	-0.7 (1.3)^{**abc}
BCM	-0.7 (1.3)*	-0.6 (1.2)^{e}	-0.9 (1.2)^{ce}	-0.6 (1.3)^{be}	-0.5 (1.1)^{e}	-1.5 (1.6)^{**abcd}
TBW	-0.3 (1.3)*	-0.3 (1.0)^{e}	-0.5 (1.2)^{ce}	-0.2 (1.1)^{be}	-0.1 (1.2)^{e}	-1.1 (1.6)^{**abcd}
ECF	-0.7 (0.7)*	-0.7 (0.6)^{e}	-0.9(0.7)^{**d}	-0.7 (0.7)^{e}	-0.7 (0.5)^{b}	-1.1 (0.8)^{**ac}

* significant difference compared to healthy controls
** significant difference compared to total patients group
$^{a\text{-}e}$ significant difference compared to patient subgroups A-E, respectively

In contrast to anthropometric data, all the parameters of bioimpedance and body composition differed significantly in the two groups (see Table 2). Importantly, the most pronounced changes occurred in the values of phase angle which is directly measurable and, hence, is independent of any assumptions accompanying BIA predictions of body composition. As opposed to Table 1, the parameters present in Table 2 showed unidirectional trends and changed only in amplitude. The subgroup E had the largest, and the subgroup A the least, extent of the parameters change (and, hence, metabolic alterations) compared to healthy controls. In the subgroups A and E, the mean values of 10 out of 13 variables presented in Table 2 were significantly different.

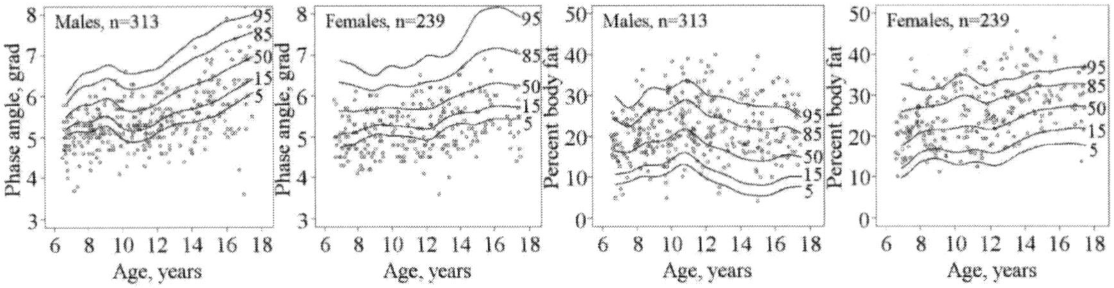

Fig. 2. Scatter plots of phase angle and percentage body fat across age in cancer patients. The smoothing reference percentile curves representing the control group are shown as continuous lines

Fig. 2 shows scatter plots of PA and %FM in cancer patients across age against the reference percentile curves for our healthy controls. It can be seen that a significant proportion of male and female patients have PA values below the 5th percentile of the reference group thus reflecting a pronounced undernutrition state. Similarly, a considerable number of patients had %FM values above the 85th and 95th percentiles regarding them as having overweight or obesity. The data on prevalence of malnutrition in the two groups according to various diagnostic criteria are summarized in Table 3.

Table 3. Prevalence of malnutrition in patients and controls (%) according to various diagnostic criteria

Malnutrition sign	Control group (n=1500)	Patients' group (n=552)	Patient subgroups A (n=64)	B (n=86)	C (n=320)	D (n=26)	E (n=56)
Overweight by %FM ($1.04<Z_{\%FM}<1.65$)	11.3	16.5	12.5	12.8	16.9	23.1	21.4
Overweight by BMI ($1.04<Z_{BMI}<1.65$)	9.8	10.6	7.8	5.8	12.8	7.7	10.7
Obesity by %FM ($Z_{\%FM}>1.65$)	5	15.4	10.9	12.8	13.8	26.9	28.6
Obesity by BMI ($Z_{BMI}>1.65$)	5.4	7.8	6.3	7.0	7.2	23.1	7.1
Normal weight obesity ($Z_{BMI}<1.04$; $Z_{\%FM}>1.65$)	0.3	4.2	1.6	3.5	3.1	3.8	14.3
Undernutrition by PA ($Z_{PA}<-1.65$)	5.1	43.5	34.4	36.0	44.4	38.5	62.5
Undernutrition by BMI ($Z_{BMI}<-1.65$)	4.7	8.2	10.9	11.6	6.6	3.8	10.7

Overweight and obesity judged from %FM occurred in 31.9% of cancer patients, twice as much as compared to healthy controls (16.3%). In spite of significant correlation between BMI and %FM in patients ($r=0.73$, $p<0.05$), the body mass index failed to detect this marked distinction between the study groups (see Table 3) and showed low diagnostic specificity for obesity (i.e. the ability to correctly identify those with the disease) at the level of 50.6% compared to bioelectric impedance analysis. This is due to the fact that an increase in body fat in patients was accompanied by a proportional decrease in fat-free mass so that BMI values were not affected significantly even in individuals with high percentage fat mass content (see Table 2). The prevalence of normal weight obesity was uncommonly sizeable, reaching a maximum of 14.3% in the subgroup E. Like the usual obesity, normal weight obesity is associated with an increased risk of metabolic syndrome and cardiovascular pathology [12].

On the basis of phase angle, reflecting a proportion of metabolically active body cell mass in fat-free mass [13], 43.5% of patients were considered malnourished (see Table 3). This number is close to the result of Murphy et al [14] who reported a proportion of malnourished of 45% among children being treated for cancer and blood-related disorders based on BCM estimates using total body potassium counting. From this, one can speculate that malnutrition, if tolerated, can persist for years after treatment of cancer thus representing a matter of concern. On the basis of BMI, only 8.2% of cancer patients were defined as undernourished (see Table 3). The sensitivity of BMI for undernutrition was only 13.4% as judged by low phase angle.

Based on meeting at least one of the following two malnutrition criteria, namely high percentage fat and low phase angle, 52.7% of subjects in the total patient group were considered malnourished with the range between 42.2% (subgroup A) and 76.8% (subgroup E of patients with CNS tumors). Because of known association of malnutrition with the reduced tolerance to chemotherapy, increased susceptibility to infections and unfavorable outcomes [15,16], it is practical to use bioimpedance analysis, as a safe, rapid, inexpensive and portable method, for monitoring of nutritional state in pediatric oncology patients. A schedule for such an assessment containing a prescription of monthly body composition screening for nutritional state and alterations with a subsequent adaptation of substrate intake according to current requirements was proposed recently in [17].

Taken together, our data show significant level of malnutrition in children and adolescents in remission of cancer depending on the type of malignancy. The results suggest an advantage of using phase angle as the most sensitive bioimpedance indicator for the assessment of metabolic alterations, associated risks, and effectiveness of rehabilitation strategies in childhood cancer patients. For adverse late effects prophylaxis, additional tracking and control of %FM could also be useful.

References

[1] Birch J M *et al* 1988 *Br. Med. J.* **296** 1372-1376
[2] Barr R D *et al* 1993 *Am. J. Pediatr. Hematol. Oncol.* **15** 284-290
[3] Tisdale M J 2003 *J. Support. Oncol.* **1** 159-168
[4] Murphy A J *et al* 2006 *Am. J. Clin. Nutr.* **83** 70-74
[5] Tseytlin G Ja *et al* 2010 *J. Phys.: Conf. Series* **224** 012116
[6] Houtkooper L B *et al* 1992 *J. Appl. Physiol.* **72** 366-373
[7] Cole T J and Green P J 1992 *Statistics in Medicine* **11** 1305-1319
[8] Stasinopoulos D M and Rigby R A 2007 *J. Stat. Software* **23** http://www.jstatsoft.org/v23/i07
[9] Starunova O A 2011 *A collection of papers of young scientists of Lomonosov Moscow State University CMC department* **8** 129-134
[10] Gupta D *et al* 2008 *BMC Cancer* **8** 249
[11] Santarpia L *et al* 2009 *Nutrition* **25** 930-931
[12] Romero-Corral A *et al* 2010 *Eur. Heart J.* **31** 737-746
[13] Lautz H U *et al* 1992 *Clin. Invest.* **70** 478-486
[14] Murphy A J, White M and Davies P S W 2010 *Am. J. Clin. Nutr.* **92** 55-60
[15] Ladas E J, Sacks N, Brophy P and Rogers P C 2006 *Pediatr. Blood Cancer* **46** 339-344
[16] Paiva S I *et al* 2011 *Support. Care Cancer* **19** 187-192
[17] Bauer J, Jurgens H and Fruhwald M C 2011 *Adv. Nutr.* **2** 67-77

First Latin-American Conference on Bioimpedance (CLABIO 2012) IOP Publishing
Journal of Physics: Conference Series **407** (2012) 012006 doi:10.1088/1742-6596/407/1/012006

Influence of regularization in image reconstruction in electrical impedance tomography

J L L Queiroz

Adjuntc Professor at State University of Santa Catarina

Email: kairos@joinville.udesc.br

Abstract. The purpose of the application of electrical impedance tomography is to obtain images of areas that are difficult to access inside of the chest and other parts of the human being. Today, this has been applied in other areas of engineering with a view to investigate phenomena for which it is difficult to obtain data for robust research. Electrical Impedance Tomography (EIT) of an inverse problem is nonlinear and ill-conditioned. This requires a careful theoretical approach and practice to get good images. To enhance the images, it is important to be sensitive to various parameters that influence the process of image reconstruction, such as the measured voltage and the current density injected into the electrodes. The impedance contact and current density are both high in a point electrode. To reduce this, a large electrode modeled with a Finite Element Method (FEM) is used. A reduced numbers iterations is found when larges electrodes are used. When FEM models are used the performance of the electrical impedance tomography reconstruction algorithm can be improved.

1. Introduction

The electric impedance tomography (EIT) techniques used in images reconstruction are divided into two parts: first the forward problem obtains a measured voltage from injecting a set of standard currents and a known object's internal conductivity (or resistivity) distribution. Through equations for electric potential and a Finite Element Method (FEM), we obtain the first estimate of the spatial variation of the conductivity within the section for those standard currents injected at the border. Then, we solve the inverse problem. The same patterns of currents are injected again into the border electrodes, using the finite element method (FEM) and the Newton-Raphson method to find the calculated voltages and compare them with the voltages measured with a desired tolerance.

EIT has numerous applications that can be categorized in three main fields: 1. Industrial. This application includes the image of fluid flow in pipeslines, measures the distribution of fluid flow in vessels of mixtures and non-destructive testing, such as the arrest of cracks (Cheney et al, 1999; Dickin et al, 1996). 2. Geophysics. These applications are geophysical, measured through the holes and surfaces (Polydorides and Lionheart, 2002). 3. Medicine. EIT is used to monitor heart and lung functions, measures of brain function (Barber 1984, Barber and Brawn 1989), detection of hemorrhage, measurement of the digestive system, detection and classification of tumors in breast tissue and functional imaging of the chest (Eyugoblu et al, 1989 by Harris et al, 1992).

EIT has some limitations that should be circumvented in its adoption in routine medical diagnostics. Its major limitation is its low spatial resolution, noise and susceptibility to errors in the electrodes. In medical applications, there is a large diversity of image characteristics. EIT is unsuitable for the anatomical images in the same manner as magnetic resonance imaging (MRI) or computed tomography (CT). However, EIT appears promising as a tool for clinical diagnosis. It has an advantage of being relatively inexpensive compared with CT and MRI. EIT is non-invasive, safe, small and easy to carry. It can be used for long periods with low energy cost and maintenance. It may be feasible in the use of continuous monitoring of air entering the lungs also avoiding risks to the patient. It can also be used in monitor gastric lavage or cerebral hemorrhage.

2. Tomography Applied in Industrial Process

The term tomography applied in industrial process (TAPI) refers to a wide range of visualization techniques that are non-invasive relatively new (since the late 80s) and still in development. The goal is to get the TAPI images of cross sections of dynamic industrial processes (Adler, 1996 and Adler, 2004; Asfaw and Adler, 2005). Tomography offers a new way of viewing the internal behavior of industrial processes. The image of the cross section produced by tomography provides an efficient process of information that can be used for viewing, monitoring, mathematical modeling and verification of possible controls. There are many forms of tomography systems such as electricity, ultrasound, radiation, magnetic resonance, microwave and optical.

The tomography methods are still under development. Some challenges remain to be focused on: improving the sensitivity of the spatial resolution of the measures, development of more precise methods of image reconstruction, improving the efficiency of data processing, and suitability designing safe and reliable electronic hardware for use in industrial environments (not just in the lab).

EIT is an ill-conditioned problem and to finds stable results if the solution incorporates some form of basic knowledge in problem behavior. In this regard, several requirements must be met to obtain the accuracy of specific images and for direct generation of models that describe more precisely the trajectory made by photon tissue models and precisely recover the inverse images of the optical properties of the tissues (Klose , 2003). Using Newton's optimization system , through the Jacobian (which combines measures of data with optical properties), these algorithms are widely used because they are easy to implement and can easily be generalized for use as the basis for simple and complex image reconstruction. (Dehghani et al, 2003, Yalavarthy, 2007). Another alternative method to reduce the number of parameters includes the use of image reconstruction base (Paulsen 1996) and also the use of mixed algorithms adapted to the environment of specific interest to be refined to higher precision numbers.

Table1. Research applications of fluid.

Applications	**Phases**	**References**
Flow in ducts	Solid -Liquid	Steverson et al, 2006; Dai, 2004; Norman et al, 2005; Pullum et al, 2006.
	Gas - Liquid	Dong et al, 2006; Wang et al, 2006.
	Liquid - Liquid	Henningson et al, 2006; Henningson et al, 2007.
Stirred tanks	Liquid – Liquid	Ricard et al, 2005; Kim et al, 2006.
	Gas –Liquid	Ricard et al, 2005; Stanley et al, 2005.
Packed bed reactor	Solid – Liquid	Bolton et al, 2004. Vijayan et al, 2007; Fransolet et al, 2001; Toye et al, 2005; Fransolet et al, 2005.

2.1. Multiphase Fluid

Industrial applications of industrial tomography with fluids are common for the different phases (liquid, gaseous or liquid particulates), and also for multiphase fluids (when different phases exist in the same fluid of an industrial process). These states are detected by different concentrations of each phase computed based on knowledge of the electrical conductivity of each phase in this case using the electric resistance tomography (ERT) which can be applied to various procedures involving the conductivity of fluids in several processes that involve conductive fluid in the continuous phase. Typical applications are the ERT to view pipes and multiphase fluids transmission in stirred tanks, for which there is commercial applications for a wide range of material types. Recent applications can be seen in Table 1 (Gigu `ere, R., et al., 2008).

3. Models

The model is characterized in two parts. The direct problem and inverse problem are presented below.

3.1. Direct Problem

If $\boldsymbol{\Omega}$ is the area subject to the presence of an electric field (Cheney et al., 1999)

$$\nabla\gamma(x,\omega)\nabla\boldsymbol{u} = 0 \tag{1}$$

Where **x** is a point inside the region $\boldsymbol{\Omega}$, and $\boldsymbol{u}$ is the electric potential and $\gamma(x,\omega)$ is the admittivity, if electric currents are injected into the boundary surface $\partial\boldsymbol{\Omega}$ then power flow density **J** is

$$\gamma\frac{\partial u}{\partial\vartheta} = J \text{ em } \partial\Omega_l \qquad l = 1,2,..L \text{ (electrode number)} \tag{2}$$

Equations (1) and (2) with the condition $\mathbf{J} = 0$ condition (charge conservation) and $\boldsymbol{u} = 0$ (reference voltage), which consists of a current pattern in electrodes is known as solid model. The total flow of current through the electrodes is equal to the integral of the current density, ie:

$$\oint_{\partial\Omega_l}\gamma\frac{\partial u}{\partial\vartheta}\,ds = V_l, \quad l = 1, 2 \ldots L \tag{3}$$

The elementary matrix equation is (Vanegas, 2002):

$$[y_c] = \iint_A \frac{1}{4A^2}[B]^t[D][B]dA \tag{4}$$

Where y_c is the local conductivity matrix, t is the thickness, B is the geometry matrix, D is the properties matrix, A is the area of each element. To calculate the total internal voltages, the global matrix is used.

$$[Y_c]*[V_i] = [C_l]\text{ . } i = 1,2\ldots n \text{ (node number)} \tag{5}$$

Where $[Y]$ is the local conductivity matrix, and $[Vi]$ is the vector of voltage at the points of the finite element mesh. And, $[Ci]$ is the pattern of current applied to the electrodes.

$$[V_i] = [C_l]*[Y_c^{-1}] \tag{6}$$

which $[Y_c^{-1}]$ is the inverse global matrix. The conductivity is a current and voltage function. In the direct problem, a set of current patterns are injected into the region and a distribution of conductivity (impedance) and voltages on the electrodes are measured.

The next step is solve the inverse problem.

3.2 The Inverse Problem

The aim of the process of image reconstruction is to get the change in conductivity occurring at a given time interval. In the first direct problem, there is the distribution of conductivity, then the measured voltages and conductivities are used to verify the accuracy of the images generated by the inverse problem.

The answers must be found within the specified tolerance. If not, a new iteration is made until the results are within the desired limit. The inverse problem is not linear and ill-conditioned. To find a satisfactory solution using good algorithm convergence, in this case we have chosen the Newton-Raphson method for its simplicity of implementation and good convergence. In the classical formulation of the Newton-Raphson method, we seek a distribution of conductivity σ^* that minimizes the differences between the measured voltages $[V_m]$ and calculated voltages $[V_c]$ to reduce the error

$$\{e\} = \{V_c\} - \{V_m\} \tag{7}$$

The finite element model for the direct problem

$$[Y_c(\sigma)] \text{ x } [Vc] = [\boldsymbol{C}] \tag{8}$$

Thus the difference between the voltages can be written as:

$$\{e\} = \{[T][Y_c(\sigma)]^{-1}\} - \{V_m\} \tag{9}$$

Where T is the matrix transformation for the equations dimensionally consistent with the voltages calculated in the electrodes,

$$\{V_m\}_{nxn} = \text{T} = \{V_m\}_{lxn} \tag{10}$$

Thus, the equation can be expressed by $\{f\} = 0$ for each standard voltage injected. To find the set of equations using the Newton-Raphson method, and using the expanded Taylor series, i.e.

$$f_i(\sigma + \delta\sigma) = f_i(\sigma) + \Sigma \frac{\partial f(\sigma)}{\partial \sigma_j} \partial\sigma_j + O(\partial\sigma_j{}^2) \tag{11}$$

The terms $\partial\sigma_j{}^2$ and higher orders do not contribute significantly in making calculations and $f(\sigma + \delta\sigma) = 0$ we have a system of linear equations in $\partial\sigma_j$

$$\sum_{j=1}^{N} J_{ij}\delta\sigma = -f_i(\sigma) \tag{12}$$

Where

$$J_{ij} = \frac{\partial f_{(i)}(\sigma)}{\partial \sigma_i} \tag{13}$$

Equation (12) defines an iterative process to calculate the homogeneous conductivity σ^* for each element. At each iteration k, a supposed conductivity σ_k is used in Equation (12) and allows you to update σ^* (Yorkey et al. 1987 a, b).

$$\sigma_i^{k+1} = \sigma_i^k + \partial\sigma_i \quad \text{i} = 1,2 \ldots \text{E} \tag{14}$$

considering the conductivity has a homogeneous value in each element of the mesh of the finite element method. When the change of the iteration produces σ^* less than the tolerance, the method converges. To find the matrix derived using a resource for finding the derivative of the matrix.

$$H_{ij} = \frac{\partial(Y^{-1}C_j)}{\partial\sigma} \tag{15}$$

Where cj = the standard currents, Y is the matrix of conductivity.
With the relation,

$$YY^{-1} = I \tag{16}$$

we derive and find the relation:

$$\frac{\partial Y^{-1}}{\partial\sigma} Y + Y^{-1}\frac{\partial Y}{\partial\sigma} = 0 \tag{16}$$

Multiplying both sides by Y^{-1} and simplifying:

$$\frac{\partial Y^{-1}}{\partial \sigma} = -Y^{-1}\frac{\partial Y}{\partial \sigma}YC_j \quad (17)$$

Finally, substituting equation (17) in Equation (15) defining the sensitivity matrix

$$H_{ij} = -Y^{-1}\frac{\partial Y}{\partial \sigma}YC_j \quad (18)$$

Where:

$\frac{\partial Y}{\partial \sigma}$ = The derivative of the conductivity with respect to matrix conductivity of the elements,

H_{ij} = the sensitivity matrix function on the elements, (i) and (j) the number of current standard,

Y^{-1} is the inverse of the global conductivity matrix,

C_j = current standards and j = the number of current standard

The next step is to find the contact model and standard current.

3.3 Electrodes Contact Model and current pattern

Besides seeking models to estimate the distribution of electrical impedance and the regularization techniques to improve the performance of the algorithms, in practice it is not enough just to model the interior of the regions which will estimate the parameters. What happens at the border becomes a more an important element for the accuracy of the results. Many authors experimentally and theoretically analyze the configuration of the electrodes and applied current patterns. What is the number and size of the electrodes? Should the contact impedance and conductivity of the electrode be modeled?

Consider the area of domain under an electric field discussed in equation 1, with the same definitions of the variables, the admittivity $\gamma(x,\omega) = \sigma(x,\omega) + i\omega\epsilon(x,\omega)$ where ϵ is the electrical permittivity and ω is the angular frequency of the applied current.

The two-dimensional model has some simplifications that do not have significantly influence. The Equation (2) together with the conservation of charge (**J**=0) reference voltage provide a possible electrode model hich is called *Continuum Model.* However, this model is inadequate, since the current density J is unknown. In practice, only the currents that are injected separately in each pair of electrodes (one takes the current in and the other moves it out) are known.

This density could be considered constant cm each electrode (Gap Model) but still be an inadequate model (Chen et al., 1989).

The important effects to be considered in the model are the discretization of the electrodes and their conductivity. This can be done as follows (Chen, et al. 1989 Samersalo E. et al., 1992): the total current passing through the electrode is equal to the current density through it as described by the Equation (3). where C_l is the current injected into the i-th electrode and $\partial\vartheta$ its corresponding area. Furthermore, this model is null the current density between the electrodes.

The conventional way of modeling the high conductivity of the electrodes is imposing the condition that u = V_i where i = 1, 2 .. L. This model is known as the *Shunt Model.*

To improve the modeling, considers the electrochemical effect that occurs in the contact area between the electrode and the contact region, which is a thin layer and has high resistance. The impedance z_l of this layer is called the effective contact impedance or impedance surface. To consider this effect, add a potential variation and replace the relation u = V_i by

$$u\,u + z_1\gamma\frac{\partial u}{\partial \vartheta} = V_1 \quad l=1,2... \quad (19)$$

resulting in the so-called *Completed Model* (Samersalo E. et al., 1992), whose unique solution has been demonstrated and is able to estimate experimental results with up to 1% accuracy.

The Completed Model is the set of equations (1), (2), (3) and (19) together with the conditions of conservation of charge $\sum_{i=1}^{L} C_l = 0$ and the definition of potential reference $\sum_{i=1}^{L} V_l = 0$. From the introduction of the first notions of visibility and sensitivity (Seager et al., 1984; Seager and Bates, 1985) an index of distinctness was defined (Gisser et al, 1987) from this introduction allowed for the quantitative measuring the ability of a standard current to distinguish between two values of conductivity. Gisser (1987) implements a set of currents that maximizes the distinguishability- *Best Pattern* (Isaacson, 1986, Valhkonen,1997) demonstrates the influence between the size and number of electrodes to distinguish between two values of conductibility.

3.4 The Model of Electrodes

The modeling of the contact impedance of the electrode-electrolyte interface has been worked out by several authors to obtain a more representative estimate of the distribution of resistivity. Along this lines, we implemented the electrodes model proposed by Hua in the paper: Finite Modeling of Skin-Contact Impedance in Electrical Impedance Tomography (Hua et al., 1993), which also represents the distribution of the electric field at the contact interface, it the dominant effect of the electrode contact impedance, such as discretization of contact impedance and shunt effect and edge. The model consists of a finite element mesh shown below:

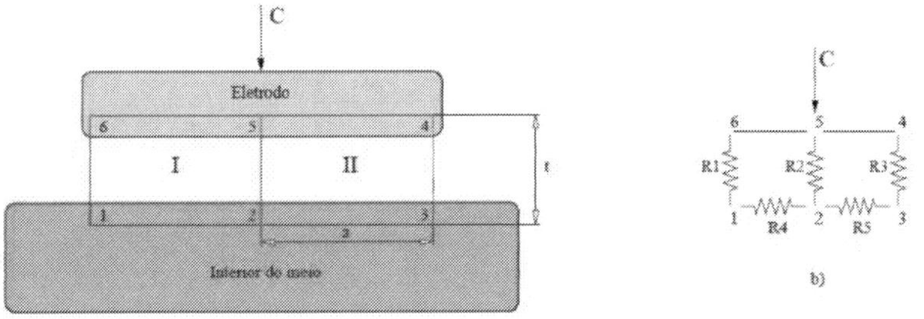

a)

Figure 1- a) Simplified model of electrode skin contact interface, b) model equivalent resistances.

submitted the following boundary conditions:

$$\int_{s4-5-6} \rho^{-1}\, \partial\phi/\partial n\, ds = \int_{s4-5-6} J ds = C_1 \tag{20}$$

Assuming that the potential across electrode is equal to $\phi_4 = \phi_5 = \phi_6$. Using the finite element method (Logan, 1986) with rectangular elements I and II, resulting in:

$$\frac{1}{3at\rho}\begin{bmatrix} a^2+t^2 & \frac{a^2}{2}-t^2 & 0 & 0 & \frac{-(a^2+t^2)}{2} & -a^2+\frac{t^2}{2} \\ & 2(a^2+t^2) & \frac{a^2}{2}-t^2 & \frac{-(a^2+t^2)}{2} & -2a^2+t^2 & \frac{-(a^2+t^2)}{2} \\ & & a^2+t^2 & -a^2+\frac{t^2}{2} & \frac{-(a^2+t^2)}{2} & 0 \\ & & & a^2+t^2 & \frac{a^2}{2}-t^2 & 0 \\ sim & & & & 2(a^2+t^2) & \frac{a^2}{2}-t^2 \\ & & & & & a^2+t^2 \end{bmatrix}\begin{bmatrix}\phi_1\\ \phi_2\\ \phi_3\\ \phi_4\\ \phi_5\\ \phi_6\end{bmatrix} = \begin{bmatrix} 0 \\ 0 \\ 0 \\ \int_4^5 J f_4 ds \\ \int_4^5 J f_5 ds + \int_5^6 J f_5 ds \\ \int_5^6 J f_6 ds \end{bmatrix} \tag{21}$$

Where ***a*** is the width of the element, ***t*** is the thickness of the interface in the *z* (thickness of the shell element), and ρ is the resistivity of the interpolation functions or shape of the elements containing geometric information. With the boundary conditions and considering that the thickness ***t*** of the interface is very small compared to half the width of the electrode, as ***t*** and ρ are unknown variables, a new variable $\rho' = t\rho$ called impedance contact with dimension $\Omega\ cm^2$ is defined. Thus, dividing ρ' by the area, Ω is obtained. While t is the order of 40µm, the value of the resistivity interface is high, therefore ρ' is in a moderate range. Hua arrives with some considerations at,

$$\frac{1}{t\rho}\begin{bmatrix} \frac{a}{2} & 0 & 0 & -\frac{a}{2} \\ & a & 0 & -a \\ & & \frac{a}{2} & -\frac{a}{2} \\ sim & & & 2a \end{bmatrix}\begin{bmatrix}\phi_1\\ \phi_2\\ \phi_3\\ \phi_4\end{bmatrix} = \begin{bmatrix}0\\0\\0\\C_i\end{bmatrix} \tag{22}$$

This model is incorporated into the matrix of the global finite element model and will be implemented in solving the inverse problem to estimate the unknown parameters like the impedances at the interface.

3.4 Regularization

The regularization method is to determine the approximate solution in a smooth and consistent way with the experimental observation data why they are mixed to a certain level of noise. The solution is adds smoother additional information, which transforms the ill-conditioned problem in a well-posed one. Among the techniques, there are the methods of Tikhonov regularization, reducing the effects of ill-conditioned systems (Bacrie Cohen et al., 1997), restoring the continuity of data solution (Cheney et al, 1990). The most widely referenced is the method of Tikhonov regularization or Tikhonov-Phillips method. It minimizes the expression of the form:

$$\min \phi_2\left(\vec{\sigma}\right) = \frac{1}{2}(\vec{V_m} - \vec{V_c}(\vec{\sigma}))'(\vec{V_m} - \vec{V}_c(\sigma)) + \alpha\left(H'H\vec{\sigma}\right) = 0 \tag{23}$$

where V_m is the vector, s, is the measured voltage value, V_c, is the initial conductivity and the operator is simulating the voltage distribution to the initial conductivity. If the ideal solution is sought iteratively, this is the distribution of conductivity that minimizes Equation (23).

3.5. Evaluation of Images

A set of image data with the ideal is possible to compare the image reconstruction algorithms. Figure 2 shows the CT reconstructed using generic iterative algorithms (GIA). The finite element method element model was used to reconstruct the images (Gigu `ere, R., et al, 2008).

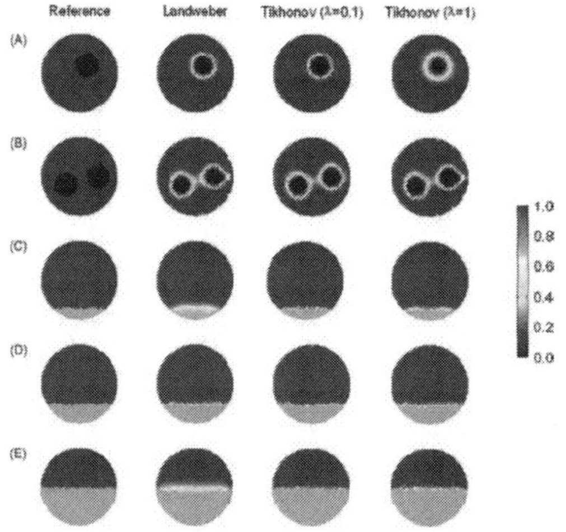

Figure 2 – Image reconstruction using generic algorithms (GIA) (Gigu ere `, R., et al, 2008).

The reconstruction in Figure 2 was performed using the Tikhonov and Landweber method with heuristic values for adjustment. The Tikhonov method requires the selection of a regularization which is heuristically found among those that reach the specified tolerance. The resulting images' associated heuristic values are selected when the comparison with the reference images are deemed satisfactory. Figure 2 shows this selection based on reference images to compare the quality of images reconstructed with the algorithms listed. Wheeler et al (2002), review several figures for use as a benchmark from those which have been proposed in the literature. Another way to compare is to do experimenting with objects of known capacitance and compare the images obtained with the figures of merit (Adler, 1996; Adler, 2004) to find the most suitable option.

3.6 Analysis of Results

The impedance problem is to find a distribution of conductivity S1 that minimizes the difference found between the voltages calculated using the finite element method and the voltages measured at the border and can be formulated as follows:

$$\{e\} = \{V_c\} - \{V_m\} \tag{24}$$

$$\min\phi_1 = \frac{1}{2}(\overrightarrow{V_m} - \overrightarrow{V_c}(\sigma))^t(\overrightarrow{V_c} - \overrightarrow{V_c}(\sigma)) \tag{25}$$

$$\min\ \phi_2 = \frac{1}{2}\alpha(\overrightarrow{\sigma^t}H^t H)\overrightarrow{\sigma^t} \tag{26}$$

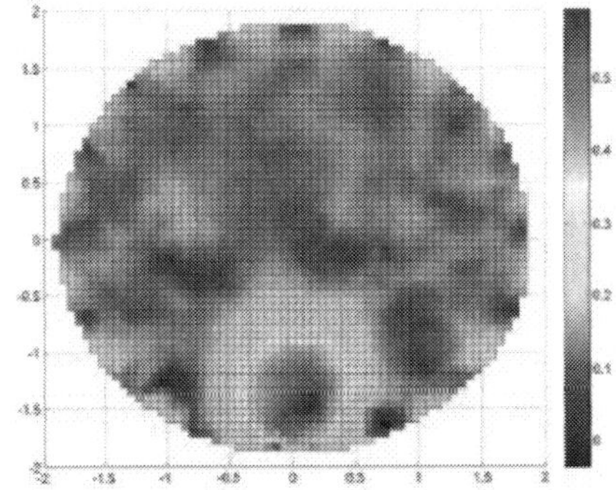

Figure 3-a)Iteration 38 with step: 0.04 and factor adjustment: λ=1e-4.

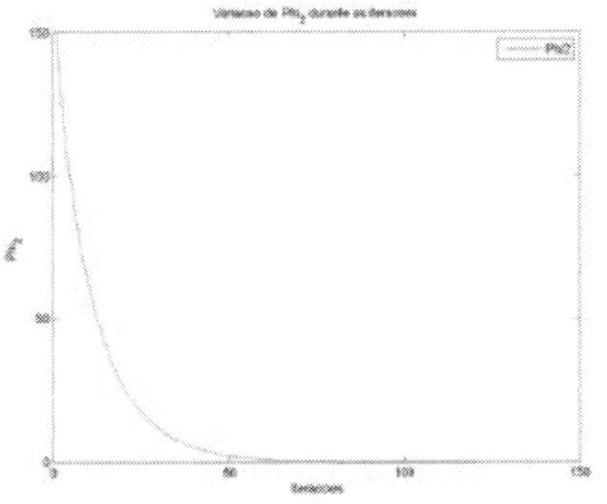

b) From approximately 50th. iteration there are no great advances in the iterative process

$\overrightarrow{V_c}$ and V_c are the calculated and measured voltages respectively. The index error (e) is obtained by solving the nonlinear system in the electrical conductivity through the Newton-Raphson method. In Figure 3, an example of the iterative process is shown, we can see that the Figure 38(a) shows an image with conditions to be adopted. In Figure 3(b), from the approximately 50th iteration there are no great advances in the iterative process. In Figure 3 (b), a graph with the performance of the iterative process is shown with the adjustment factor (1e-4), phi_1 is the first term of Equation (23) and is defined in equation (25). In each iteration, comparing the two terms of equation (23) may evaluate the effect of a greater or lesser regularization factor. In the present study, the iterative method of Newton-Raphson was used.

The capacitances obtained were tested to ensure that an adequate reconstruction of the original image maintained. The high values of factor adjustment influencing the image can produce solutions that reduce the data scale to maximum. The reducing adjustment factor scale preserves as much as possible the image to obtain a proper reconstruction. This approach allows for a theoretical interpretation of the image reconstruction algorithm that takes into account the various sources of prior information, such as the magnitude of electronic equipment that can introduce noise and its spatial resolution obtained for the maximum number of electrodes used.

4. Conclusion

The objective of this study was to review the techniques of electrical impedance tomography, highlighting the relevant topic of the impedance of contact electrodes. The contact impedance is high and generates an interference in the readings of voltages and the conductivities of the elements near the electrodes. The control for this effect is accomplished through the use of the Hua model (1993). Their use contributes to the reduction of noise and artifacts in the image. Use the Hua model of electrode (1993) contributes to model the impedance contact to get more accurate conductivity and voltages. The inclusion of this type of electrode improves the results, contributing to find a more appropriate distribution of conductivity images. The Newton-Raphson method used in the solution of this system with nonlinear behavior is quite effective. The method was used in its classical formulation with convergence in a small number of iterations.

5. REFERENCES

[1] Adler A and Guardo R Electrical Impedance Tomography: Regularized Imaging and Contrast Detection 1996 *IEEE Trans Med Imaging*, **170-9** 15
[2] Adler A Accounting for Erroneous Electrode Data in Electrical Impedance Tomography, 2004 *Physiol Meas*, **227-238**.
[3] Asfaw Y and Adler A Automatic Detection of Detached and Erroneous Electrodes in Electrical Impedance tomography, 2005 *Physiol Meas*, **175-183** 26
[4] Barber D C and Brawn B N Applied Potential Tomography, 1984 J. *Phys. E. Sci. Instrum* 723- **733** 17
[5] Barber D C A Review of Image Reconstruction Techniques for Electrical Impedance Tomography 1989 *Med. Phys*. **162-9** 16
[8] Bolton G T, Hooper C W, Mann R and Stitt E H 2004 Flow Distribution and Velocity Measurement in a Radial Flow Fixed Bed Reactor Using Electrical Resistance Tomography, 1997 *Chem. Eng. Sci*. **59** 1989.
[9] Borsic A 2002 *Regularisation Methods for Imaging from Electrical Measurements*, PhD Thesis, Oxford: Brooks University)
[10] Cheney M, Isaacson D and Nevell J C. *Electrical Impedance tomography*, 1999 *SIAM J. Applied Math. Review*. 85-101.
[11] Cohen-Bacrie C, Goussard Y and Guardo R Regularized Reconstruction in Electrical Impedance Tomography Using a Variance Uniformization Constraint, 1997 *IEEE Trans.Medical Imaging*, **16** 562-571.
[12] Dehghani H, Pogue B W, Jiang S, Brooksby B and Paulsen K D Three dimensional optical tomography: resolution in small object imaging, 2003 *Appl. Opt*. **42**, 3117-3128.
[13] Dai, Y. Pachowko, A.D. Wang, M. The Control of Polymer Addition to Raw Waste Sludge Using Electrical Resistance Tomography-a Preliminary Study, 2004 *Part. Syst. Charact*. **21** 228–233.
[14] Dickin F and Wang M, Electrical resistance tomography for process tomography, 1996 *Meas Sci Tech*, **7** 247-260.
[15] Eyuboglu B M, Brown B H and Barber D C, In Vivo Imaging of Cardiac Related Impedance Changes, 1989 *IEEE Eng Med Biol Mag* **8** 3945.
[16] Fransolet E Crine M, L'Homm G, Toye D and Marchot P Analysis of Electrical Resistance Tomography Measurements Obtained on a Bubble Column, 2001 *Meas. Sci. Technol* **12** 1055–1060
[17] Fransolet E, Crine M, Marchot P and Toye D Analysis of gas holdup in bubble columns with non-Newtonian fluid using electrical resistance tomography and dynamic gas disengagement technique 2005 *Chem. Eng. Sci*. **60** 6118–6123.
[18] Gigu`ere R, Fradette L, Mignon D and Tanguy P A ERT Algorithms for Quantitative Concentration 2008 *Measurement of multiphase flows Chemical Engineering Journal* **141**305–317
[19] Harris N D, Brown B H and Barber D C Continuous Monitoring of Lung Ventilation with Electrical Impedance Tomography, 1992 *Med Biol Socv* 1754-1755.
[20] Henningsson M, Ostergren K, Dejmek and P Plug flow of Yoghurt in Piping as Determined by Cross-Correlated dual-Plane Electrical Resistance Tomography, 2006 *J. Food Eng*. **76.** 163–168.
[21] Henningsson M, Regner M. Ostergren K, Tragardh C and Dejmek P CFD simulation and ERT Visualization of the Displacement of Yoghurt by Wter on Industrial Scale, 1997 *J. Food Eng*. **80** 166–

175
[22] Hua P J, Woo J G, Webster and Tompkins, Finite Element Modeling of Electrode Skin Contact Impedance in electrical Impedance Tomography 2001 *IEEE trans. Biomed.* **40(4)** 335-343.
[23] Karl WC Regularization in Image Restoration and Reconstruction, in Handbook of Image and Video Processing 2000 (Academic Press: Bovik A, Ed. chapter **3**, 141-161).
[24] Klose A D and A H Hielscher Quasi Newton Methods in Optical Tomography Image Reconstruction 2003 *Inverse Probl.* **19** 387-409
[25] Kim S N and Dyakowski A N T Measurement of Mixing of two Miscible Liquids in a Stirred Vessel with Electrical Resistance Tomography 2006 *Int. Commun. Heat Mass Transfer* **33**1088–1095.
[26] Miranda D A A Algoritmo da Tomografia por Impedância Elétrica Baseada em Newton Raphson 2000 (São Paulo: Ed. USP).
[27] Norman J T and Bonnecaze R T Measurement of Solids Distribution in Suspension Flows using Electrical Resistance Tomography 2005 Can. *J. Chem. Eng.* **83** 24–36.
[28] Paulsen K D and Jiang H 1996 Enhanced Frequency-Domain Optical Image Reconstruction in Tissues Through Total Variation Minimization *Appl. Opt.* **35**, 3447-3458.
[29] Polydorides N, Lionheart W R B A MATLAB Based Toolkit for Three-Dimensional Electrical Impedance Tomography: A contribution to the EIDORS project, 2002 *Measurement Science and Technology* **13** 1871-1883
[30] Pullum L, Graham L, Rudman M and Hamilton R High concentration suspension pumping 2006 *Miner. Eng.* **19** 471– 477.
[31] Ricard F, Brechtelsbauer C, Xu Y, Lawrence C and Thompson D Development of an electrical resistance tomography reactor for pharmaceutical processes, 2005 *Can. J. Chem. Eng.* **83** 11–18.
[32 Soleimani M, G´omez-Laberge Cand Adler A Imaging of Conductivity Changes and Electrode Movement in EIT 2006 *Physiol. Meas.* **27** S103–S113.
[33] Stanley S J, Mann R and Primrose K Interrogation of a Precipitation Reaction by Electrical Resistance Tomography *(ERT), AICHE J.* **51** (2005) 607–614.
[34] Stevenson R, Harrison S T L, Miles N and Cilliers J J 2006 Examination of swirling flow using electrical resistance tomography *Powder Technol.* **162** 157–165.
[35] Toye D, Fransolet E, Simon D, Crine M, L'Homme P and Marchot G Possibilities and Limits of Application of Electrical Resistance Tomography in Hydrodynamics of Bubble Columns 2005 *Can. J Chem. Eng.* **83** 4–10.
[36] Vanegas Molina, N.A, 2002, 2002 "Redução do Erro Numérico do Filtro Extendido de Kalman Aplicado a Tomografia de Impedância Elétrica", São Paulo, 108p.
[37] Vauhkonen M Electrical Impedance Tomography and Prior Information 1997 PhD thesis(Kuopio University: Department of Applied Physics)
[38] Vauhkonen M, Lionheart W R B, Heikkinen L M, Vauhkonen P J and Kaipio J P, A Matlab Package for the EIDORS project to reconstruct two-dimensional EIT images", 2001 *Physiol Meas*, 2001, **22** 107-111.
[39] Vijayan M, Schlaberg H I and Wang M Effects of Sparger Geometry on the Mechanism of Flow Pattern Transition in a Bubble Column" 2007 *Chem. Eng. J.* **130** 171–178.
[40] Wheeler J, Wang W and Tang M, A Comparison of Methods for Measurement of Spatial Resolution in two Dimensional Circular EIT Images, 2002 *Physiol Meas*, **23** 169-76.
[41] Yalavarthy P K, Pogue B W, Dehghani H and Paulsen K D Weight-matrix Structured Regularization Provides Optimal Generalized Least-Squares Estimate in Diffuse Optical Tomography, 2007 *Med. Phys.* **34**, 2085-2098.

Journal of Physics: Conference Series **407** (2012) 012007 doi:10.1088/1742-6596/407/1/012007

Comparison of body fat in Brazilian adult females by bioelectrical impedance analysis

Vaz Altamir[1,2], Slaets Annie France Frère[2], López Ramírez Leonardo[3]
[1] Member of the Brazilian Society of Nutrology, Abran, São Paulo, Brazil.
[2]Technological Research Center, Mogi Das Cruzes University, São Paulo, Brazil.
Rua Candido Xavier de Almeida Souza, 200. CEP 08780-911
[3]Telemedicine Research Group, Nueva Granada Military University, Bogotá, Col.

E-mail: dr.altamirvaz@uol.com.br Phone: +55 11 81010333

Abstract. Body-fat is essential for human body, provided that its amount is at healthy levels. If in-excess body-fat is deleterious, its lack is otherwise also harmful. Estimated percent body-fat performed with commercially available devices measuring bioimpedance have many advantages, such as easy measurement and low cost. However, these measurements are based on standard models and equations that are not disclosed by manufacturers, and this leads to questioning the validity of these estimates for Brazilian females. The aim of this study was to compare electrical tetrapolar and octapolar impedance results obtained with commercially available equipment: Maltron BF-906 and OMRON 510-W. Data analysis involved descriptive and inferential statistics. Devices used in this study to estimate body fat quantity have not shown any significant differences in results; this is a major issue when selecting equipment based on three factors: study focus, available financial resources, and target population. Results obtained from the two devices have not shown any significant differences, which lead to the conclusion that either device may be reliably used.

1. Introduction

Assessment of body composition is the focus of sciences in the search for the ideal human body composition; however, to date there is not an absolute standard parameter for this "ideal body" both for males and females [1]. The Body Mass Index (BMI) valid for adults above 20 years old, also known as Quetelet's index, is accepted by the World Health Organization as the official index for calculation of obesity [2], even though this index does not show body composition related to body fat rate.

Among the existing methods to assess body composition, the direct method is the most accurate one, though impracticable, as it requires dissection of the body, and is therefore used for research only. Indirect methods may be mentioned where there is no separate handling of components, but the application of chemical-physical principles to estimate fat and lean mass quantity such as the chemical methods, namely the radioactive potassium count (K40 and K42), deuterium oxide dilution, and urine creatinine clearance, in addition to ultrasound, X-rays, computed tomography, nuclear magnetic resonance, and densitometry [3]. The less precise but easily performed double-indirect methods are the ones routinely used, such as the skinfold measurement or the anthropometric technique, near infrared interactance, and bioimpedance or bioelectrical impedance, validated from indirect methods [4].

Bioelectrical Impedance
Owing to their low operating costs and relatively simple use, the double-indirect methods have been preferably used by Physical Education, Nutrition, and Medicine area professionals. Bioimpedancemetry or bioelectrical impedance (BIA) is an easy, low-cost, non-invasive method, not requiring skilled operator, which may be used in any patient [5]. There is a wide range of commercially available devices to measure bioimpedance, which vary depending on the segment, number, type and location of electrodes used [6]. Among these devices, the tetrapolar one with four electrodes, two of which placed on upper limbs and two on the lower limbs at the right-side of the body in order to measure all body segments, has been validated by several scientific studies and is widely used in clinical practice [7,8]. In addition to the tetrapolar, there are devices that use bipolar, scale-type, foot-to-foot measurement techniques with a metal plate as electrode, but that measure the lower body segment only. Hand-to-hand bipolar devices, also with electrodes, measure specifically the upper segment. More recently, devices called octapolar, with eight body contact points through plates are available. Nevertheless, results are frequently discrepant [9-12], due to methodological differences as for target populations, devices used, and different data collection methodologies [13]. For this reason, more research is required to test the different devices in the market and specificity of their predictive equations for different populations, in order to obtain evidences on result validity, accuracy, and reliability [14]. With this in mind, this trial evaluated two percent body fat measurement techniques in *menacme* females. The tetrapolar electrical bioimpedance has been first evaluated by using BIAt (hands and feet) and then percent fat results (%BF) obtained with octapolar electrical bioimpedance have been compared.

1.1.1. Preliminay:
Human body may be thought as a cylinder divided into five smaller cylinders full of fluids and with electric activity. Biological tissues act as current conductors or isolators, and this current applied to the body will tend to follow the lowest-resistance pathway.

- Tetrapolar method is based on the measurement of total body endurance to the passage of an electric current [15]. This method is considered as the standard in BIA but it does not define body fat distribution, which is of major relevance due to the several changes in lipid profile observed in its distribution in viscera.
- Octapolar: The full-body method calculates the percent fat through segmental measurements, by measuring individual values for arms and legs plus body. For such, eight surface electrodes are used to measure weight, BMI, %BF, percent viscera fat, and percent muscle mass.

Main characteristics of both devices are shown in Table 1.

Table 1. Device characteristics.

Device	**Octapolar**	**Tetrapolar**
Frequency [Hz]	50	50
Current [μA]	500	700
Technique	BI	BI
Variable	%BF	%BF

2. Method

2.1. Subjects

A total of 55 volunteer healthy females not on any medications that might affect the results were enrolled to participate in this study. Each signed the informed consent form of IRB's process 134/2011 protocol CAAE: 0134.0.237/000-11. Previously to the tests, personal data and anthropometric measures were recorded and results are shown in Table 2.

Table 2. Sample characteristics.

Variables	Mean	Standard Deviation	Minimum	Maximum
Age (years)	35.8	11.1	13	55
Weight (kg)	72.7	16.0	40.8	121.8
Height (cm)	159.3	6.0	146	175
BMI	28.6	5.7	15.4	45.3

2.2. Materials

- Matron tetrapolar bioimpedance apparatus, model BF-906 (Maltron, UK), with placement of adhesive electrodes in silver solution gel as follows: right foot, distal electrode at middle toe base and proximal electrode between middle and side malleoli, and for right hand, distal electrode on middle finger base and proximal electrode matching styloid process. Above 5-cm distance was kept between electrodes, with the subject lying supine with right foot and hand slightly apart from the body.
- Omron octapolar bioimpedance apparatus, model HBF-510W (Omron Inc., USA), with the subject standing up with two plate electrodes attached to the hands and two electrodes to the feet.
- A 300 kg x 100g scale (Balman, Brazil) was used to determine subject body mass and height.
- Portable estadiometer (Balman, Brazil) accurate to 1.0mm, respectively.
- Data collected was recorded in proper form and tabulated in a worksheet prepared by the author using MS-Excel (Microsoft, USA).
- Correlation analysis as well as t-Student test were applied to results for statistical analysis.

2.3. Experimental Protocol

Bioimpedance measurements were carried out once with each apparatus, first lying down, in supine position on a stretcher following a 5-minute rest. Before placement of electrodes on the subject skin, a 70% alcohol solution was applied and then measurement with tetrapolar device was carried out. Following tetrapolar data collection, subjects undergone measurement with octapolar device standing up and contacting the device's eight metal plates.

3. Results

Table 3 shows the results of percent body fat estimate provided by the devices used in this trial. These values are the result of the pooled sample analysis (n=55) with mean, standard deviation, maximum and minimum values calculated for each equipment.

Table 3. Measurement results

Variables	Mean	Standard Deviation	Minimum	Maximum
Tetrapolar	37.0	7.9	14	52.3
Octapolar	38.2	8.1	14.4	54.8

Figure 1 shows regression analysis between standard equipment (Maltron) and target equipment (Octa) indicating high correlation $r=0.92$ for $p<0.05$ with $n=55$.

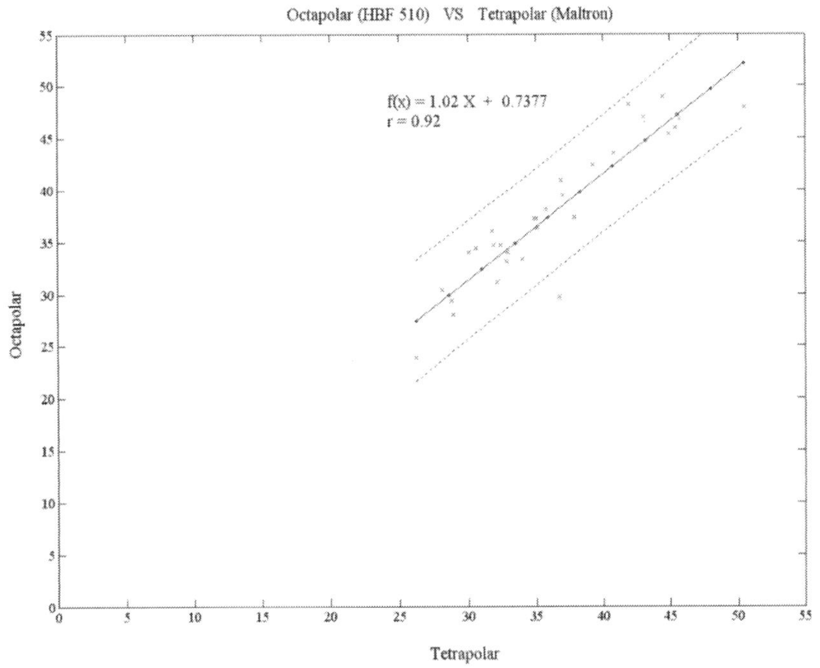

Figure 1. Correlation analysis between Tetra and Octa

4. Discussion

Tetrapolar equipment has been chosen as standard in this trial based on scientific papers [7,8].

Another study argues that bioimpedance measurement equipment vary based on the segment under analysis, however, other factors have been analyzed in this study such as number, type and location of electrodes used to measure percent body fat [6].

The comparison of equipment results is also important due to the differences in the price of the devices, as the tetrapolar is much more expensive (almost fivefold) than the octapolar one. Are these devices so different in their results? May cheaper equipment be used in clinical practice?

Correlation values obtained indicate a correlation deemed as high $r=0.92$ for $p<0.05$. Error estimation obtained from a data longitudinal analysis showed that the volunteers with higher measurement values were measured with lower error by Maltron equipment as they presented with higher body fat quantity at the lower body segment. Volunteers with lower measurement values showed higher conformance and lower error with octapolar equipment.

5. Conclusion

Anthropometric measures are essential for the evaluation of clinical treatment progress. The body composition evaluation method based on bioimpedance shows to be effective in clinical follow-up. Devices used in this trial to estimate body fat quantity have now shown significant differences in results; this issue is important when choosing the equipment based on three factors: study focus, available financial resources, and target population. Results obtained from the two devices have not shown any significant differences, which leads to the conclusion that either device may be reliably used.

Journal of Physics: Conference Series **407** (2012) 012007 doi:10.1088/1742-6596/407/1/012007

References

[1] Ozenoglu, A; Ugurlu, S; Can, G; Hatemi, H. 2009. Reference values of body composition for adult females who are classified as normal weight, overweight or obese according to body mass index. Endocrine Regulations, v.43, p.29-37.

[2] World Health Organization. 1995. Physical status: The use and interpretation of anthropometry. WHO Technical Report Series, p.854.

[3] Katch FI & McArdle WD. 1996. Nutrição, exercício e saúde. 4ª.ed. Rio de Janeiro, Medsi.

[4] Roche, A; Heymsfield, SB; Lohman, T. 1996. Human Body Composition. Human Kinetics, Champaign, IL.

[5] Lukaski, HC; Bolonchuk, WW. 1988. Estimation of body fluid volumes using tetrapolar impedance measurements. Aviation, Space, and Environmental Medicine, V.59, p.1163-1169.

[6] Flegal, KM; Shepherd, JA; et al. 2009. Comparisons of percentage body fat, body mass index, waist circumference, and waist-stature ratio in adults. Am J Clin Nutr, v.89, p.500-8.

[7] Rodrigues, MN; Silva, SC; Monteiro, WD; Farinatti, PT. 2001. Estimativa da gordura corporal através de equipamentos de bioimpedância, dobras cutâneas e pesagem hidrostática. Rev Bras Med Esporte, V.7, n.4, p.125-131.

[8] Rodrigues, PN; et al. 2008. Composición corporal en niños preescolares: comparación entre métodos antropométricos simples, bioimpedancia y absorciometría de doble haz de rayos X. Arch. argent. Pediatr, V.106, n.2, p.102-109.

[9] Lukaski, HC; Johnson, PE; Bolonchuk, WW; Lykken, GI. 1985. Assessment of fat-free mass using bioelectrical impedance measurements of human body. Am J Clin Nutr, V.41, p.810-7.

[10] Lukanki, HC; Bolonchuk, WW; Hall, CB; Siders, WA. 1986. Validation of tetrapolar bioelectrical impedance method to asses human body composition. J Appl Physiol, V.60, n.4, p.1327-1332.

[11] Segal, KR; Gutin, B; Presta, E; Wang, J; Van Itallie, TB. 1985. Estimation of human body composition by electrical impedance methods: a comparative study. J Appl Physiol, V.58, p.1565-71.

[12] Paijmans, IJ; Wilmore, KM; Wilmore, JH. 1992. Use of skinfolds and bioelectrical impedance for body composition assessment after weight reduction. J Am Coll Nutr, V.11, n.11, p.145-51.

[13] Kushner, RF. 1992. Bioelectrical impedance analysis: a review of principles and applications. J Am Coll Nutr, V.11, p.199-209.

[14] Sun, SS; Chumlea, WC; Heymsfield, SB; Lukaski, HC, et al. 2003. Development of bioelectrical impedance analysis prediction equations for body composition with the use of a multicomponent model for use in epidemiologic surveys. Am J Clin Nutr, V.77, p.331-40.

[15] Steijaert M, Deurenberg P, van Gaal L, De Leeuw I. 1997. The use of multi-frequency impedance to determine total body water and extracellular water in obese and lean female individuals. Intern J Diab; 21:930-4.

Use of electrocardiogram (ECG) electrodes for Bioelectrical Impedance Analysis (BIA)

J C Caicedo-Eraso[1,2], C H González-Correa[1], C A González-Correa[1]
[1] Research Group on Electrical Bioimpedance, University of Caldas, Manizales, Colombia, South America
[2] Department of Systems and Informatics, University of Caldas, Manizales, Colombia, South America

E-mail: julioc.caicedo@ucaldas.edu.co

Keywords: Bioelectrical Impedance Analysis (BIA), body composition, electrocardiogram (ECG) electrodes.

Abstract: BIA is a safe, noninvasive, portable and relatively inexpensive method of estimating body composition that is practical and suitable for individual use and large-scale studies. However, the cost of the electrodes recommended by some BIA manufacturers is too high for developing countries; where very often the long and complicated process of importation reduces the time they can be used. The purpose of this study was to evaluate the use of two types of ECG electrodes (2290 and 2228 by 3M®) in BIA measurements to decrease the costs of the test. The results showed that the 2228 ECG electrodes can be used in BIA measurements for adult's body composition assessment. These electrodes are available in the domestic market and their costs are 92% lower than the electrodes recommended by manufacturer. The results show a new cost-benefit relation for BIA method and make this a more accessible tool for individual tests, large-scale researches and studies in the community.

1. Introduction

According to WHO, the global prevalence of overweight and obesity in adults in 2008 was 30%, with a projection of reaching 43% by 2015 [1]. Colombia's situation is even worse as 51% of adults are overweight or obese [2]. There are several methods to determine body composition but greater precision and accuracy require a greater complexity and higher costs. Total body water by isotope dilution, hydrodensitometry, magnetic resonance imaging (MRI) and computerized tomography (CT) provide precise and reliable measurements of body composition and can be considered as reference methods for in vivo measurements. However, cost, accessibility, complexity and the radiation exposure with MRI and CT limit the use of these techniques [3,4]. Recently, dual-energy X-ray absorptiometry (DEXA) has become an attractive alternative method; nevertheless, all methods described above are poorly accessible for large-scale research in the community [4]. BIA is based on measuring tissue conductivity and the relation between the volume of conductor and its electrical impedance applied in the study of human body composition [5]. BIA is a simple, safe, noninvasive, relatively inexpensive method that is practical and suitable for individual use, large-scale researches and field studies in the community. [6]. However, the cost of the electrodes recommended by some manufacturers of bioimpedance analyzers is high to disseminate the use of BIA in developing countries, where usually the long and complicated process of importation reduces the time they can be used. The purpose of this study was to evaluate the use of two types of ECG electrodes in BIA measurements to reduce the costs of the test.

2. Materials and methods

2.1. Subjects

The methods, classified as minimum risk by the Colombian Ministry of Health, were approved by the Bioethics Committee of the University of Caldas. Variations in body composition related to gender, age, ethnicity and genetic influence that affect BIA measurements [7] were minimized by using an adequate measurement protocol and selecting a homogeneous group of young adult females from a specific region in Colombia [1,8]. A sample of 12 young adult females of the University of Caldas (Colombia) was evaluated. The purpose and procedures of the study were explained to the volunteers and the inclusion/exclusion criteria were verified after completing a questionnaire. Afterwards, the volunteers signed an informed written consent. Inclusion criteria were to be female, aged between 18 and 24 years, with body mass index (BMI) between 18.5 and 29.9 kg/m^2 and without co-morbidities. Exclusion criteria were smoking, having any metal implant (including copper T) or a cardiac pacemaker, be pregnant, using diuretics, had undergone surgery for weight reduction or had silicone implants in the breasts. Subject's characteristics are shown in table 1.

Table 1. Subject's characteristics (n=12).

Variables	Mean	SD
Age (years)	21.7	4.05
Height (cm)	157.5	6.06
Weight (kg)	56.9	7.45
BMI (kg/m^2)	23.5	4.87
Hand electrodes distance (cm)	8.5	0.78
Foot electrodes distance (cm)	12.7	1.09

2.2. Data acquisition

Measurements were performed in one session early in the morning to minimize environmental [9] and biological [7] variations. Relative humidity and environment temperature were controlled with an electric heater (BFH416 by Bionaire™) and a dehumidifier (BMD100 by Bionaire™). Relative humidity (RH) (73.5 ± 1.9%) and environmental temperature (18.8 ± 0.7 °C) were measured with a thermo-hygrometer (13307 by DeltaTrak®, ±0.1 °C / ±1% RH) and atmospheric pressure (787.4 ± 0.6 mmHg) was measured with a barometer (K4 by Konustar®,±0.1 mmHg). Volunteers were asked to comply with the following requirements before the test: 48 hours of not drink alcohol, 12 hours of not exercising vigorously and 12 hours of fasting but keeping normal water hydration. All were asked to evacuate her bladder and colon 30 minutes before the test and all wore underwear and a hospital gown during the test. All volunteers were evaluated out of menstrual cycle [8,10,11]. Other possible sources of error like hydration status, subject position, limb abduction, electrode position, wearing jewellery and skin alcohol use, were also controlled in this study [8,10,11,12].

2.3. Anthropometric measurements

Height (S) (Heightronic-235 by Seca®, ±0.01 cm) and weight (W) (PP2000 by Icob-Detecto®, ±0.1 kg) were measured twice, and a third measurement was taken if it was found a difference greater than 0.5 cm or 0.1 kg respectively [13].

2.4. BIA measurements

Bioelectrical impedance was measured (Hydra 4200 by Xitron Technologies©) on the dominant side of body for three times at the end of an exhalation on a nonconductive surface. Raw resistance (R), capacitive reactance (Xc) and impedance (Z) data were stored in a computer [14,15]. Subjects remained in a supine position by 5 minutes and BIA measurements were made between minutes 6 to 10, with the arms comfortably separated from the body 15 degrees, and the legs comfortably separated about 45 degrees (Figure 1) [8,16]. Dorsal hand and anterior foot surfaces were cleaned with alcohol and dried with a paper towel [8]. For reproducible measurements, four landmarks were made for the placement of the electrodes: midline between prominent ends of radius and ulna of wrist, midline of third metacarpal-phalangeal joint on dorsal hand surface, midline between the medial and lateral malleolus of ankle and midline of third metatarsal-phalangeal joint on anterior surface of foot [8,16]. Current was applied at distal electrodes and voltage was measured at proximal electrodes (Figure 2) [8,16]. Minimum distance between current and voltage electrodes must be 5 cm [8]; and this condition was met with all participants in this study. A blanket was used for comfort and homogeneous skin temperature [8,17].

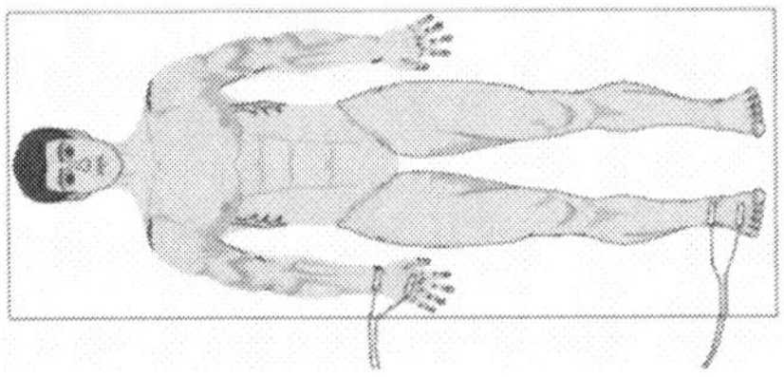

Figure 1. Position of subject.

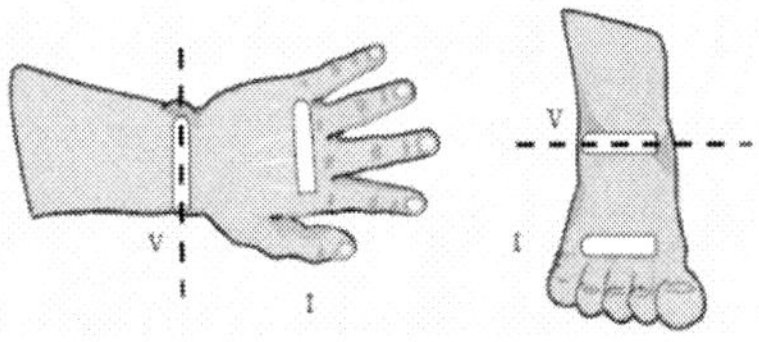

Figure 2. Location of electrodes.

Results obtained with two types of ECG electrodes (2290 and 2228 by 3M®) were compared to the results obtained with the electrodes (292-STE by Impedimed©) recommended by manufacturer. The 292-STE electrodes were used as the reference BIA measurement. Table 2 shows some characteristics of the three different types of electrodes tested in this study.

Table 2. Electrodes.

Electrode	Image	Use	Backing	Gel	Skin sensor contact area (cm^2)	Cost[c] (US$)
292-STE		BIA	Printed film[a]	Solid adhesive[b]	5.75	6.0
2290		ECG	Micropore	Dried	2.55	0.75
2228		ECG	Foam sticky	Wet	3.14	0.50

[a]Latex free [b]Impedimed© proprietary [c]Cost for testing in one person (four electrodes)

The 12 subjects were divided into three groups (A, B and C) of 4 individuals each. The measurement protocol used for each type of electrode is shown in figure 3 [8, 16]. A variation in the order of using the reference electrodes was made to evaluate BIA variations generated by test time and change in position of subjects (figure 4) [8]. Biological BIA variation by change of subject position and time of the tests was controlled adopting a stand position to restore the distribution of the body liquids between measurements with different types of electrodes [6,8].

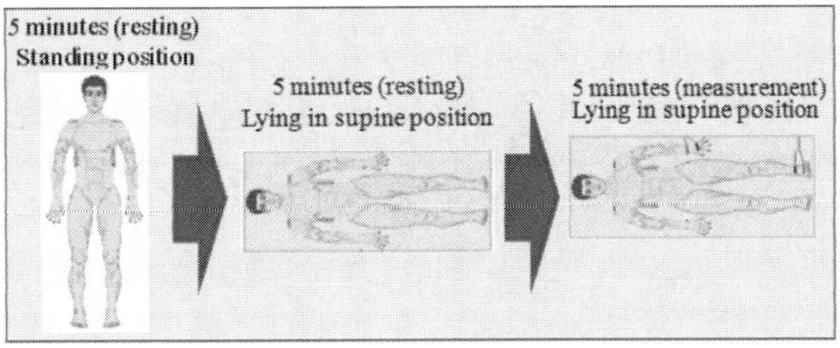

Figure 3. BIA measurement protocol for each type of electrodes.

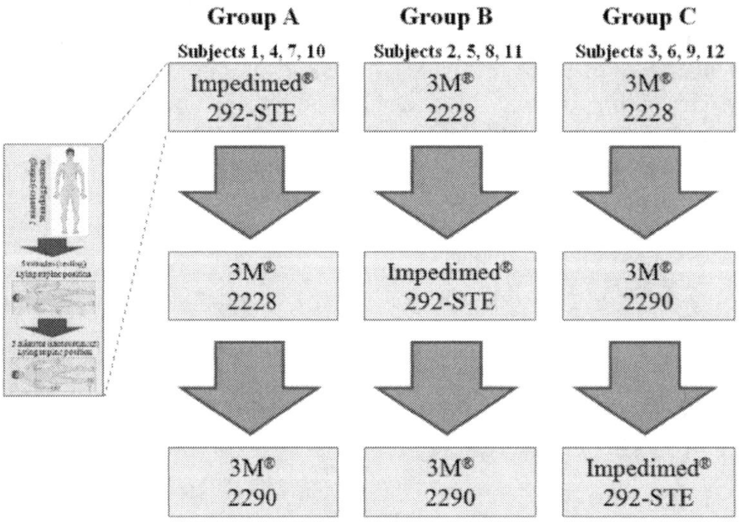

Figure 4. Variation in the order of using the reference electrodes in the groups A, B and C.

2.5. Statistical methods

The mean and the standard deviation (SD) were used to evaluate the characteristics of subjects and laboratory conditions. The paired Student's t-test ($p<0.01$) and the graphical method of Bland and Altman were used to evaluate the significant differences in all measurements [18].

Low and high frequencies were represented by 5 and 500 kHz respectively. Raw data (R, Xc and Z) at 5 and 500 kHz were analyzed to determine the effect of changing the order of using the reference electrodes. Raw data, model data (extracellular resistance Rec, intracellular resistance Ric, capacitance C and infinite resistance Rinf), volume data (extracellular water ECW and intracellular water ICW) and body composition data (fat-free mass FFM, body fat BF and percent body fat %BF) were analyzed to determine significant differences between electrodes. The 292-STE Impedimed© electrodes were used as the reference BIA measurement. Fat-free mass (FFM) was calculated with the equation (1) [16]. Body fat (BF) was obtained by subtracting FFM from total body mass (W). Percent body fat (%BF) was calculated with (BF/W)*100.

$$\begin{gathered} FFM = d_{ECW}V_{ECW} + d_{ICW}V_{ICW} \\ d_{ECW} = 1.106, d_{ECW} = 1.521; d: \text{density}, V: \text{volume} \end{gathered} \tag{1}$$

The Hydra_S_Acquisition Utility® Version 1.0 (2003) was used to capture the raw data and the Hydra_Data_Model_Vol Utility® Version 2.2 (1997) was used to obtain the model data and volumes. Body composition data were calculated with Microsoft® Excel® Version 2007 and the statistical analysis was made with XLSTAT® Trial version (2010). Also, the standard error of estimate (SEE) was use to evaluate if the body composition data were acceptable (table 3) [19].

Table 3. Acceptable SEE.

ECW, ICW (lt)	FFM, BF (kg)	%BF (%)
1.0 to 1.5	2.0 a 3.5	3.5 a 5

3. Results

Comparison of results of raw data (R, Xc and Z at 5 and 500 kHz) obtained with the reference electrodes (292-STE) in the three groups A, B and C (4 subjects each) showed that all *p* values were greater than 0.01, then the paired Student's t-test found no significant differences when using the strategy of changing the order of using the reference electrodes (Table 4). The Bland and Altman analysis [18] showed similar results.

Table 4. Comparison of raw data between groups A, B and C (n=4).

Raw	A vs. B		A vs. C		B vs. C	
data	p_t	B&A	p_t	B&A	p_t	B&A
R_5 (Ω)	0.4	Not	0.6	Not	0.5	Not
Xc_5 (Ω)	0.5	Not	0.9	Not	0.3	Not
Z_5 (Ω)	0.4	Not	0.6	Not	0.5	Not
R_{500} (Ω)	0.4	Not	0.6	Not	0.5	Not
Xc_{500} (Ω)	0.5	Not	0.8	Not	0.4	Not
Z_{500} (Ω)	0.4	Not	0.6	Not	0.5	Not

p_t: *p* value by paired Student's t-test (Significant difference if p_t<0.01)
B&A: significant difference by Bland & Altman method (Yes or Not)

Comparison of results of raw data (R, Xc and Z at 5 and 500 kHz) obtained in 12 subjects with the ECG electrodes (2290 and 2228) against results obtained with the reference electrodes (292-STE), showed significant differences in most cases with 2290 electrodes; while 2228 electrodes showed a significant difference only in Xc at 500 kHz (Table 5).

Table 5. Comparison of raw data with ECG electrodes against reference electrodes (n=12).

Raw	2290		2228	
data	p_t	B&A	p_t	B&A
R_5 (Ω)	0.451	Not	0.023	Not
Xc_5 (Ω)	0.000	Yes	0.168	Not
Z_5 (Ω)	<0.0001	Yes	0.023	Not
R_{500} (Ω)	0.001	Yes	0.026	Not
Xc_{500} (Ω)	<0.0001	Yes	<0.0001	Yes
Z_{500} (Ω)	0.015	Not	0.085	Not

pt: p value by paired Student's t-test (Significant difference if pt<0.01)
B&A: significant difference by Bland & Altman method (Yes or Not)

Comparison of results of model data (Rec, Ric, C and Rinf) obtained in 12 subjects with the ECG electrodes (2290 and 2228) against results obtained with the reference electrodes (292-STE) showed significant differences in Ric and Rinf with 2290 electrodes and none significant difference with 2228 electrodes (Table 6). Volume (ECW and ICW) and body composition (FFM, BF and %BF) data showed significant differences in most cases with 2290 electrodes and none significant difference with 2228 electrodes.

Table 6. Comparison of model data with ECG electrodes against reference electrodes (n=12).

Model	2290		2228	
data	p_t	B&A	p_t	B&A
Rec (Ω)	0.130	Not	0.029	Not
Ric (Ω)	0.024	Yes	0.274	Not
C (nF)	0.403	Not	0.076	Not
Rinf (Ω)	0.632	Yes	0.098	Not

p_t: *p* value by paired Student's t-test (Significant difference if $p_t<0.01$)
B&A: significant difference by Bland & Altman method (Yes or Not)

The SEE values obtained for volume and body composition data with ECG electrodes (2290 and 2228) showed acceptable SEE in all cases with 2228 electrodes (Table 7). Percentages of error for FFM and BF with the 2228 ECG electrodes were 0.3% and 0.0%; while the 2290 ECG electrodes produced errors of 3.6% and 6.7% respectively.

Table 7. SEE obtained for volume and body composition data.

Volume and body composition data	2290	2228
ECW (lt)	0.2	0.2^*
ICW (lt)	1.6^*	0.3
FFM (kg)	2.3^*	0.4
BF (kg)	2.3	0.4^*
%BF (%)	3.8	0.7^*

* underestimation

Bland & Altman plots for volume and body composition data obtained in 12 subjects with the ECG electrodes (2290 and 2228) against the reference electrodes (292-STE) showed that 2228 ECG electrodes had a good response to the prediction of ECW, ICW, FFM, BF and %BF because the average is at zero or very close to it and the scatter data were ever within ±2SD (Figure 5).

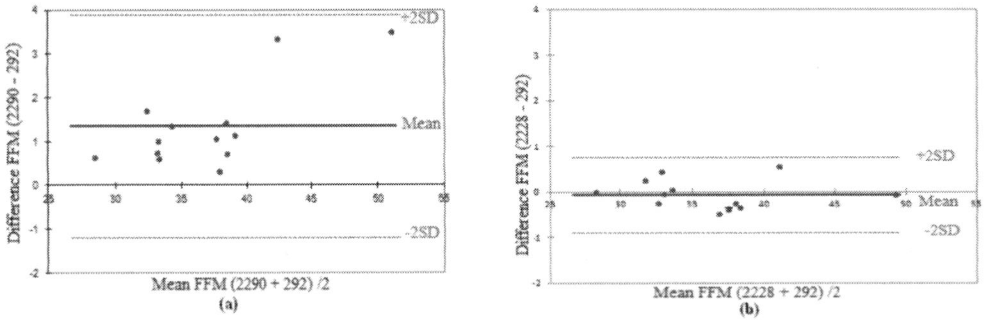

Figure 5. Bland & Altman plots of FFM (kg) with ECG electrodes against the reference electrodes (292-STE), (n=12). **(a)** 2290 ECG electrodes, **(b)** 2228 ECG electrodes.

4. Discussion and conclusion

Electrodes are critical for the accuracy of BIA measurements and researchers have studied the skin–electrode impedance, type of gel (dried or wet gel), shape and size electrode, inter-electrode distance, electrode material, electrode construction, electrode placement and material of the interface [20,21]. Additionally, there have been significant efforts for developing systems to characterize and compare textile electrodes against conductors [22] and compare different textile electrodes against standardized electrolyte electrodes [23]. Some authors have showed that ECG electrodes can be useful for reducing costs when monitoring anesthetic depth with bispectral index [24]. For BIA measurements, foam sticky, wet gel and Ag/AgCl ECG electrodes have been used [25], however, there has been no validation of the ECG electrodes in BIA measurements. The purpose of this study was to fill this gap.

A good system of electrodes for bio-potential measurements should have low coupling electrode-gel-skin impedance, high adhesion, good physical stability, large effective area and high flexibility [26]. Three types of electrodes used in this study met these characteristics. Although dried gel electrodes have more advantages over wet gel electrodes, dried gel electrodes have a weakness: they are less conductive and less humid than wet gel electrodes. Wet gel is able to get into the pores of the skin which ensures better interface gel-electrode-skin and it is easier to clean. For this reason wet electrodes are the most commercial ones [20]. In this study it was found that 2228 wet-gel electrodes are more suitable for BIA measurements than the 2290 dried-gel electrodes.

Recommended area for electrodes for measuring BIA is greater than 4 cm^2 [8] since a smaller area decreases the capacitance, increases the impedance [21] and deteriorates the coupling of the gel-electrode and skin interface. However, area of 3.14 cm^2 was apparently enough in this case. Some authors have shown that this variable only plays a minor role [27]. We suggest use standardized techniques for BIA measurements for minimize environmental and biological variations [28]. Other ECG electrodes with foam backing, wet gel and a minimum skin sensor contact area of 3.14 cm^2 could be used in BIA measurements but they should be validated with a similar process as in this study.

This study suggests that the 2228 ECG electrodes can be used in BIA measurements with the bioimpedance analyzer Hydra 4200 for adult's body composition assessment. The 2228 ECG electrodes are available in the domestic market and their costs are 92% lower than the electrodes recommended by manufacturer of this bioimpedance analyzer. The results show a new cost-benefit relation for BIA method and make this a more accessible tool for individual tests, large-scale researches and studies in the community.

5. References

[1] World Health Organization 2011 Country estimates of NCD mortality and selected risk factors 2008 Overweight and obesity: 2008 comparable estimates of prevalence of overweight and obesity *Global status report on non-communicable diseases 2010* Annex 4

[2] Ministry of Social Protection of the Republic of Colombia 2010 *National Survey of the nutritional situation in Colombia*

[3] Ellis KJ 2001 Selected body composition methods can be used in field studies *J Nutr* May **131** (5) pp 1589-1595

[4] Chien MY, Huang TY, Wu YT 2008 Prevalence of sarcopenia estimated using a bioelectrical impedance analysis prediction equation in community-dwelling elderly people in Taiwan *J Am Geriatr Soc* Sep **56** (9) pp 1710-1715

[5] Houtkooper LB, Lohman TG, Going SB, Howell WH 1996 Why BIA should be used for estimating adiposity *Am J Clin Nutr* Sep **64** (3 Suppl) pp 436-448

[6] Kyle et al, 2004. BIA--part I: review of principles and methods *Clin Nutr* **23** (5) pp 1226-1243

[7] Deurenberg P, Deurenberg-Yap M 2003 Validity of body composition methods across ethnic population groups *Forum Nutr* **56** pp 299-301.

[8] Kyle et al, 2004 BIA-part II: utilization in clinical practice (Review) *Clin Nutr* **23** (6) pp 1430-1453

[9] Cornish BH, Thomas BJ, Ward LC 1998 Effect of temperature and sweating on bioimpedance measurements Appl Radiat Isot May-Jun 49 (5-6) pp 475-476.

[10] Gleichauf CN, Roe DA 1989 The menstrual cycle's effect on the reliability of bioimpedance

measurements for assessing body composition *Am J Clin Nutr* Nov **50** (5) pp 903-907
[11] Mitchell CO, Rose J, Familoni B, Winters S, Ling F 1993 The use of multifrequency bioelectrical impedance analysis to estimate fluid volume changes as a function of the menstrual cycle *Basic Life Sci* **60** pp 189-191.
[12] Deurenberg P, Weststrate JA, Paymans I, Van Der Kooy K 1988 Factors affecting bioelectrical impedance measurements in humans *Eur J Clin Nutr* Dec **42** (12) pp 1017-1022.
[13] Lohman TG, Roche AF, Martorell R 1988 Stature, Recumbent Length, and Weight *Anthropometric standardization reference manual* **Chapter 1** (Champaign-IL, Human Kinetics Ed)
[14] Nyboer J 1959 Electrical impedance plethysmography (Springfield, IL, Charles C. Thomas Publisher)
[15] Thomasset A 1962 Bioelectrical properties of tissue impedance measurements *Lyon Med* **207** pp 107-118
[16] Xitron Technologies 2007 Hydra ECF/ICF (Model 4200) Bio-impedance Spectrum Analyzer Operating Manual (San Diego-CA, Xitron Technologies Inc.)
[17] Liang MT, Su HF, Lee NY 2000 Skin temperature and skin blood flow affect bioelectric impedance study of female fat-free mass *Med Sci Sports Exerc* Jan **32** (1) pp 221-227
[18] Bland JM, Altman DG 1986 Statistical methods for assessing agreement between two methods of clinical measurement *Lancet* Feb **8**, 1 (8476) pp 307-310.
[19] Baumgartner RN 1996 Electrical Impedance and Total Body Electrical Conductivity *Human Body Composition* **Chapter 5** (Champaign-IL, Human Kinetics Ed)
[20] Eggins BR 1993 Skin contact electrodes for medical applications *Analyst* Apr **118** (4) pp 439-442
[21] Abdurrahman A, Price D, Bhansali S 2007 Effect of electrode geometry on the impedance evaluation of tissue and cell culture *Sensors and Actuators B: Chemical* Vol **127** Issue 1 pp 89-96
[22] Beckmann L, Neuhaus C, Medrano G, Jungbecker N, Walter M, Gries T, Leonhardt S 2010 Characterization of textile electrodes and conductors using standardized measurement setups. *Physiol Meas* Feb **31** (2) pp 233-47.
[23] Márquez JC, Seoane F, Välimäki E, Lindecrantz K 2010 Comparison of Dry-Textile Electrodes for Electrical Bioimpedance Spectroscopy Measurements. *Journal of Physics: Conference Series* **224** 012140
[24] Hemmerling TM, Harvey P 1994 Electrocardiographic electrodes provide the same results as expensive special sensors in the routine monitoring of anesthetic depth *Anesth Analg* Feb **94** (2) pp 369-37116
[25] Tengvall M, Ellegård L, Malmros V, Bosaeus N, Lissner L, Bosaeus I 2009 Body composition in the elderly: reference values and bioelectrical impedance spectroscopy to predict total body skeletal muscle mass *Clin Nutr* Feb **28** (1) pp 52-58
[26] Luo S, Afonso VX, Webster JG, Tompkins WJ 1992 The electrode system in impedance-based ventilation measurement *IEEE Trans Biomed Eng* Nov **39** (11) pp 1130-1141
[27] Gonzalez-Correa CH, Evans JA, Smye S, Holland P 1999 Variables affecting BIA measurements of body water *Medical & Biological Engineering & Computing* Vol **37** No 2 pp 106-107
[28] Gonzalez-Correa CH, Caicedo-Eraso JC Bioelectrical impedance analysis (BIA): a proposal for standardization of the classical method in adults *J. Phys.: Conf. Ser.* (In press 2012)

6. Conflict of interest
The authors declare that there is no conflict of interest for this work and that any sponsorship from the manufacturers or brands mentioned in this study was received.

7. Acknowledgement
The authors would like to thank volunteers for their valuable participation in this study.

Bioelectrical impedance analysis of bovine milk fat

E A Veiga, P Bertemes-Filho
Department of Electrical Eng., State University of Santa Catarina, Joinville, Brazil.

E-mail: dee6eav@joinville.udesc.br

Abstract. Three samples of 250ml at home temperature of 20°C were obtained from whole, low fat and fat free bovine UHT milk. They were analysed by measuring both impedance spectra and dc conductivity in order to establish the relationship between samples related to fat content. An impedance measuring system was developed, which is based on digital oscilloscope, a current source and a FPGA. Data was measured by the oscilloscope in the frequency 1 kHz to 100 kHz. It was showed that there is approximately 7.9% difference in the conductivity between whole and low fat milk whereas 15.9% between low fat and free fat one. The change of fatness in the milk can be significantly sensed by both impedance spectra measurements and dc conductivity. This result might be useful for detecting fat content of milk in a very simple way and also may help the development of sensors for measuring milk quality, as for example the detection of mastitis.

1. Introduction

Over a decade the Bioelectrical Impedance Analysis (BIA) is being investigated as a potential technique for tissue characterization [1,2]. BIA technique has also been widely used to evaluate body composition and the quality of food, water and milk [3,4,5]. This technology is relatively simple, quick and noninvasive way for obtaining the impedance spectrum of the material under study. The impedance can be related to an equivalent electrical model of the material in order to fit to the measured data [6].

In order to obtain the most biological information, multi-frequency excitation signal should be required. It can be done by using multiple Direct Digital Synthesizer (DDS) in order to produce discrete frequencies in the time domain, and then sine waves are obtained by filtering the high-frequency harmonics [7]. Digital Signal Processing (DSP) integrated circuits (IC) are also widely used for this purpose, but they suffer from speed, power consumption and limiting number of storage points [8]. Furthermore, the number of bits available in its internal D/A converter is limited and so the precision of the signal generator is reduced. Multi-frequency systems use multiple DDS but the synchronization is also difficult [8]. Field-Programmable Gate Array (FPGA) has been proposed as a hardware technology for DSP systems. It can develop the most suitable circuit architectures in terms of computing, memory storage and power consumption in a similar way as DSP [8]. The FPGA Spartan 3 from Xilinx Inc. provides a DDS that can be used to generate multi-frequency sine waves. Though, bioelectrical impedance analyser can be easily developed by using a FPGA circuit.

Bovine milk is one of the most consumed food in the entire world. The fat present in the milk is a parameter that has great importance for dairy industry. For many reasons, consumers ask for information about the food they are purchasing, especially for people who have clinical recommendations regarding the consumption of saturated fat. Electrical measurement is a simple tool for material characterization and it is been of most interest for checking quality and fat content [9,10].

Journal of Physics: Conference Series **407** (2012) 012009 doi:10.1088/1742-6596/407/1/012009

The main objective of this work is to investigate the conductivity spectra of three types of bovine milk by measuring the electrical impedance spectra.

2. Methodology

Three samples of 250ml at about 20ºC were obtained from free fat (skimmed), low-fat (semi-skimmed) and whole UHT milk. Analyses were performed by measuring both impedance spectra and conductivity. The system setup is shown in figure 1.

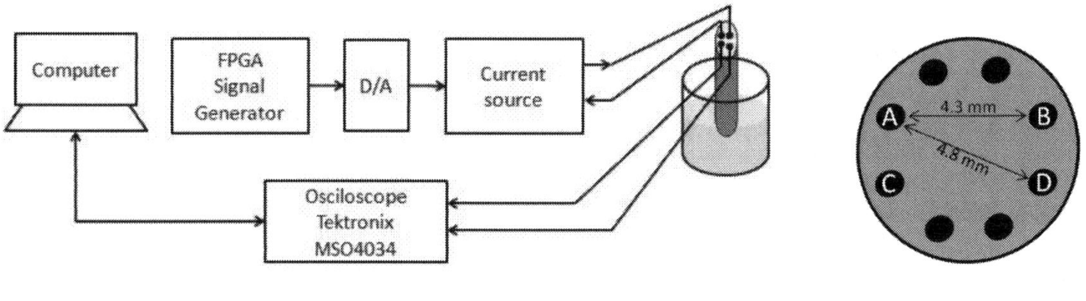

(a) (b)

Figure 1. (a) Block diagram of the implemented BIA system. **(b)** Transversal view of the impedance probe tip, where A,B are injecting electrodes and C,D are receiving electrodes, where the electrode diameter is 1 mm.

2.1 Signal Generator

The sine wave generator is based on FPGA Spartan 3 from Xilinx (model XC3S200 Starter kit board), which contains a DDS module with a phase generator and a sine/cosine lookup table [10]. The output frequency is controlled by a Phase Increment (PINC_IN). The equation 1 shows the relationship between the output frequency F_{out} and the PINC_IN.

$$F_{out} = \frac{F_{clk}\, PINC\_IN}{2^L} \quad (1)$$

where F_{clk} is the system clock frequency and L represents the number of bits in the phase accumulator.

The data was converted to digital by a DAC (Digital to Analog Converter) of 8 bits at a sampling rate of 667 kHz. In practice, 10 discrete frequencies were generated in the frequency range from 1 kHz to 100 kHz.

2.2 Current Source

The voltage signal is converted into a current by a modified Howland bipolar current source [11], which converts an input voltage of 2 Vpp (peak-to-peak) to an output current of 1mApp over the frequency range. The current is injected by two electrodes of the probe, which contains 8 gold electrodes (see figure 1b).

2.3 Measuring System

The signal acquisition was done by using the MSO 4034 oscilloscope from Tektronics [12], which has a bandwidth of 350 MHz and a sample ratio of 2.5GS/s. The differential voltage was measured cross other two electrodes of the probe by using the oscilloscope. The oscilloscope was connected to a PC based on the HTTP communication protocol. Data was recovered by using a Matlab interface and then both impedance and conductivity σ (=1/ρ) spectra were calculated (see equation 2). In order to calculate the impedance at dc frequency, a conductivity meter (model CD4303) from Digital Instruments was used to measure the dc conductivity of the samples (see equation 2).

Journal of Physics: Conference Series **407** (2012) 012009 doi:10.1088/1742-6596/407/1/012009

$$Z_{calc} = \frac{\rho}{2\pi} \cdot \gamma \tag{2}$$

where ρ is the medium resistivity and γ is the probe factor given by equation 3.

$$\gamma = \left(\frac{1}{R_{AC}} - \frac{1}{R_{AD}} + \frac{1}{R_{BD}} - \frac{1}{R_{BC}}\right) \tag{3}$$

where the electrode distances AC, AD, BD and BC are given in figure 1.

The bovine milk samples are UHT type from milk manufacturer. The contents of sodium, calcium, proteins and carbohydrates are similar for the three types of milk used, as shown in table 1.

Table I. Milk contents for a volume of 200ml, according to the manufacturer.

Milk type	**Carbohydrate (g)**	**Protein (g)**	**Fat (g)**	**Calcium (mg)**	**Sodium (mg)**
Whole	9	6	6	230	138
Low-fat	9	6	2	230	138
Fat free	9	6	0	230	138

3. Results

The impedance was calculated by measuring both injecting current and resulting voltage. The measurements were collected 10 times for each type of milk at each discrete frequency. The time estimated for taking each measurement in specified frequency was 8.6 seconds. Table 2 shows the mean impedance modulus for each type of milk and its respective conductivity obtained from CD4303.

Table II. Measured milk conductivity from CD4303 at 20°C.

Milk type	**Conductivity (mS/cm)**
Whole	3.62
Low-fat	3.91
Fat free	4.63

Figure 3 shows a variation of approximately ± 1.3 % in the milk impedance modulus below 50 kHz whereas ± 4 % at 100 kHz. The impedance modulus increases as increasing fat content.

Table 3 shows the mean values of the measured impedance (Z_{meas}) and the calculated impedance (Z_{calc}) from equation 2 by using the data from table 2.

Table III. Both measured and calculated milk impedance at 20°C

Milk type	**Fat (%)**	**Mean Z_{meas} (Ω)** (figure 3)	**Z_{calc} (Ω)** (equation 2)
Whole	3	219.56	168.38
Low-fat	2	203.57	155.89
Free fat	0	170.08	131.65

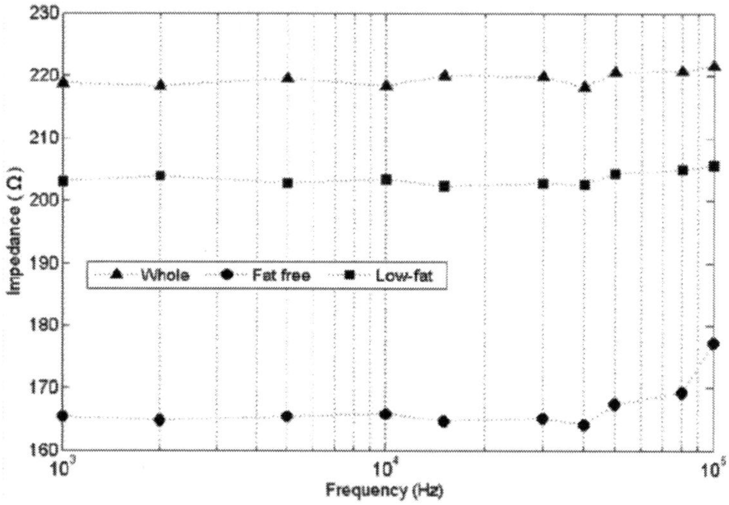

Figure 3. Impedance modulus of the milk in the frequency range from 1 kHz and 100 kHz.

Figure 4 shows the relationship between the calculated (continuous lines) and measured (bullet lines) conductivity over the frequency range of 1 kHz to 100 kHz. It can be seen that the conductivity data are different and the maximum error is approximately 23.3 %. This might be explained by the fact that the conductivity meter probe has a different geometry (ring electrodes) and then the probe factor is different.

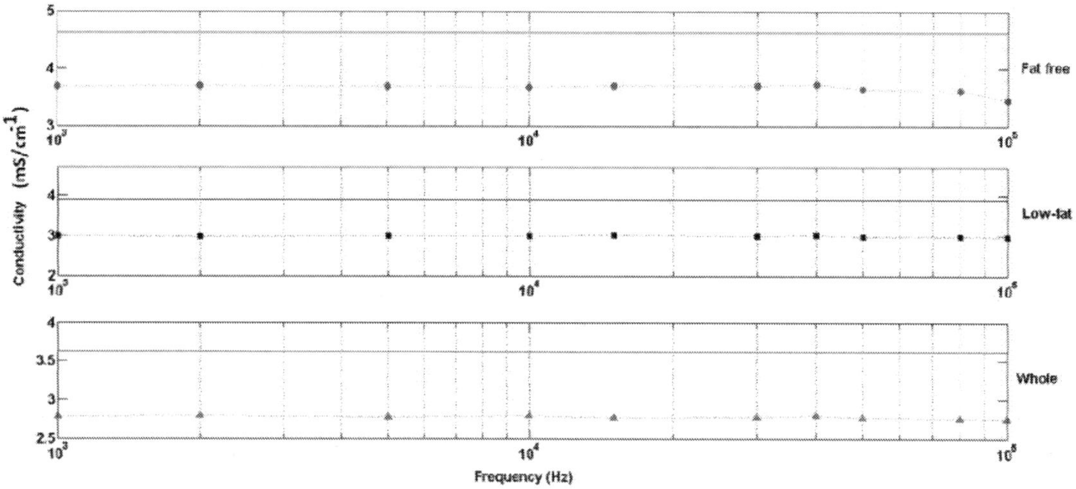

Figure 4. Bovine milk conductivity spectra from 1 to 100 kHz.

4. Discussions and Conclusions

The electrical conductivity spectra of bovine milk by using impedance measurements over the frequency range from 1 to 100 kHz were performed. It was showed that there was a conductivity difference of approximately 7.9 % between whole and low-fat milk whereas 15.9 % between low-fat and free fat one.

Although the concentration of calcium is similar for each type of milk, it was found that there is a change in the impedance modulus as fat concentration changes. There might be explained by the fact that there are other types of ions in the milk which may contribute for the current flow.

The maximum difference between calculated conductivity from impedance spectra and measured conductivity from CD4303 was approximately 23.3 %, which can be corrected by changing the probe factor. It must be emphasized that the conductivity meter uses the two electrode technique whereas the impedance spectra used the four electrode method.

It can be concluded that the change in the fat content of milk can be significantly sensed by both impedance and conductivity spectra measurements. This results might be useful for detecting milk fatness in a very simple way and also it might help the development of sensors for measuring milk quality, as for example mastitis.

References

[1] Abdul S, Brown B H, Milnes P and Tidy J A Int J. Gynecological Cancer **16** 1823-1832. 2006

[2] Keshtkar A, et al. Virtual bladder biopsy using bio-impedance spectroscopy at 6.500 Hz-1.024 Mhz. Measurement. **40** p.585-590. 2007

[3] Piton C , Dasen A and Bardoux I. Assessment of the impedance measure for a rapid evaluation of the bacteriological quality of raw milk Lait **68** 467-484. 1988

[4] McDougalla S, Murdough P, Pankey W, Delaney C, Barlow Jand Scruton D. Relationship among somatic cell count. **40** 245-254. 2001

[5] Yang L and Bashir R. Biotech Advanced **26** 135-150. 2008

[6] Kyle U G etal. Clinical Nutrition **23** 1226-1243. 2004

[7] Zhang X, Wang H, Chen S and Zhang Y FPGA-based Multi-frequency Excitation and Modulation Technology in EIT System 2010 IEEE

[8] Woods R, McAllister, Lightbody G and YI Y FPGA-based Implementation of Signal Processing Systems. 2008

[9] Pliquett U Bioimpedance: A Review for Food processing. 2010 Springer

[10] Bertemes-Filho P, Valicheski R C, Pereira R M and Paterno A S. Bioelectrical impedance analysis for bovine milk: preliminary results. Journal of Physics. Conference Series (Print), v. 224, p. 1-4, 2010

[11] Xilinx Inc. LogicCore IP DDS Compiler v4.0 – Applications Note 2011

[12] Bertemes-Filho P, Paterno A S and Pereira R M. Multichannel Bipolar Current Source used in Electrical IMpedance Spectroscopy: Preliminary Results. IFMBE **25** 657-660. 2009

[13] Tektronix. Mixed Signal Oscilloscope Series. Available on http://www2.tek.com/cmswpt/psdetails.lotr?ct=PS&cs=psu&ci=13296&lc=EN

Journal of Physics: Conference Series **407** (2012) 012010 doi:10.1088/1742-6596/407/1/012010

Early Detection of Cervical Intraepitelial Neoplasia in a Heterogeneos Group of Colombian Women Using Electrical Impedance Spectroscopy and the Miranda-López Algorithm

David A. Miranda[1], Sandra P. Corzo[1] and Carlos-A. González-Correa[2]

[1]CIMBIOS, Escuela de Física, Universidad Industrial de Santander, Cra. 27 Cll 9, Bucaramanga, Colombia

[2] GRUBIE (Grupo de Bio-Impedancia Elctrica), Universidad de Caldas, Manizales, Colombia

E-mail: dalemir@uis.edu.co

Abstract. Electrical Impedance Spectroscopy (EIS) allows the study of the electrical properties of materials and structures such as biological tissues. EIS can be used as a diagnostic tool for the identification of pathological conditions such as cervical cancer. We used EIS in combination with genetic algorithms to characterize cervical epithelial squamous tissue in a heterogeneous sample of 56 Colombian women. All volunteers had a cytology taken for Papanicolau test and biopsy taken for histopathological analysis from those with a positive result (9 subjects). ROC analysis of the results suggest a sensitivity and specificity in the order of 0.73 and 0.86, respectively.

1. Introduction

Cervical cancer is the leading malignancy among Colombian women, where the situation is worse when compared to South America and Europe [1, 2, 3]. Therefore, cervical cancer in Colombia is an issue of high concern and the Colombian government is interested in strengthening the prevention programs implemented in the country [4].

Electrical characterization of cervical tissue has been suggested as a tool for the screening of cervical cancer [5, 6]. Electrical tissue properties studied through EIS depend upon the amplitude and frequency of the electrical excitation. For low amplitudes (i.e. electric field and current density below $1V/cm$ and $1mA/cm^2$ respectively) electrical properties of tissues have a linear behavior. At low frequencies (alpha dispersion zone) electrode-tissue properties, relaxation and polarization of charged spaces dominate. Between frequencies around 1 kHz and 1 MHz (beta dispersion zone) polarization and interfacial relaxation characterizes the electrical properties of tissue. At higher frequencies (gamma dispersion zone) the dipolar relaxation of water and other molecules take relevant importance. In the beta dispersion zone it is possible to study the properties of biological tissues by the induced polarization effect that takes place at these frequencies [7, 8].

Biologic tissues are characterized by aqueous solution compartments separated by membranes of cells attached together. These membranes limit the free movement of ions through the extracellular compartment, determining its macroscopic behavior as a heterogeneous dielectric [7]. As such, they can be modeled by the Cole-Cole semi-empirical equation [9]. Pelton et al [10] suggested an alternative form of Cole-Cole equation that better describes the induced polarization effect:

$$\rho = \rho_0 \left\{ 1 - m \left[1 - \frac{1}{1 + (j\omega\tau)^c} \right] \right\} \tag{1}$$

Where ω is the frequency of the excitation, ρ_0, the resistivity at low frequency, τ, the time parameter, c, the relaxation parameter and m, the intrinsic chargeability (also called limited polarizability). The intrinsic chargeability, m, characterizes the intensity of the induced polarization effect [11].

In this article, we use the algorithm proposed by Miranda and López-Rivera [12] to obtain the Cole-Cole parameters from real part of the electrical impedance cervical measurements taken in a heterogeneous sample of 56 Colombian women. The results suggest good sensitivity (0.73) and specificity (0.85) when using two different parameters: low frequency resistivity and intrinsic chargeability, respectively.

2. Methods and Materials

The study was approved by the institutional Ethics Committee and women included in it signed a written informed consent. In this study we present the results obtained from a heterogeneous sample of 56 women. Electrical impedance spectrum were measured in cervical tissue in the eight zones showed in figure 1. On each zone impedance spectrum for seven different frequencies between 9.6kHz and 614kHz was measured using the Mark III bioimpedanciometer [5, 13]. A cervical sample for the Papanicolau test was taken from each woman after the electrical measurements were registered. A biopsy and histopatological study of the cervix were performed to women with a positive Papanicoalu test (9 women).

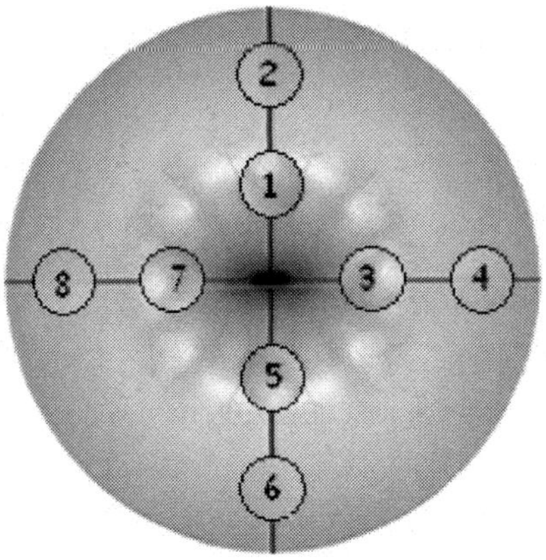

Figure 1. Zones in cervix were electrical impedance spectrum were measured.

Electrical impedance spectra were processed using the Miranda-López algorithm [12] to obtained the following four parameters of the spectra: intrinsic chargeability (m), low frequency resistivity (ρ_0), central relaxation time (τ) and relaxation parameter (c) for tissue.

In accordance with Bethesda standard for cervical cytology, four classification groups were defined to the screening: normal (NO), low-grade squamous intraepithelial lesion (LSIL), high-grade squamous intraepithelial lesion (HSIL) and cancer (CA). ANOVA and ROC curves were used to analyze the results.

3. Results and Discussion

The real part of the electrical impedance spectrum for measures classified as NO, LSIL, HSIL and CA are showed in figure 2. This figure shown differences in the spectra due to tissue condition. For normal tissue (NO) low frequency resistivity is greater than that of abnormal tissue (LSIL, HSIL and CA), i.e. as tissue become abnormal, a conductive behavior is predominant with a decrease in low frequency resistivity.

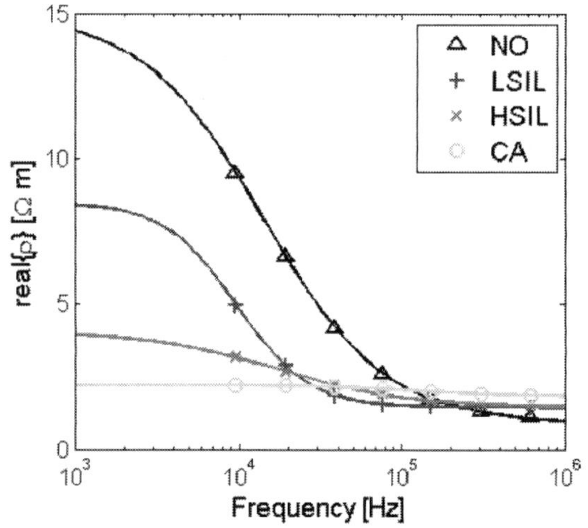

Figure 2. Real part of the electrical impedance spectrum for NO, LSIL, HSIL and CA tissues.

From the four Cole-Cole parameters only the low frequency resistivity (ρ_0) and intrinsic chargeability (m) show a good behavior as an classifier. ANOVA analysis revealed a statistical differentiation between classification groups and numerical values for ρ_0 and m ($p < 0.001$). The sensitivity and specificity were studied by ROC curves for m and ρ_0 (figures 3 and 4, respectively). The cut-off point of ROC curve was selected in the form that the best sensibility and specificity were simultaneously achieved, i.e. the point near to the left up corner in the ROC curve. A sensitivity and specificity over 70% and 80%, respectively, were suggested by ROC curves. Because measures were made on a heterogeneous population, the sensitivity and specificity obtained suggest that electrical properties of tissue are not significantly influenced by population heterogeneity.

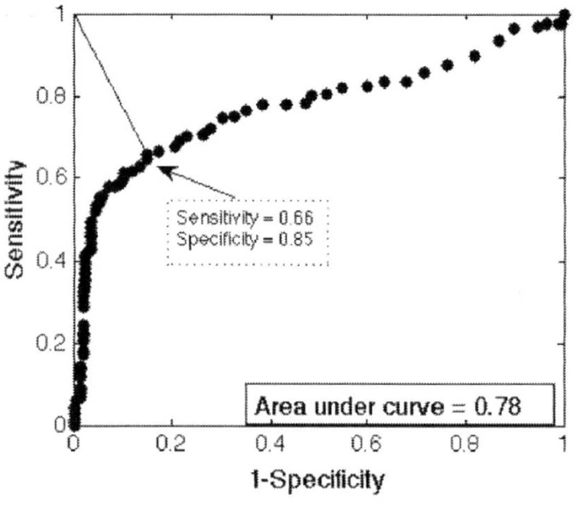

Figure 3. ROC curve for intrinsic chargeability (m).

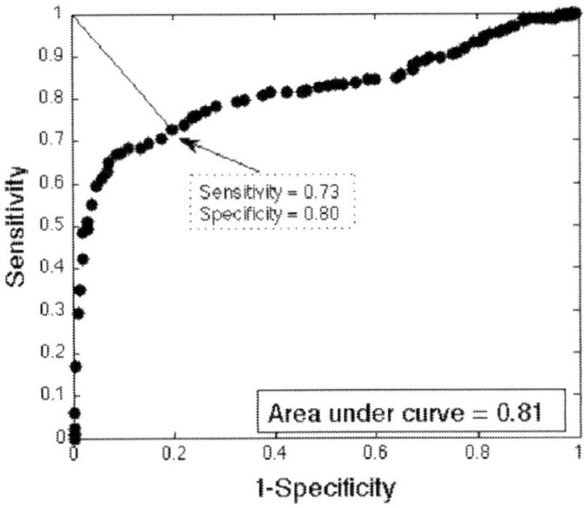

Figure 4. ROC curve for low frequency resistivity (ρ_0).

4. Conclusions

Electrical impedance spectroscopy was used to study electrical properties of cervical tissue in a heterogeneous population of Colombian women. Cole-Cole parameters were studied as screening tool. Statistical differences between numerical values for parameters m and ρ of Cole-Cole model for normal and abnormal tissue were observed ($p < 0.001$). The area under ROC curves ($A_{ROC} > 0.78$) suggest that these two parameters (m and ρ) could be used, in a complementary way, for the screening of cervical cancer in Colombian women.

5. Acknowledgment

We gratefully acknowledge the support of "Liga de Lucha Contra el Cancer, seccional Bucaramanga", "Hospital Universitario de Santander (HUS)", "Hospital de Floridablanca", "Universidad Industrial de Santander (UIS)" and COLCIENCIAS.

References

[1] Lewis, MJ. 2004 *A Situational Analysis of Cervical Cancer in Latin America and the Caribbean.* (Washington, DC: PAHO/WHO).
[2] WHO/ICO 2010 Information Centre on HPV and Cervical Cancer. Summary report on HPV and cervical cancer statistics in Colombia. 2010. `www.who.int/hpvcentre`. Accessed the 26/03/2012.
[3] Howlader N. et. al. 2011 SEER Cancer Statistics Review, 1975-2008, National Cancer Institute. Bethesda, MD, `http://seer.cancer.gov/csr/1975_2008/`. Based on November 2010 SEER data submission, posted to the SEER web site, 2011. Accessed the 26/03/2012.
[4] Palacio-Betancourt D. et. al. 2010 "*Plan Nacional para el Control del Cáncer en Colombia 2010-2019, Republica de Colombia, Minsterio de la Protección Social, Instituto de Cancerología*" (`http://www.cancer.gov.co/documentos/Plannacionalparaelcontroldelcancer/PlanParaControlCancer.pdf`).
[5] Brown B. H. et. al. 2000 Lancet 355 892.
[6] Abdul S et. al. 2005 Gynec. Oncol. 99 S64.
[7] Grimnes S. and Martinsen Ø. G. 2008 *Bioimpedance and bioelectricity basics, second edition* (Great Britain: Academic Press).
[8] Schwan H. 1957 *Electrical properties of tissue and cell suspensions* In: Advances in Biological and Medical Physics, editors: J. H. Lawrence and C. A. Tobias (New York: Academic Press) 5 1929-1931.
[9] Cole K. and Cole H. 1971 J. Chem. Phys. 9 341.
[10] Pelton W. H. et. al. 1978. Geophysics 43 788.
[11] Zhdanov M. 2008 Geophysics 73 F197.
[12] Miranda D. and López-Rivera 2008 Physiol. Meas. 29 669.
[13] Wilkinson B. A. et. al. 2002 J. Urology 168 1563.

Design and characterization of a multi-frequency bioimpedance measurement prototype

O E Mattia Neto, R W Porto and J C C Aya

Biosignals Laboratory, University of Caxias do Sul. Alameda Joao dal Sasso, Bento Goncalves, Brazil.

E-mail: {oscar.mattia, wolffporto, jccaya} @gmail.com

Abstract. A multi-frequency bioimpedance measurement prototype is proposed, validated and characterized. It consists of an Improved Howland Current Source controlled by voltage, a load voltage sensing scheme through a discrete 3-opamp instrumentation amplifier, a phase and quadrature demodulation setup through analog multipliers, and digitization and processing of the signals using a digital benchtop multimeter. The electrical characterization of the measurement channel was done for resistive loads only, on four different circuits. Measurements were made on 10 frequencies, from 100 kHz to 1 MHz, with 10 load resistances, from 100 Ω to 1 kΩ, to obtain linearity, absolute error and frequency response. The best performance among the four circuits was a maximum absolute error of 5.55 %, and -1.93 % of load current variation at the worst case scenario.

1. Introduction

Many papers have already discussed the difficulties regarding the instrumentation used in bioimpedance analysis and electrical impedance tomography, like the studies of [7], [4] and [3], just to mention a few. Also, special attention was given to current sources, as in the works of [5], [8], [9], [11] and [10]. The present paper builds upon and intends to contribute to the knowledge of the afore mentioned works by detailing the design, implementation and characterization of a multi-frequency bioimpedance measurement prototype. It starts with the proposed measurement channel, followed by a description of each implemented block and a brief discussion on the printed circuit boards used. Then, its functioning is shown through waveforms measured at specific points. A methodology is proposed for characterizing linearity, absolute error and frequency response of the whole channel, as well as output impedance of the current source. In the end, a thorough discussion of the results is presented, concluding the paper with future works and possibilities.

2. Proposed measurement channel

The block diagram of Figure 1 illustrates the proposed measurement channel. The actual implemented blocks are: voltage controlled current source (VCCS), instrumentation amplifier (INA), demodulation and filtering, while the function generator and the data acquisition system are benchtop ones. The following sections will detail the design and implementation of each block.

Journal of Physics: Conference Series **407** (2012) 012011 doi:10.1088/1742-6596/407/1/012011

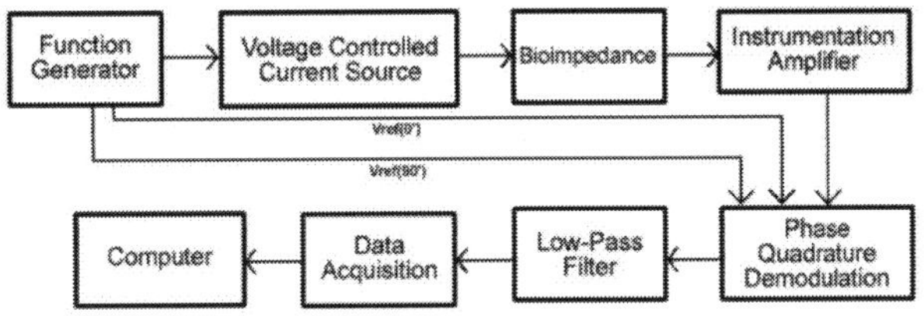

Figure 1. Block diagram of the proposed measurement channel.

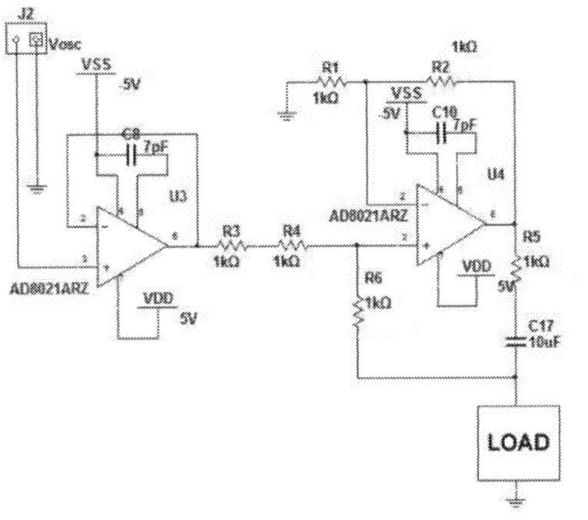

Figure 2. Implemented Improved Howland current source.

2.1. Current source

The chosen current source was the Improved Howland, with a DC decoupling capacitor on the feedback network, as proposed by [8]. The implemented circuit can be seen in Figure 2.

The input buffer is used to reduce the output impedance of the function generator, connected to $J2$. Capacitors $C8$ and $C10$ are gain compensation capacitors, used *only* with the AD8021 opamp, since the circuit tends to oscillate in higher frequencies. The ideal transfer function of the current source is $I_{LOAD} = V_{osc}/R1$.

2.2. Instrumentation amplifier

The instrumentation amplifier used is based on three discrete operational amplifiers, with FET inputs, like the AD8065/AD8066, from Analog Devices. This is due mainly to the fact that INAs with a higher bandwidth have digitally programmable gain, which would add complexity of hardware and software to the project. Besides, these amplifiers have a strong attenuation on their CMRR, reaching approximately 20 dB at 1 MHz for the AD8253, for example [1]. The topology is presented on figure 3, which also allows the implementation of a guardshield. According to [6], this technique allows a compensation of the degradation of the CMRR of the amplifier due to the different impedances of the source, and the different capacitances from the cable signals to the shielding. The signal transmission is balanced, since both cables are affected by the same noise sources, which shows up to the INA as a commom mode voltage. The shielding is used to reduce differential noise pick up.

The ideal transfer function is unitary. Connectors $J2$ and $J3$ are the inputs for the sensing

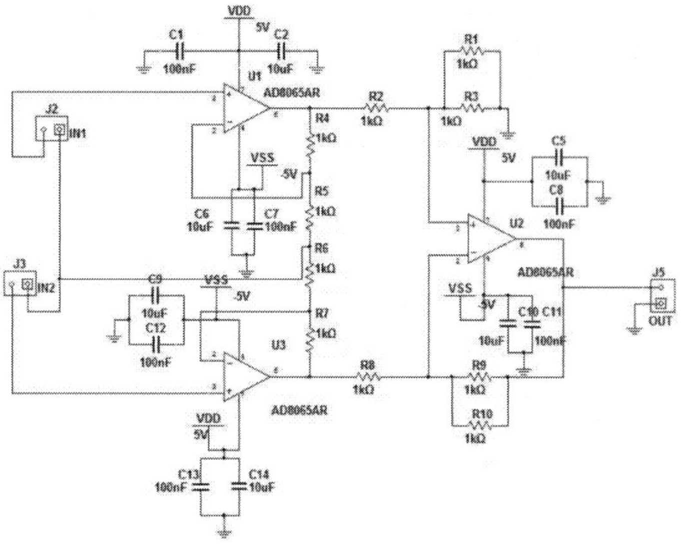

Figure 3. Implemented Instrumentation Amplifer.

electrodes, while connector *J5* is the output for the demodulator. All capacitors shown are used for power supply decoupling.

2.3. Phase and quadrature demodulation

The phase and quadrature demodulation is done using the AD835, an analog multiplier from Analog Devices. It generates the linear product of the inputs X and Y, with a -3 dB bandwidth of 250 MHz. This kind of demodulation allows the use of low-cost data acquisition systems, since the sampled data will be a DC voltage. It can also be used for higher frequencies of operation, as long as the analog multiplier's bandwidth is respected.

2.4. Low-pass filter

The choice of filter was a 4th-order Butterworth, for maximum bandwidth flatness, and of Sallen-Key topology, allowing gain adjustment independently of the filter's frequency of operation. Two independent filters are needed, one for the phase demodulation, and one for the quadrature demodulation.

2.5. Printed circuit board

The matching of the resistances, both in the current source and in the instrumentation amplifier, demand careful design of the printed circuit boards to keep these traces as short as possible, and also symmetric. Special attention was given to the parasitic capacitances of the main high frequency nodes, the output of the current source and the input of the instrumentation amplifier, by keeping the power planes as far away as possible. Due to tight project schedule, the first prototypes were manually built, and are single sided.

3. Validation

The following figures present the waveforms on different points of the measurement channel. The amplitude of the function generator was 1 V_p, at 1 MHz, and the load resistance was 500 Ω.

The stable functioning of the Improved Howland circuit can be seen on Figure 4, showing its input and output waveforms. Next, figure 5 shows, on channel 1, the output of the current source, and on channel 2, the output of the instrumentation amplifier, with its unitary gain.

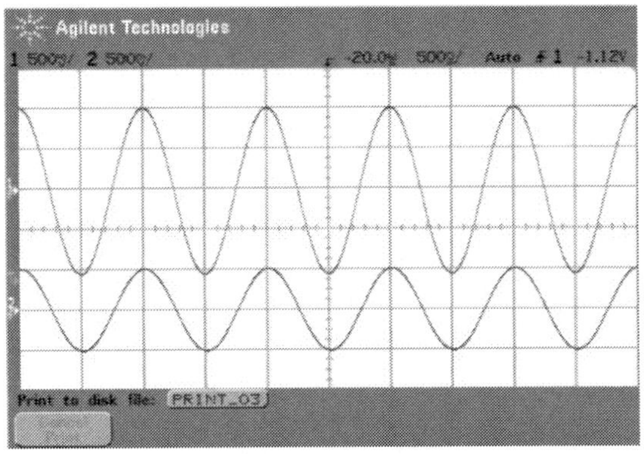

Figure 4. Improved Howland waveforms. Channel 1: Function generator output; Channel 2: Improved Howland output.

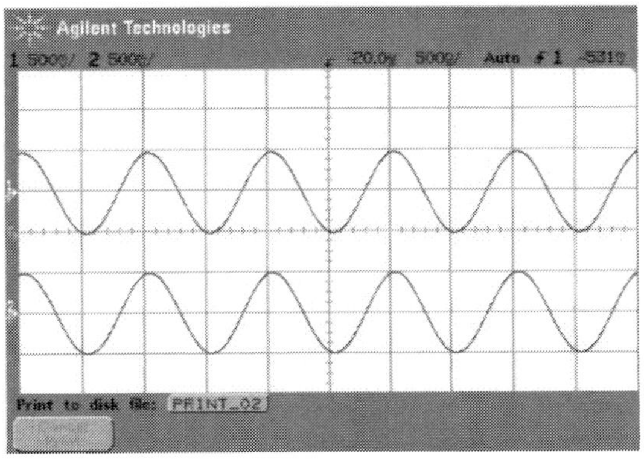

Figure 5. Improved Howland and INA waveforms. Channel 1: VCCS output; Channel 2: INA output.

On figure 6's channel 1, the output of the instrumentation amplifier is shown against the output of the phase demodulation (channel 2). It is clear that the signal's amplitude is halved, and a DC component of the same value shows up as the result of the demodulation procedure. This signal is then low-pass filtered and acquired digitally acquired as a DC voltage.

4. Characterization

The output of the phase demodulation was measured for 10 different frequencies, from 100 kHz to 1 MHz, for 10 purely resistive loads, from 100 Ω to 1 kΩ, totalling 100 acquisition procedures. For each procedure, 200 samples were averaged using a 10 Samples/s sampling rate, measured with an Agilent 34401A digital multimeter. The excitation signal was obtained from an Agilent 33210A function generator.

Characterizing the equipment for purely resistive loads only, any attenuation of the output

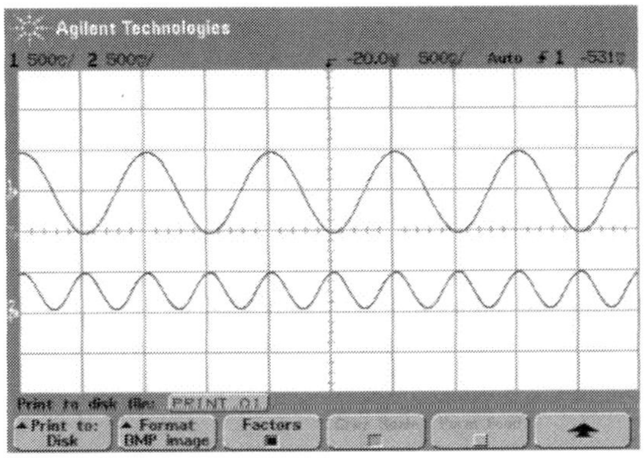

Figure 6. Instrumentation amplifier and analog multiplier waveforms. Channel 1: INA output; Channel 2: Phase demodulation output.

voltage, in function of the frequency, will be due to static elements of the measurement channel, that are:

- parasitic capacitances of the printed circuit board;
- input parasitic capacitance of the instrumentation amplifier;
- degradation on the frequency response of the operational amplifiers;
- phase delay of the current source and INA circuits.

The injected sinusoidal waveform has an amplitude of 1 V_p, leading to a load current of 1 mA_p. We assumed that R1 = 1 kΩ, and that the amplitude of the function generator does not vary with frequency. Also, that the transfer function of the instrumentation amplifier, the analog multiplier and the low pass filter remains the same over the whole range of frequencies.

The block diagram of the experiment is shown in figure 7, followed by its benchtop implementation, on figure 8.

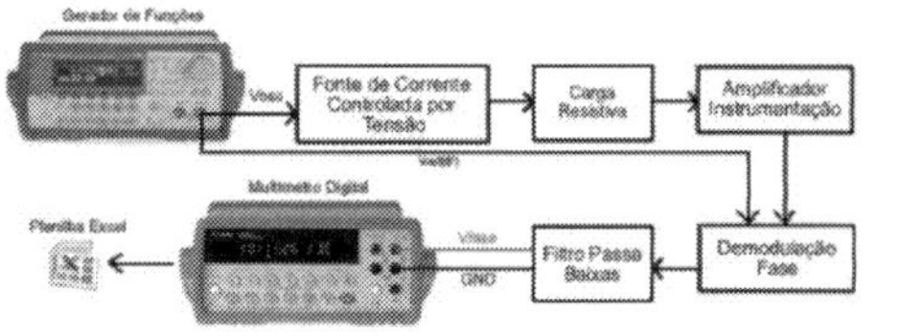

Figure 7. Block diagram of the experiment.

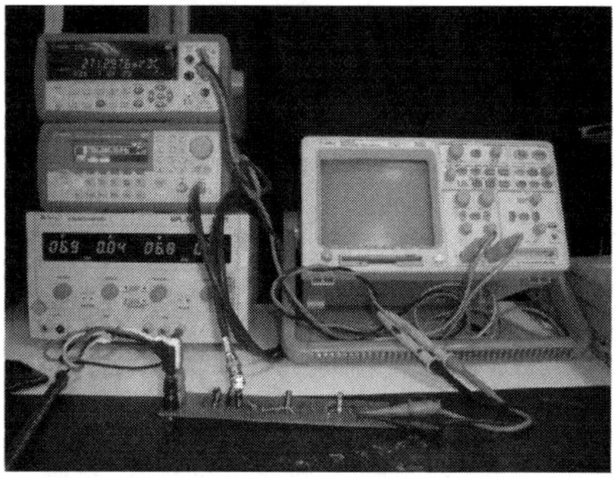

Figure 8. Bechtop implementation.

5. Results

The main reason for this characterization was to determine a minimum value for the output impedance of the current source, usually presented in literature as a means of comparison between different circuits and equipments. The first analysis made regarded linearity of the prototype, for all frequencies and load resistances. On figure 9 it can be seen that the equipment is linear over all its operating range. The fitting curve is only shown for the 100 kHz frequency to avoid crowding the graphic.

Load current can be seen on figure 10. Different colors represent different load resistances, presented with their ideal values, for clarity. Real values can be found on the tables at the appendix.

As expected, the load current shrinks with increasing frequency and load resistance, mainly due to parasitic capacitances that shunt it away from the load. Its value is always less than the ideal 1 mA_p, due to systematic errors of the measurement channel, like phase delays influencing the demodulation and the tolerance of the resistances. To clarify this errors, figure 11 shows the absolute error, as a function of frequency and load resistance. It is given by the difference between the ideal value and the experimental one, divided by the ideal value.

The absolute error is relatively high, varying from 4 % up to around 6 %. But, once this absolute error has been characterized, it can be software compensated, since it depends only on

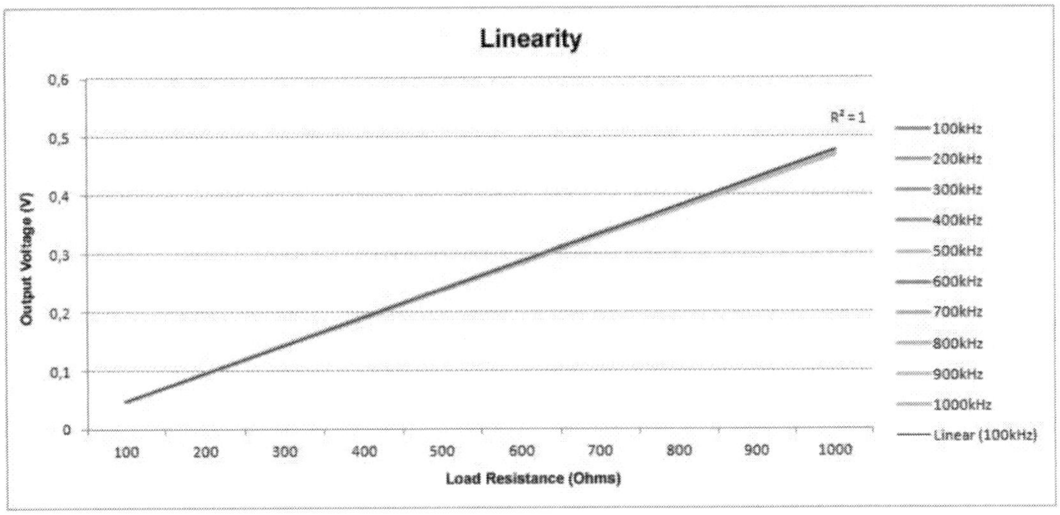

Figure 9. Linearity graphic of the output voltage, as a function of frequency and load resistace.

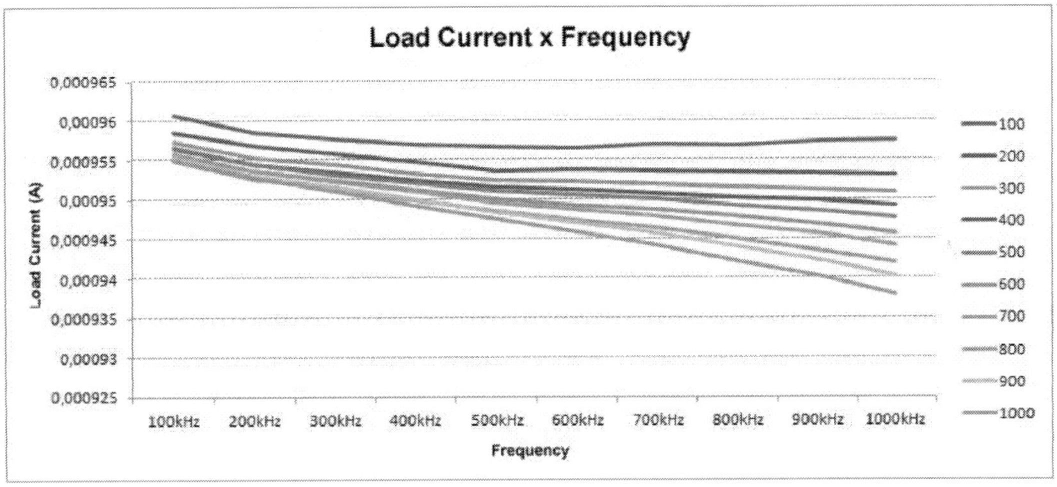

Figure 10. Output current graphic, as a function of frequency and load resistance.

static elements of the measurement channel. Under this perspective, an analysis of the variation of the load current becomes more interesting, and is shown in figure 12.

The reference for calculating the variation is the load current value for a 100 Ω load, at 100 kHz, and it considers that the rest of the circuit is not affected by the operating frequency or by the load resistance. It is known that other factors will influence the degradation of the load current measured, such as phase delay of the current source and the instrumentation amplifier, for example. Anyway, a tendency of increasing variation, and consequently increasing error can be seen, as we increase load resistance and frequency. For the worst case, that in fact is for the highest frequency and load resistance, the maximum variation is inferior to -2,5 %.

This maximum variaton could lead to the conclusion that the output impedance of the current source is greater than 40 kΩ, since the greatest load resistance is 1 kΩ. But, for that to be true, the current variation would have to be constant for all frequencies, resulting in a constant

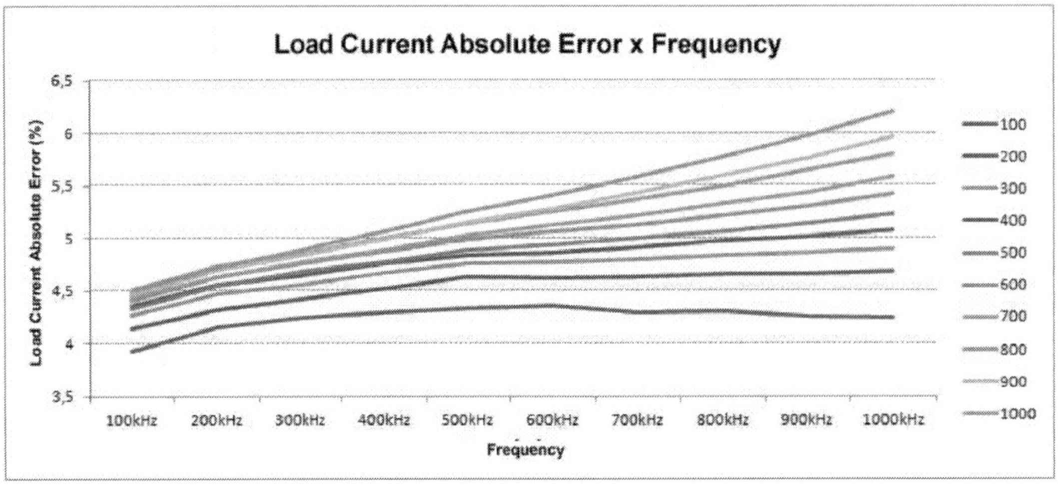

Figure 11. Absolute error graphic, as a function of frequency and load resistance.

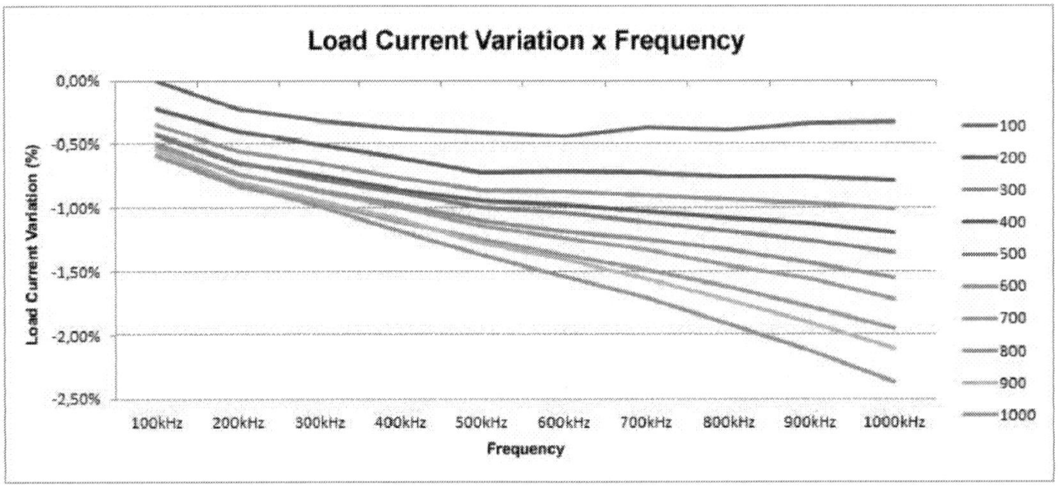

Figure 12. Load current variation graphic, as a function of frequency and load resistance.

inclination to the curve and an equal spacing between curves. But the output impedance varies with frequency, since it depends on factors such as the differential and common-mode open loop gain of the operation amplifier, as can be seen on figure 13.

In figure 13, it becomes clear that the output current shows different variations for the same load resistance variation, depending on the operating frequency. But, maybe not as explicit in the graphic is the fact that, for the same frequency, the circuit presents different values of output impedance, depending on the load resistance.

For example, between load resistances of 900 Ω and 1000 Ω, we have almost zero load current variation on the 200 kHz frequency. On the same frequency, with the load varying from 700 to 800 Ω, the load current variation is greater, around 0,25 %. The presence of this different derivatives on the curve could suggest a dependency of the output impedance of the current source with the load resistance, assuming that the other blocks of the circuit are not affected by

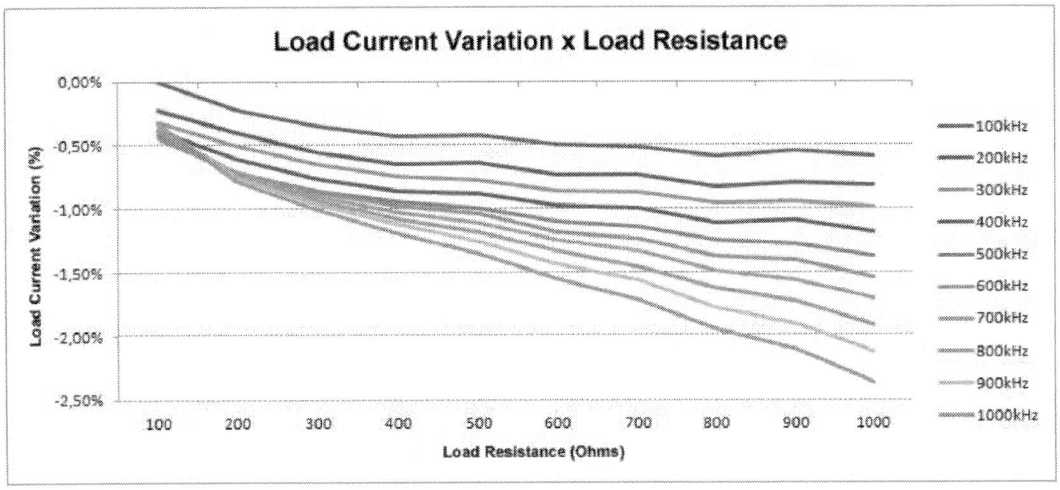

Figure 13. Load current variation graphic, as a function of load resistance and frequency.

it.

It is important to highlight that the output impedance being analysed is not only that of the current source, since the input capacitance C_i of the instrumentation amplifier, tipically 2 pF for the AD8065/AD8066 [2], is also present, besides the parasitic capacitances of the PCB. Since traditional methods of current measurement consist actually of measuring a voltage developed across a known resistance, any data acquisition system that is connected in parallel to the current source will inevitably alter it's output impedance, shunting current away from the load. The effective characterization of the output impedance of the current source *only*, could be made measuring the current in an indirect form, with the use of magnetic field sensors, for example, disconnecting the instrumentation amplifier from the load. This approach presents its difficulties due to the few milliamp amplitude of the current used on bioimpedance systems.

5.1. Comparison

The graphics exposed first were obtained from Circuit 1. Characterization data was gathered on four different circuits, in an attempt to verify the influence of the following parameters on the overall performance of the measurement prototype:

- matching of resistances;
- open loop gain of the operational amplifiers;
- bandwidth of the operational amplifiers;
- printed circuit board.

The difference between the circuits can be seen on table 1.

Two prototytpes of printed circuit boards were made, due to different component packages. Basically, prototype A was built for three AD8065, and individual 0805 packaged resistors, while prototype B was built for one AD8066 plus one AD8065, using matched resistors on a SO16 package [12]. The VCCS OpAmp mentioned is the one used on the current source.

6. Conclusion

This paper presented the design, implementation and characterization of a multi-frequency bioimpedance measurement prototype. It was the author's undergratuate Electrical Engineering

Table 1. Four different characterized circuits.

	Circuit 1	Circuit 2	Circuit 3	Circuit 4
Printed Circuit Board	A	B	B	A
VCCS OpAmp	AD8038	AD8021	AD8038	AD8038
Resistor Matching	1%	0.05%	0.05%	0.05%
Min. Absolute Error	3,93%	4,05%	3,80%	3,68%
Max. Absolute Error	6,20%	6,14%	6,15%	5,55%
Max. Load Current Var.	-2,37%	-2,04%	-2,34%	-1,93%

final project, and also the first step on the construction of an electrical impedance tomograph at the Biosignals Laboratory at UCS, where it can complement the works already being made on EEG.

During the implemetation, we could prove the functioning of the measurement channel on the desired frequency and load ranges, demodulating the high frequency signals through analog multipliers. Besides that, the characterization procedure showed some interesting results that need further analysis, specially on the behavior of the current source. The best results were obtained on a single sided PCB, using the AD8038 op-amp, 0.05 % tolerance resistors, resulting in a maximum absolute error of 5,55 % and a maximum load current variation of -1,93 %. We beleive that the main limitation for improving these results reside on the input parasitics of the instrumentation amplifier that will always shunt some current away from the load.

The degrading effect of higher frequencies on the measurement channel was clear, mainly due to the parasitic capacitances of the current source and the input of the instrumentation amplifier. Future works on this project may include the integration of a Direct Digital Synthesizer as function generator, and the construction of an aquisition system and a graphical user interface that allows the equipment to be operated by lay users. Since many biological tissues have known conductivity values, it could be used on those tissues as another way to validate the measurements.

Another interesting point to verify would be the sensibility of the output impedance of the current source against the open loop gain of the operational amplifier, and the variation of each resistance. Other analysis that could be done is the possible dependency of the current source and the load resistance. The characterization of each circuit block individually could contribute to answer this questions.

Appendix

Experimental data gathered on the four implemented circuits can be seen on Figures A1, A2, A3, A4. The first two lines are the ideal and real values of the resistances, in Ω. The following lines are all voltage measurements, in V. The last line represents the DC offset of the analog multipliers' output, in V.

References

[1] Analog Devices Inc. 2008 AD8253 Datasheet (Programmable Gain Instrumentation Amplifier).
[2] Analog Devices Inc. 2010 AD8065/AD8066 Datasheet (High Performance, 145 MHz Fast*FET* Op Amps) Norwood, MA 02062-9106, U.S.A.
[3] Bayford H 2006 Bioimpedance tomography (electrical impedance tomography) *Annual Review of Biomedical Engineering* v. 8, p. 63-91.
[4] Boone K G and Holder D S 1996 Current approaches to analogue instrumentation design in electrical impedance tomography *Physiological Measurement*, v. 17, p. 229-247.

Ideal	100	200	300	400	500	600	700	800	900	1000
Real	100,0307	200,3005	299,971	400,335	500,3895	600,0975	700,53	800,2475	900,2235	1000,275
100kHz	0,0827678	0,13071775	0,1783001	0,2261745	0,274064	0,321531	0,369472	0,416821	0,464762	0,5123425
200kHz	0,0826588	0,13054125	0,17800745	0,225768	0,273526	0,320855	0,368737	0,4159125	0,463669	0,511248
300kHz	0,0826156	0,1304488	0,1778716	0,225588	0,273204	0,320486	0,368259	0,415414	0,463048	0,510432
400kHz	0,0825871	0,13034665	0,1777015	0,225362	0,272956	0,320151	0,367858	0,414834	0,462395	0,509494
500kHz	0,0825716	0,13023145	0,1775685	0,225204	0,272679	0,319786	0,367357	0,41429	0,461604	0,508598
600kHz	0,0825569	0,1302384	0,1775491	0,225145	0,272571	0,319561	0,367031	0,413824	0,461034	0,507802
700kHz	0,0825881	0,13023185	0,17750795	0,225044	0,272409	0,319371	0,366714	0,413367	0,460385	0,506956
800kHz	0,08257895	0,1302071	0,17745435	0,22493	0,272235	0,319115	0,3662995	0,4128645	0,459653	0,505962
900kHz	0,0826033	0,1301989	0,1774087	0,22485	0,272042	0,318829	0,36594	0,412241	0,458877	0,5049935
1000kHz	0,0826107	0,1301742	0,1773564	0,224716	0,271831	0,318504	0,365436	0,411624	0,458004	0,503814
Offset	0,03472	0,03472	0,03472	0,03472	0,03472	0,03472	0,03472	0,03472	0,03472	0,03472

Figure A1. Data results from characterization 1.

Ideal	100	200	300	400	500	600	700	800	900	1000
Real	100,43915	200,2505	300,6105	400,909	500,5505	600,567	700,2305	800,2955	900,5895	1000,2
100kHz	0,0835209	0,1312572	0,1792333	0,227399	0,275157	0,323042	0,370763	0,4186675	0,466647	0,514372
200kHz	0,083446	0,1310793	0,17894975	0,227007	0,274639	0,322393	0,369989	0,417751	0,465604	0,513128
300kHz	0,08341225	0,13099695	0,17880365	0,226782	0,274332	0,322011	0,369494	0,417138	0,464832	0,512204
400kHz	0,0833933	0,1309386	0,1786789	0,226588	0,274061	0,321638	0,369013	0,416512	0,464034	0,511203
500kHz	0,0833814	0,13087755	0,17858275	0,226435	0,273811	0,3212915	0,368542	0,41586	0,463218	0,5102085
600kHz	0,0833987	0,1308843	0,1785371	0,226336	0,2736775	0,321044	0,368165	0,415336	0,462515	0,509214
700kHz	0,08343305	0,1308982	0,1785352	0,226285	0,273538	0,3208195	0,367789	0,414812	0,46175	0,508233
800kHz	0,0834688	0,130922	0,1785256	0,226225	0,273398	0,320581	0,3673795	0,414219	0,46091	0,507111
900kHz	0,0835239	0,1309705	0,1785408	0,226197	0,273298	0,320354	0,367001	0,413633	0,460074	0,505957
1000kHz	0,08358585	0,1310493	0,1785866	0,226177	0,273232	0,320099	0,366626	0,413047	0,459223	0,504676
Offset	0,03540295	0,03527475	0,03527475	0,03527475	0,03527475	0,03527475	0,03527475	0,03527475	0,03527475	0,03527475

Figure A2. Data results from characterization 2.

Ideal	100	200	300	400	500	600	700	800	900	1000
Real	100,23645	200,269	300,619	399,9555	500,345	600,646	700,0945	800,217	900,206	999,948
100kHz	0,0834364	0,1313584	0,1793847	0,226944	0,275013	0,32304	0,37058	0,41855	0,46637	0,5140295
200kHz	0,0833562	0,13119905	0,1791171	0,226572	0,274532	0,322426	0,369876	0,417672	0,465348	0,512891
300kHz	0,08332045	0,1311166	0,17897945	0,226368	0,274253	0,322058	0,369391	0,417067	0,464607	0,512036
400kHz	0,08329535	0,1310481	0,17885255	0,226181	0,273968	0,321669	0,368882	0,4164395	0,463808	0,511044
500kHz	0,08328935	0,13100075	0,1787567	0,226022	0,273746	0,321327	0,368438	0,415825	0,462982	0,510021
600kHz	0,08330155	0,1309918	0,1787154	0,2259155	0,273562	0,321051	0,368019	0,415267	0,462249	0,509059
700kHz	0,08333225	0,13102115	0,17871865	0,225867	0,27345	0,320843	0,367674	0,414725	0,4615	0,508042
800kHz	0,0833661	0,1310425	0,1787119	0,225806	0,273304	0,320572	0,367262	0,4141025	0,460631	0,50689
900kHz	0,08342255	0,13109225	0,1787296	0,225776	0,2731855	0,320334	0,366862	0,413498	0,45976	0,505714
1000kHz	0,08349005	0,1311618	0,1787674	0,2257615	0,273081	0,320152	0,36648	0,41288	0,4589235	0,504499
Offset	0,03527475	0,03527475	0,03527475	0,03527475	0,03527475	0,03527475	0,03527475	0,03527475	0,03527475	0,03527475

Figure A3. Data results from characterization 3.

[5] Bertemes-Filho P, Brown B H and Wilson A J A A 2000 A comparison of modified howland circuits as current generators with current mirror types circuits *Physiological Measurement*, v. 21.
[6] Franco S 2002 *Design with operational amplifiers and analog integrated circuits* New York, U.S.A: McGraw-Hill.
[7] McEwan A, Cusick G and Holder D S 2007 A review of errors in multi-frequency eit instrumentation *Physiological Measurement*, v. 28, n. 7, p. S197-S215.
[8] Nowicki D J and Webster G J 1989 A one op-amp current source for electrical impedance tomography *Annual International Conference of IEEE Engineering in Medicine Biology Society* S1.
[9] Porto R W 2008 Projeto e Avaliacao de um Canal de Medicao de Bioimpedancias *Master Thesis* Universidade Federal do Rio Grande do Sul, Brazil.
[10] Rafiei-Naeini M and McCann H 2008 Low-noise current excitation sub-system for medical EIT *Physiological*

Ideal	100	200	300	400	500	600	700	800	900	1000
Real	100,0661	200,537	300,0935	400,6675	501,01	600,317	700,084	800,3855	900,085	1000,27
100kHz	0,08280355	0,13103625	0,1787786	0,227127	0,275247	0,322994	0,37085	0,418627	0,4667765	0,51484
200kHz	0,08272885	0,13084265	0,1784951	0,226756	0,2747955	0,322421	0,370171	0,41783	0,465843	0,51374
300kHz	0,08269525	0,13075885	0,17835305	0,226564	0,274544	0,3220935	0,36975	0,417359	0,465244	0,51303
400kHz	0,08267295	0,13068135	0,17823295	0,226403	0,274306	0,321768	0,36934	0,416804	0,464562	0,512258
500kHz	0,0826585	0,1306289	0,1781238	0,22623	0,274085	0,321484	0,368956	0,416317	0,463948	0,511378
600kHz	0,0826564	0,1306275	0,17807655	0,226171	0,273964	0,321233	0,368647	0,415912	0,4634045	0,510709
700kHz	0,0826648	0,1306225	0,17805325	0,226099	0,27384	0,321061	0,368335	0,4155185	0,462873	0,509913
800kHz	0,0826699	0,1306151	0,17801545	0,226005	0,273673	0,320805	0,367974	0,414994	0,462207	0,508982
900kHz	0,0826797	0,13061155	0,1779763	0,225935	0,273542	0,320578	0,367621	0,414483	0,461513	0,50803
1000kHz	0,0826923	0,13062665	0,17795315	0,225855	0,273372	0,3203	0,367243	0,413937	0,4607065	0,506997
Offset	0,03461495	0,03461495	0,03461495	0,03461495	0,03461495	0,03461495	0,03461495	0,03461495	0,03461495	0,03461495

Figure A4. Data results from characterization 4.

Measurement, v. 29, p. S173-184.

[11] Ross A S, Saulnier G J, Newell J C and Issacson D 2003 Current source design for electrical impedance tomography *Physiological Measurement* v. 24, p. 509-16.

[12] Vishay 2011 NOMCT16-1KACT-ND Datasheet (Molded, 50 mil Pitch, Dual-In-Line Thin Film Resistor, Narrow Body, Surface Mount Network) Document Number 60007.

Effect of osmotic pressure to bioimpedance indexes of erythrocyte suspensions

A A Melnikov[1], D V Nikolaev[2], M V Malahov[1], A V Smirnov[2],

[1] State Pedagogical University, Respublikanskaya str. 108, 150000 Yaroslavl, Russia
[2] Scientific Research Centre 'Medass', Moscow, Russia

Email: a.melnikov@yspu.yar.ru; dvn@medass.ru

Abstract. In the paper we studied effects of osmotic modification of red blood cells on bioimpedance parameters of erythrocyte suspension. The Cole parameters: the extracellular (Re) and intracellular (Ri) fluid resistance, the Alpha parameter, the characteristic frequency (Fchar) and the cell membranes capacitance (Cm) of concentrated erythrocyte suspensions were measured by bioimpedance analyser in the frequency range 5 – 500 kHz. Erythrocytes were incubated in hypo-, hyper- and isoosmotic solutions to achieve changes in cell volume. It was found that Re and Alpha increased in the suspensions with low osmolarity and decreased in the hypertonic suspensions. Ri, Fchar and Cm were higher in the hyperosmotic and were lower in the hypoosmotic suspensions. Correlations of all BIS parameters with MCV were obtained, but multiple regression analysis showed that only Alpha parameter was independently related to MCV (β=0.77, p=0.01). Thus Alpha parameter may be related the mean corpuscular volume of cells.

1. Introduction

The electrical properties of the erythrocyte suspensions with standard hematocrit are dependent to three factors: the volume of single red blood cells, their number in the measured sample of blood, as well as the concentration of ions in the suspending solution [1]. Osmotic pressure affects both the electrical resistance of the extracellular water, and the size and shape of cells [2]. It is known that the impedance parameter Alpha may depend on those parameters of red blood cells [1, 3]. However, the relationship between the osmotic modification of erythrocyte suspensions and bioimpedance parameters of the suspensions is not fully understood. In this report, we tested the proposition that the osmotic effects of suspension medium on red cell volume will be relationships to the bioimpedance parameter Alpha.

2. Materials and methods

2.1. Preparation of hypotonic, isotonic, and hypertonic erythrocyte suspensions

The blood samples with volume 9 ml (n=8) were divided into three aliquots by 3 ml. First aliquot was mixed with 6 ml of the hypotonic 0.7% NaCl solution (osmolarity – 240 mosmol/l), second – with 6 ml of the isotonic 0.85% NaCl solution (osmolarity – 290 mosmol/l) and third – with 6 ml of the hypertonic 1.1% NaCl solution (osmolarity – 375 mosmol/l). Then the erythrocytes were washed in these solutions by triple centrifugation at 3000 rpm for 10 minutes. Then washed erythrocytes were incubated for 1 hour in solutions with different osmolarity. After incubation concentrated erythrocyte suspensions with 92% hematocrit were prepared.

2.2. Estimation of mean corpuscular volume, hematocrit and red blood cells concentration in suspensions with standardized hematocrit

Hematocrit (Ht) was derived by centrifugation of the suspensions in the microcapillaries at 5000 rpm for 15 minutes. Red blood cells concentration (RBC) was measured by the photocolorimetric method. Mean corpuscular volume (MCV) was calculated by the formula: MCV=Ht/RBC.

2.3. Bioimpedance measurements of the erythrocyte suspensions

Prepared concentrated erythrocyte suspensions with standardized hematocrit (Ht=92%) were analyzed with the BIS method. The electric measurements were performed at room temperature on the BIA analyzer ABC-01 "Medass" (Russia) at the frequency range 5 – 500 kHz (31 frequencies). The erythrocyte suspensions of 1 ml were drawn into conductivity cell. The cell is a plastic tube with two pairs of potential and current electrodes. The electrodes were connected to the BIA analyzer. Data from the analyzer were processed on the PC with special software. The method of electrical measuring is described in our previous work in details [4]. The impedance locus was built by bioimpedance analyzer software. The example of the hodograph is shown in the figure 2.

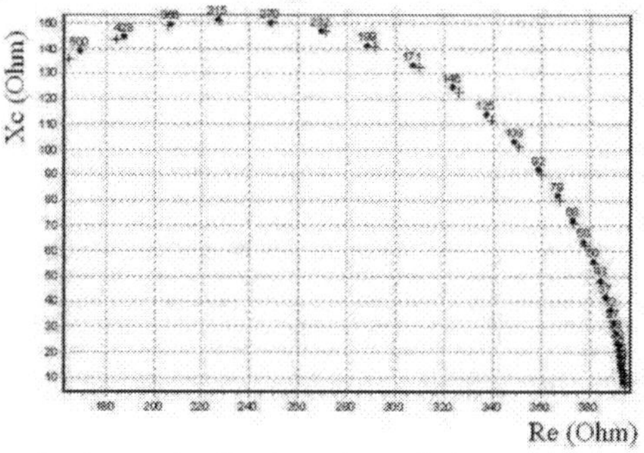

Figure 2. The example of the hodograph. Each point of the graph shows the ratio of reactance to resistance at the appropriate frequency marked with the number.

2.4. Statistics

Results are expressed as mean + SD. Paired Student's t-tests were used for comparison of BIS parameters in suspensions with different osmolarity. Pearson's correlation (r) was utilized to study the relations between the bioimpedance spectroscopy parameters and MCV. Multiple regression analyze was used to establish independent correlations between the BIS parameters and MCV. The multiple standardized regression coefficient (β) was calculated.

3. Results

The values of hematocrit, red blood cells concentration and mean corpuscular volume are presented in the table 1.

Table 1. The values of hematocrit, red blood cells concentration and mean corpuscular volume in solutions with different osmolarity

	Osmolarity, mosmol/l		
	240	290	375
Ht, %	92,4±2,5	93,6±2,5	91,9±2,0
MCV, fl	106,3±7,6***	94,7±7,1	85,8±6***
RBC, 10^{12}/l	13,1+0,8***	14,9+1	16,1+1,2***

(*** – p <0.001 compared to 290 mosmol/l suspension)

Ht of the erythrocyte suspensions was equal. MCV was higher in the hypotonic and lower – in the hypertonic solutions. Vice versa, RBC decreased in the hypoosmotic and increased in the hyperosmotic solutions.

The BIS parameters of the erythrocyte suspensions in solutions with different osmolarity are shown in the table 2.

Table 2. The BIS parameters of erythrocyte suspensions in solutions with different osmolarity

	Osmolarity, mosmol/l		
	240	290	375
Re, Ohm	660,0±102,5**	562,9±72,6	450,7±28,9**
Ri, Ohm	64,3±4,6*	68,7±1,5	73,0±6,1*
Fchar, kHz	219,5±31,1*	245±31,3	291,4±19,3**
Alpha	0,318±0,002*	0,316±0,002	0,314±0,002*
Cm, pF	47,4±5,6**	53,4±4,6	59,7±3,4**

(*, ** – $p<0.05$, <0.01 respectively compared to 290 mosmol/l suspension)

We found that Re and Alpha increased in the suspensions with low osmolarity and decreased in the hypertonic suspensions. On the contrary, Ri, Fchar and Cm were higher in the hyperosmotic and were lower in the hypoosmotic suspensions.

The results of correlation analysis of RBC and MCV with the BIS parameters are presented in the table 3.

Table 3. Correlations of RBC and MCV with the BIS parameters

	Re	Ri	Fchar	Alpha	Cm
RBC	-0,55**	0,50*	0,52**	-0,56**	0,59**
MCV	0,62**	-0,46*	-0,60**	0,65**	-0,68**

(*, ** – $p<0.05$, <0.01 respectively)

As it's shown, RBC was positively related to Ri, Fchar and Cm and was negatively related to Re and Alpha. We found that the correlations of MCV with Re and Alpha were positive, and with Ri, Fchar and Cm were negative.

The results of multiple regression analysis of MCV with the BIS parameters are presented in the table 4. One can see that only correlation of the Alpha parameter with MCV was statistically independent.

Table 4. The results of multiple regression analysis of MCV with the BIS parameters

	Re, Ohm	Ri, Ohm	Fchar, kHz	Alpha	Cm, pF
β	0,70	-0,35	0,60	0,77	0,12
p	0,56	0,07	0,40	0,01	0,86

4. Discussion

The main results of our study showed that all BIS parameters differed in suspensions with different osmolarity. The BIS parameters were related to MCV and RBC values, but only correlation of Alpha parameter with MCV was statistically independent.

The changes of RBC and MCV in hypertonic and hypotonic suspensions were determined by the movement of water on the osmotic gradient. As soon as in the solutions with low osmolarity water entered the erythrocytes they swelled, their volume increased and their shape became more spherical. Since total cell volume of the suspensions was equal, RBC in the hypotonic solutions decreased. On the contrary, in the hypertonic solutions water leaved the erythrocytes, so the cells shrunk and their volume decreased, and RBC of the suspensions increased [2].

Obviously in our experiment condition, corpuscular volume of erythrocytes, red blood cells concentration and ions concentration in suspending solution influence the electric properties of measured suspensions.

It's known that Re reflects extracellular and Ri – intracellular fluid volume [5]. Since hematocrit of the erythrocyte suspensions was equal, total extracellular and intracellular fluid volume didn't change in the suspensions with different osmolarity. Evidently, the Re and Ri changes associated with difference in the ions concentration in the hypotonic and hypertonic solutions. Electrical conductivity was higher and Re value was lower in the hyperosmotic solution with high NaCl concentration, in the hypoosmotic solution with low ion concentration – vice versa. However changes in the osmotic concentration of the suspending media cause flow of water into or out of the erythrocyte until osmotic equilibrium is restored [2]. In the solutions with high osmolarity water leaves red blood cells and hemoglobin concentration in the erythrocytes increases. Since hemoglobin reduces ion mobility in the erythrocytes, high hemoglobin concentration causes a decrease of conductivity and an increase of Ri [6]. In low osmolarity suspensions hemoglobin concentration is lower, and Ri decreases. Thus in our experiment conditions the Ri changes are related to the hemoglobin concentration changes in red blood cells and doesn't depend on MCV and RBC.

Fchar value is inversely proportional to resistance and capacitance of measured object [1]. In biological objects capacitance is determined by cell membranes quantity or, in other words, by a number of cells. Since in the hypertonic solutions RBC is higher, it may be expected that Fchar increases. On the contrary, in the hypotonic solutions RBC is lower, so Fchar should decrease. However we found that Fchar increased in the hyperosmotic and decreased in the hypoosmotic suspensions (table 2). Obviously, resistance influenced the Fchar value in this case. The resistance, in turn, depends on the ions concentration. Hence, we suggest that Fchar was determined by the ions concentration, like Re. In the hypotonic suspensions the resistance increased so that Fchar decreased, in the hypertonic solutions – vice versa. This is confirmed by close correlation of Fchar with Re in our study ($r=-0.94$, $p<0.001$) and in the article [7].

It was found that Cm increased in hypertonic and decreased in hypotonic suspensions (table 2). Obviously, since in the hypertonic solutions RBC was higher and total cell membranes area increased, Cm value grew. In the hypotonic solutions – vice versa, RBC and total cell membranes area decreased, so Cm value diminished. In the papers [4, 8] a correlation between Cm and Ht was also obtained. The authors also believe that a gain of the cell membranes number in the blood samples with high hematocrit causes a Cm increase. Thus it is possible that Cm depended on RBC and didn't reflect MCV.

The Alpha parameter and osmotic effects. All BIS parameters were related to MCV (table 3) but the multiple regression analysis showed that only correlation of MCV with Alpha was statistically independent (table 4). This indicates that Alpha parameter may be determined by erythrocyte corpuscular volume. Besides, Grimnes S. et al [1] postulated that the Alpha depends on the cell size. Obviously cell volume can also influence Alpha. However, it's known that the Alpha parameter is also related to the cell size and shape heterogeneities [3]. Alterations of the heterogeneity of erythrocytes in our experiments were induced by changes in solutions osmolarity. In the hypotonic solutions erythrocytes shape became more spherical, suspensions heterogeneity decreased, so Alpha should be increased. In the hypertonic solutions the erythrocytes shrunk, cell heterogeneity in suspensions increased, and Alpha should be decreased. Thus Alpha parameter may be related both to the cell heterogeneity and mean corpuscular volume.

5. Conclusions

Thus, the results showed that all investigated bioimpedance parameters changed in according to the osmotic modifications of erythrocyte suspensions. Changes in Re, Ri, Fchar were caused concentrations of NaCl ions, Cm led to the red blood cell concentrations, however, the Alpha parameter is related to mean corpuscular volume in concentrated erythrocyte suspensions.

References

[1] Grimnes S, Martinsen O G 2008 *Bioimpedance and bioelectricity basics* (L.: Academic Press) pp 311 – 312
[2] Lang F and Busch G 1998 *Physiol. Rev.* **78** 247-306.
[3] Ivorra A et al 2004 *Proc. XII Int. Conf. on electrical Impedance & V electrical Impedance Tomography (Gdansk, Poland, 20 – 24 June 2004)* vol 1 pp 87-90.
[4] Malahov M V et al 2010 *J. Phys.: Conf. ser.* **224** (1). 012130.
[5] Cornisht B H et al 1993 *Phys. Med. Biol.* **38** 337-46.

[6] Pauly H and Schwan H P 1966 *Biophys. J.* **6** 621–39.
[7] Poon C S and Choy T T 1981 *Biophys. J.* **34** 135-47.
[8] Ulgen Y Sezdi M 2007 *Med. Biol. Eng. Comput.* **45** 653-660.

Journal of Physics: Conference Series **407** (2012) 012013 doi:10.1088/1742-6596/407/1/012013

Humidity in sandy soil measured by passive, wireless, and resonant sensor with bifilar coil

S F Pichorim, V A Marcis and G T Laskoski

CPGEI/UTFPR-Federal University of Technology-Paraná. Av. Sete de Setembro, 3165 CEP(ZIP): 80230-901 Curitiba, Paraná, Brazil.

pichorim@utfpr.edu.br

Abstract. In this paper a passive, wireless, and resonant (PWR) sensor for monitoring soil humidity is presented. This sensor employs a bifilar coil and it is compared with the classical PWR sensor with interdigital capacitor. Besides that, a theoretical model is developed to evaluate the influence of medium in sensor's response. A resonant circuit is used to detect the humidity variation. In a setup for 300 g of fine sand, where sensor's responses were evaluated for dry and wet sand, tests show a linear correlation of 0.7195 and a sensitivity of 252 kHz / % between resonant frequency and fresh water-to-sand mass ratio.

1. Introduction

Passive sensors, that do not require internal battery, generally are simple, small, and cheap. In general, passive sensors are wireless (communication by telemetry) and resonant (their frequency is modulated as a function of a measured parameter) using capacitive and inductive components. There are several applications for PWR (passive, wireless, and resonant) sensor, for example, pressure, humidity, gas concentration, pH, elasticity, radiation, and food quality measurement [1]-[6]. The modulated resonant frequency of PWR can be detected remotely by a simple excitation coil and reading circuit [7]-[8].

Humidity PWR sensors have the ability to detect water content, where it can be used in civil engineering, environmental biology, or agriculture to investigate the drying rate of seeds, wood, sand, concrete or soil [9]. The soil humidity is also important in estuarine and coastal environment, where the amount of sand and water is significant. These regions are characterized by high concentration of nutrients, temperature gradients, variable salinity and other physical-chemical parameters (including humidity) that are very important for the breeding and feeding of most species that inhabit the oceans [10]. Therefore, PWR sensor for humidity is important in bioimpedance area to characterize bacterial growth in presence of water [6], biological environment behavior [10], wood drying quality, plant properties and soil quality [11].

Ong *et al.* developed a PWR sensor to measure the water content (or moisture content) in sand and concrete [9]. This recent work was fundamental to the development of this sensor for sandy soils. Thus, it is described in more details and its results will be compared to the sensor proposed here. The sensor (see figure 1a) is made of a single-sided printed circuit board with square spiral inductor with 7 turns (inner and outer lengths of 25 and 40 mm, respectively). An interdigital capacitor with 19 sets of electrodes of 20 mm in length is used as sensitive element. All tracks have a width of 0.5 mm and the same spacing of 0.5 mm. This PWR sensor presented a resonant frequency of 22.5 MHz for dry sand and about of 20 MHz for sand with 21.4 % of mass ratio of water to sand [9].

2. Development

In this work, a new technique to build up a PWR sensor for humidity was developed. This new sensor employs bifilar coils, which inductors are wound with two parallel wires (i.e., bifilar coil), where higher self-capacitances were obtained [12]. In this technique, the end of first wire is connected to the beginning of the second one by a jumper wire. A PWR sensor with bifilar winding is shown in figure 1(b). It can be seen that bifilar turns work as an inductor and a capacitor, simultaneously. As a consequence of an inductor with high self-capacitance, the interdigital capacitor can be omitted in the new PWR sensor [7], [13]. This bifilar arrangement reduces the number of components in PWR sensor and increases the useful area for humidity when compared with interdigital capacitor area (figure 1(a)). For example, figures 1(a) and 1(c) show the conventional sensor and its useful area for the capacitive sensor (interdigital) in area R (purple). The area V (green in figure 1(d)) is the capacitive area of bifilar PWR sensor. Also, in area A (blue) is the useful area for the inductor in the conventional sensor compared with the area V (green), which is also the inductive element of the bifilar sensor. A larger area makes possible greater number of turns, increasing the distance for telemetry, that means a better inductive link.

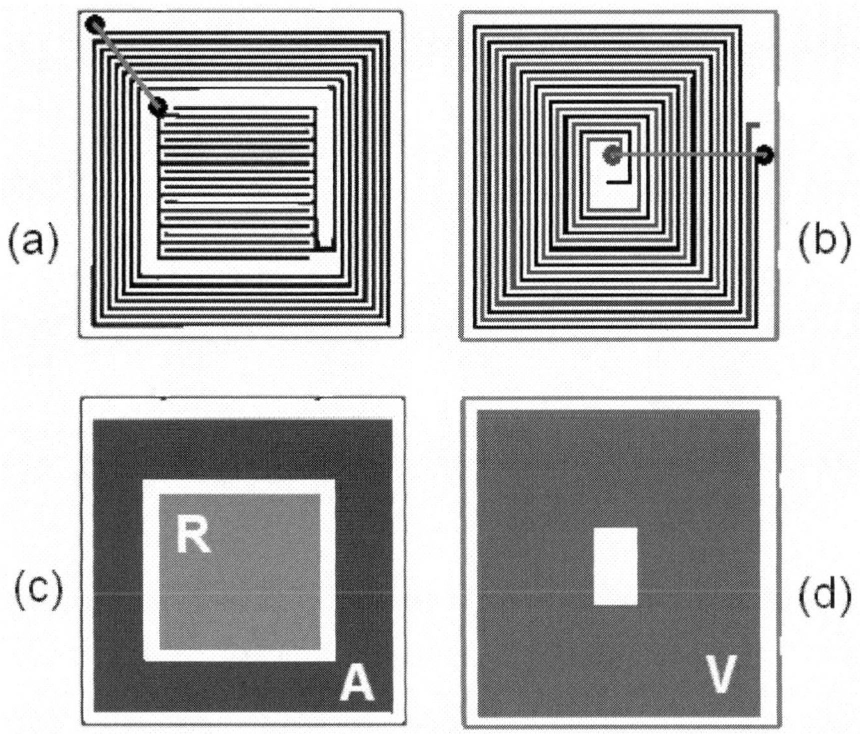

Figure 1. The conventional PWR (a) and bifilar winding PWR (b) sensors. Both are built on a printed circuit board. In (a), there are a square spiral inductor, the interdigital capacitor, and the jumper (red) completing the LC parallel circuit [9]. In (b), there are the bifilar inductor (two wires are painted with black and blue for clarity the explanation) and jumper (red), that completes the bifilar coil. In (c) and (d), a comparison of the useful areas for the capacitive element R (purple) and inductive element A (blue) for conventional and bifilar sensor. It is important to point out that bifilar sensor presents both (inductive and capacitive) elements in the area V (green).

2.1. Electrical model of PWR sensor

The capacitance between adjacent tracks of bifilar coil is used as a capacitive sensor. The total capacitance of sensor (figure 2) is composed by a substrate (permittivity ε_S), the medium under test (with permittivitty ε_M and conductance σ_M), and a thin enameled layer of polyester or polyamide-imide (permittivity ε_E about $4.\varepsilon_0$), which covers the copper tracks and protects them against oxidation caused by the medium.

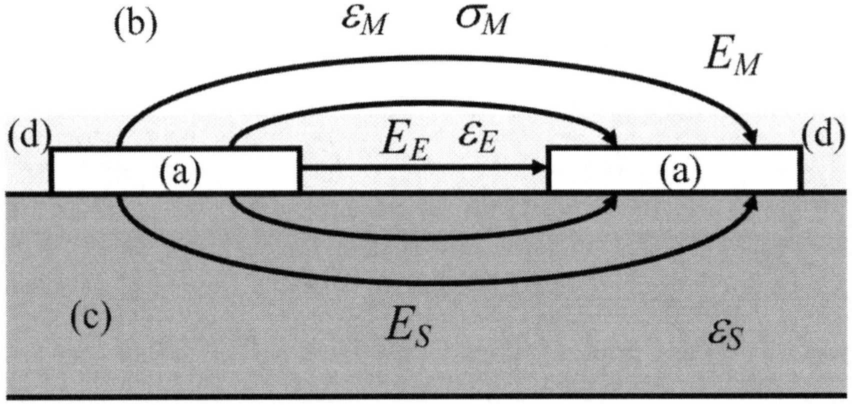

Figure 2. Electric fields between two adjacent tracks (a), crossing the medium (b), the substrate (c), and enameled layer (d).

The admittance of the medium (Y_M) is capacitive and can be calculated by

$$Y_M = j\omega C = j\omega\varepsilon K \quad , \tag{1}$$

where K is a constant that represents the geometry of sensor, and ε is the complex permittivity of the medium. The electrical permittivity is complex number due to dispersive effects caused by conductivity (σ_M), which it can be written as

$$\varepsilon = \varepsilon_M - j\frac{\sigma_M}{\omega} \quad . \tag{2}$$

Applying equation (2) in (1), yields

$$Y_M = j\omega\varepsilon_M.K + \sigma_M.K \quad . \tag{3}$$

Defining C_M and R_M as total capacitance and resistance of the medium, respectively, which can be written as

$$C_M = \varepsilon_M.K \quad \text{and} \quad R_M = \frac{1}{\sigma_M.K} \quad , \tag{4}$$

the admittance of the medium is

$$Y_M = j\omega C_M + \frac{1}{R_M} \quad . \tag{5}$$

Thus, rewriting equation (5) as the impedance of the medium (Z_M), a parallel association of the resistance R_M and capacitive reactance $X_{CM}=(2.\pi.f.C_M)^{-1}$, can be written as

$$Z_M = -jX_{CM} \ // \ R_M \quad , \tag{6}$$

where it is possible to see that impedance of medium, due to the electric fields between two adjacent tracks, depends on capacitance and resistance of medium. As can be observed in figure 2, electric fields E_S and E_E, that cross only the substrate and enameled layer, respectively, cause a constant capacitance (called here C_S), which is independent of the medium influence. The electric field thought the medium also crosses the enameled layer. The influence of this layer is represented by a serial capacitance C_E. Therefore, a representation of an electrical circuit equivalent for the sensor is shown in figure 3. In this circuit, the total capacitance (C_T) of the PWR sensor can be determined by

$$C_T = C_S + \frac{C_E.C_M}{C_E + C_M} \quad . \tag{7}$$

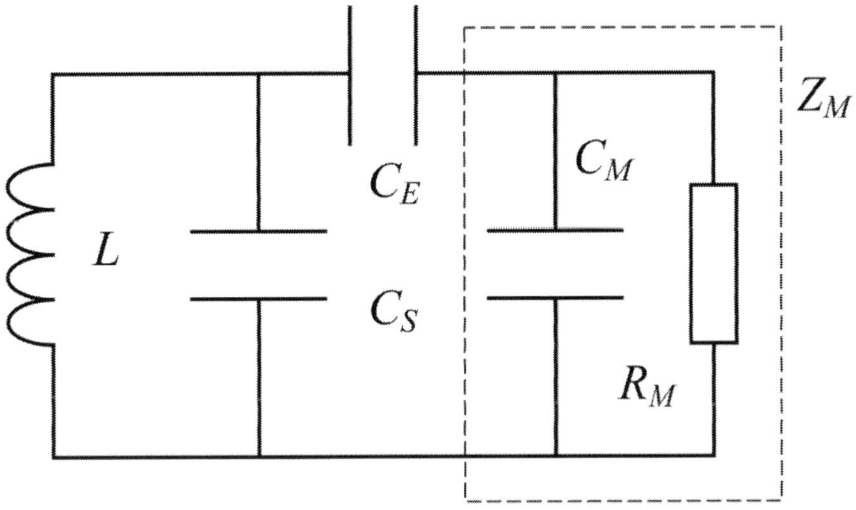

Figure 3. Electrical circuit for PWR sensor.

As the enameled layer is very thin, the capacitance C_E is greater than C_M , but it cannot be neglected in the electrical model analysis. The humidity increases the capacitance C_M due to high relative permittivity of water (about 80) in comparison with the air or dry sand (about 4) [9]. When the PWR is inserted in a sandy soil, the capacitance C_M increases about 6.25 times, since dry sand and wet sand have relative permittivity of 4 and 25, respectively [14]. It is important point out that constant capacitances of sensor (C_S and C_E according to equation 7) reduce the variation of total capacitance C_T. However, the change of humidity in soil significantly modulates the capacitance C_M and C_T, where it can be measured as a function of resonant frequency (fo) by the classical equation

$$f_0 = \frac{1}{2\pi\sqrt{L.C_T}} \quad . \tag{8}$$

As can be observed in figure 3, the conductivity of water in the soil can modify R_M and the response of PWR sensor. Although typical fresh water is much more conductive (range of few mS/m) [14] than deionized water, the resistance R_M is high enough to be neglected in sensor analysis ($R_M \to \infty$). However, the presence of sea water, which has high conductivity (some S/m) caused by its salinity, reduces drastically the resistance R_M, and, consequently, decreases the quality factor of resonant sensor. This effect causes a reduction in signal amplitude, getting more difficult to read remotely *fo*. For low values of R_M, the sensitivity of C_M is also reduced.

2.2. Sensor construction

The PWR bifilar sensors were made of a single-side printed circuit board (PCB) with 9 pairs of turns with a spacing of 0.5 mm between tracks. The figure 4 shows the sensor with dimension of 37.5 × 40 mm^2. External dimensions, tracks width, and tracks separation were chosen based on a previous work [9] to be possible comparisons. To complete the bifilar coil a jumper is soldered between points 1 and 2, resulting a spiral coil with 18 turns.

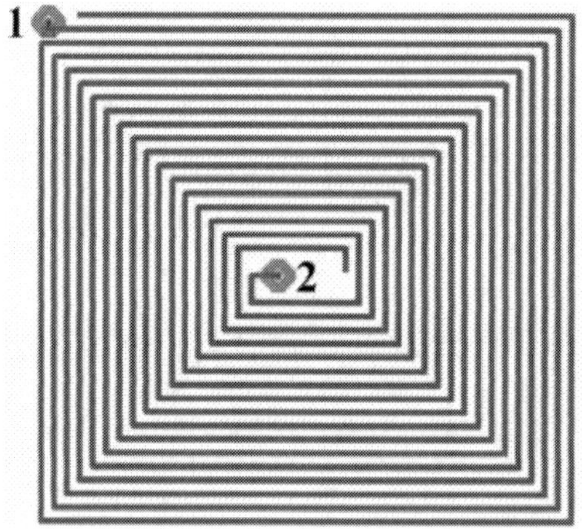

Figure 4. The new PWR sensor with 9 pairs of bifilar turns.

3. Tests and results

In general, the resonant frequency is obtained by the response frequency analysis using an RF signal generator and an oscilloscope to measure the maximum amplitude. Meanwhile, for wireless sensors, other technique must be employed. Resonant frequency (*fo*) can be detected remotely using a dip-meter, where the absorption of an RF oscillator signal is measured by a meter pointer deflection when a resonant LC circuit is placed near to its antenna coil. In this experiment, a Leader LDM-815 dip-meter was used. However, the value of *fo* is shown in an analog dial with poor precision and resolution. Thus, a second coil (loop coil in figure 5) connect to a digital oscilloscope (Agilent model MSO 6034A) was used to measure, with higher precision, the oscillation frequency of dip-meter.

The sensor must be placed close to dip-meter antenna coil (few centimeters). Turning the dial of dip meter, it changes the oscillation frequency. When the oscillation frequency corresponds to the PWR resonant frequency, there is a maximum amplitude reduction, where it can be easily measured in a pointer of dip-meter. The dip point (minimum for *fo*) also can be observed in the digital oscilloscope by monitoring signal amplitude reduction. When dip point is found, the resonant frequency of PWR sensor is read with an oscilloscope.

Tests of PWR sensor for humidity measurement were made by insert the sensor in a sample of sandy soil with a volume of water. The sand was collected in Pontal do Paraná beach, where it was verified a particle size distribution with 82% of fine sand, about 10% of medium sand, about 7% of very fine sand, and 1% of fine sediments (silt and clay). The sand was inserted in a box with dimensions of 90 × 50 × 75 mm^3. A procedure to dry and wet the sand changes the humidity values, where the weight of water was used to determine the humidity of sample soil. Initially, the sample of sand was heated up to 200°C for two hours, and a mass of 300 g was measured with a precision weighing balance (iPlena with resolution of 1 g). This dry sample is denominated as *md*. Afterwards, the sample was saturated with the addition of fresh water and it was obtained the mass of wet sand mass, denominated *mw*. The drying process was natural evaporation of water. Some hours were necessary to complete the process. Periodically, the *fo* of PWR sensor was measured and the sample weighed to determine the new mass of water in soil (*mw*). This process was repeated until the mass of sand is equal to initial value (*md*). The humidity of soil was related with mass ratio of water to sand (*h*), calculated by

$$h = \frac{mw - md}{md}.100\% \quad . \tag{9}$$

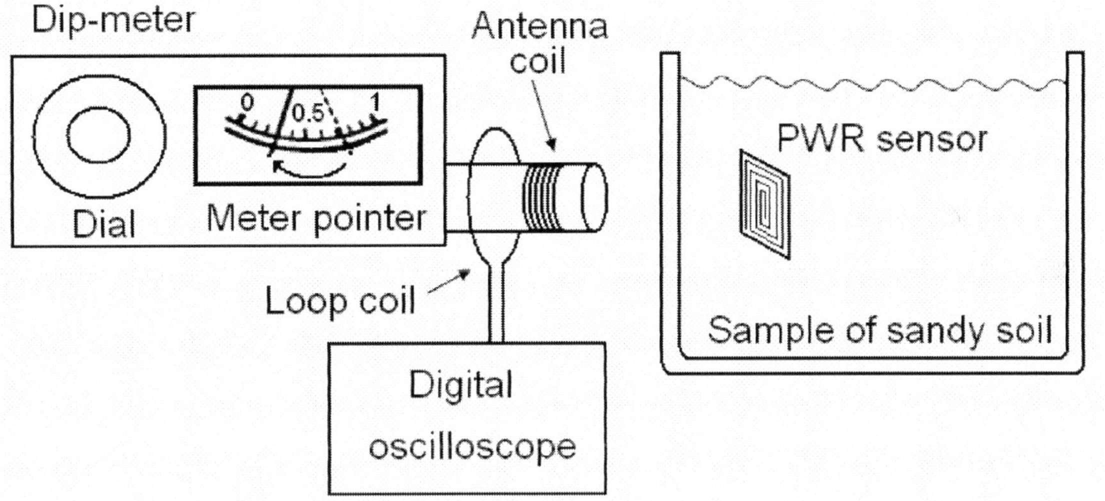

Figure 5. Setup for experiment.

In figure 6 is shown the relation between resonant frequency of sensor and the water-to-sand mass ratio (soil humidity). Frequencies of about 19 MHz and 14 MHz were measured for dry and wet sandy soil (dots), respectively. A linear-curve fitting measured points was calculated, presenting a slope of 190.9 kHz per each percent of mass ratio and coefficient of determination about 0.7195.

Other measurements were done to evaluate the electrical model of sensor (figure 3). When the PWR sensor was removed from soil, its *fo* and self-inductance *L* were measured by precision vector impedance analyzer (Agilent model 4294A). Using measured values of *fo* and *L*, the electrical model of PWR sensor could be solved by equation 7, yielding the capacitances C_S, C_E, and C_M. These capacitances values are presented in table 1.

Journal of Physics: Conference Series **407** (2012) 012013 doi:10.1088/1742-6596/407/1/012013

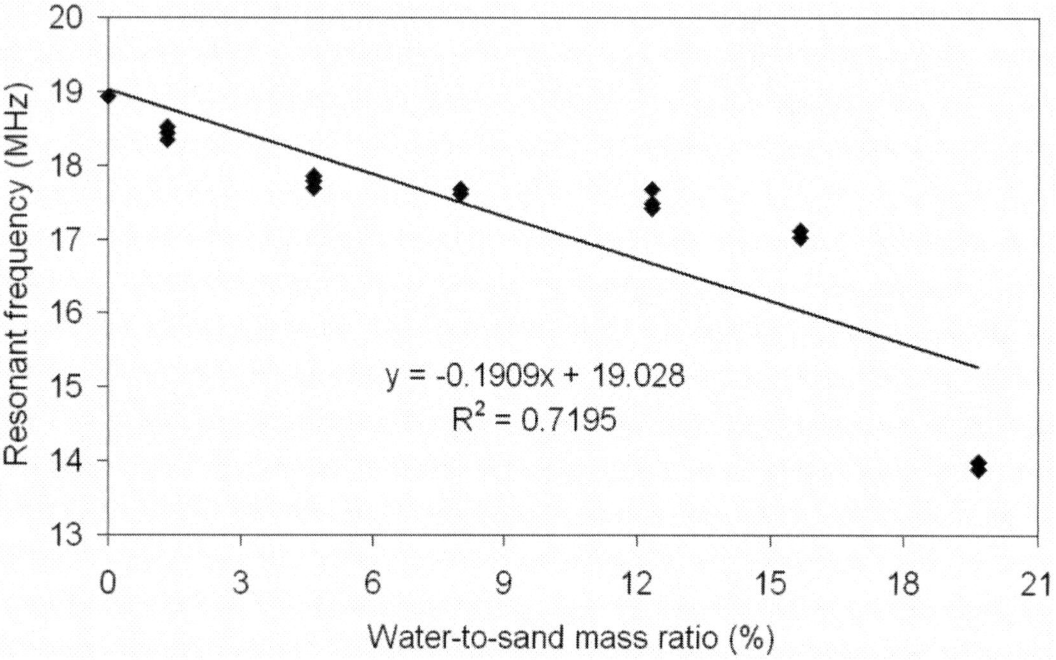

Figure 6. Resonant frequency of PWR bifilar sensor as a function of mass ratio of water to sand (dots) and a linear-curve fitting measured points.

Table 1. Measured parameters of PWR bifilar sensor and capacitances of electrical model obtained by equation 7.

Parameter	Situation	Value
fo	Air	19.6 MHz
fo	Dry sand	18.94 MHz
fo	Wet sand	13.9 MHz
L	at 1 MHz	6.58 μH
Sensitivity	Full range	252 kHz / %
C_S		8.6 pF
C_E		62 pF
C_M	Dry sand	2.2 pF
C_M	Wet sand	13.7 pF

4. Discussion and Conclusion

In this work the development and implementation of a PWR sensor for monitoring soil humidity was presented. The developed sensor employs a bifilar coil that presents a high self-capacitance. This intrinsic capacitance changes as a function of medium and it is used to monitoring the humidity. The main advantage of this sensor is that bifilar arrangement reduces the components and increases the useful area for humidity when compared with the PWR with interdigital capacitor. Besides that, it increases the useful area for the inductor, improving the inductive link between PWR sensor and reading circuit. In this work it was obtained 123% of more inductance (6.58 μH versus 2.95 μH[9]) than a PWR sensor with interdigital capacitor, same dimension, and similar humidity sensitivity.

The theoretical model shows that sensor's capacitance presents an impedance that depends on capacitance and resistance of medium. In this work, a resonant circuit was used to detect the variation of capacitance as a function of the humidity of medium. However, it is necessary to ensure that medium presents a low conductivity due to the dependence of the resistance between the PCB tracks in capacitor's response.

In a setup for 300 g of fine sand, tests shows a linear correlation between resonant frequency and fresh water content. Analyzing figure 6, the points below 17% for mass ratio present a better linear correlation (R^2=0.8548) and a sensitivity of 90 kHz/%. However, including the points of mass ratio up to 21%, the sensitivity increases to 190.9 kHz/%, but linear correlation is reduced (R^2=0.7195). This sensitivity is higher than obtained by Ong *et al.* (114 kHz per each percent of mass ratio)[9]. The problem of non-linearity may be related to the sand drying process. During the process it was observed that some regions of soil sample were drier than others. It means that balance measured total mass of water and sensor analyzed the humidity in its neighborhood. For greater volumes of soil the problems of sand drying could be reduced. Thus, for dry and water-saturated sand, the water distribution is more homogeneous and these values of humidity and resonant frequency are more reliable. Thus, the full range sensitivity of 252 kHz / % can be accepted.

The electrical model of sensor developed in equation 7 and figure 3 could describe the behavior of total capacitance of sensor with the influence of medium humidity, and could be used to predict, with acceptable precision, the resonant frequency of PWR sensor.

Future works can be developed to improve sensor's response (linearity and sensitivity). Besides that, it is possible to analyze sensor with other dimensions and frequencies. Despite of the good response of theoretical model, it must be improved to verify the influence of medium. Especially in medium with high conductivity. Finally, the sensor may have its application extended to other areas, such as monitoring of moisture in the soil for agriculture, construction, farming, and other possible applications.

Acknowledgment

Authors would like to thank to Brazilian Council for Scientific and Technological Development (CNPq) and Brazilian Coordination of Improvement of Personnel of Superior Level (CAPES) for their financial support.

References

[1] Ong K G Zeng K and Grimes C A 2002 A wireless, passive carbon nanotube-based gas sensor *IEEE Sensors Journal* **2** 82-88

[2] Harpster T J Stark B and Najafi K 2002 A passive wireless integrated humidity sensor *Sens. Actuators A, Phys* **95** 100-107

[3] Pichorim S F and Abatti P J 2006 Biotelemetric passive sensor injected within tendon for strain and elasticity measurement *IEEE trans. Biomedical Engineering* **53** 921-925

[4] Baldi A Choi W and Ziaie B 2003 A self-resonant frequency-modulated micromachined passive pressure transensor *IEEE Sensors Journal* **3** 728-733

[5] Son C and Ziaie B 2008 A wireless implantable passive microdosimeter for radiation oncology *IEEE Trans. on Biomedical Engineering* **55** 1772-1775

[6] Tan E L Ng W N Shao R Pereles B D and Ong K G 2007 A wireless, passive sensor for quantifying packaged food quality *Sensors* **7** 1747-1756

[7] Pichorim S F and Destefani F A 2010 Estudo de bobina bifilar de Tesla como sensor para engenharia biomédica *XXII Congr. Bras. de Eng. Biomédica* 359-362

[8] Pichorim S F and Abatti P J 2008 A novel method to read remotely resonant passive sensors in

Journal of Physics: Conference Series **407** (2012) 012013 doi:10.1088/1742-6596/407/1/012013

biotelemetric systems *IEEE Sensors Journal* **8** 6-11
[9] Ong J B You Z Mills-Beale J Tan E L Pereles B D and Ong K G 2008 A wireless, passive embedded sensor for real-time monitoring of water content in civil engineering materials *IEEE Sensors Journal* **8** 2053-2058
[10] Lana P Bianchini A Ribeiro C Niencheski L Filmann G and Santos C 2006 Avaliação ambiental de estuários brasileiros: diretrizes metodológicas *Série Livros* **22** (Rio de Janeiro: Museu Nacional)
[11] Grimnes S and Martinsen Ø G 2008 *Bioimpedance & Bioelectricity Basics* (London: Elsevier) 2nd ed., p.5
[12] Tesla N 1894 Coil for electro-magnets *US patent office* 512.340
[13] Miranda C M and Pichorim S F 2012 Equacionamento e modelagem da bobina bifilar de Tesla *Revista Uniandrade* **13** **1** 42-59
[14] Liu L and Li Y 2001 Identification of liquefaction and deformation features using ground penetrating radar in the New Madrid seismic zone, USA *Journal of Applied Geophysics* **47** 199-215

First Latin-American Conference on Bioimpedance (CLABIO 2012) IOP Publishing
Journal of Physics: Conference Series **407** (2012) 012014 doi:10.1088/1742-6596/407/1/012014

The detection of chronic cerebral hemorrhage in rabbits with magnetic induction

Jian Sun[1], Gui Jin[1], Mingxin Qin[1*], Zibing Wan[2], Jinbao Wang[3], ChaoWang[4], Wanyou Guo[4], Lin Xu[1], Xu Ning[1], Jia Xu[1], Xianjie Pu[1], Mingsheng Chen[1] and Hongmei Zhao[2]

[1] College of Biomedical Engineering and Medical Imaging,
[2]Experimental Animal Center,
Third Military Medical University, Gaotanyanzheng street 30,
Shapingba district, 400030 Chongqing, China
[3]College of Life Sciences and Technology, Xidian University,
[4]College of Electronic Engineering, Xidian University,
Taibai South Road 2, 710126 Xi'an, China
Email: nksunj@126.com, qmingxin@tmmu.edu.cn
[*] Corresponding author.

Abstract. Chronic cerebral hemorrhage (CCH) in the brain is an important clinical problem that is often monitored and studied with expensive devices such as MRI and PET, which are not readily available in low economical resource parts of the world. We have developed a less expensive tool for non-contact monitoring of CCH in the brain. The system measures the phase shift between the electromagnetic signals on the two coils. CCH was induced in the brain of rabbits by stereotactic method. Intracranial pressure (ICP) and Electrocardiograph (ECG) of subjects were monitored for 1.5h. Signals were continuously monitored up to t=1.5h at exciting frequency 10.7MHz. From 0.8 to 2.4 ml of autologous blood was injected (each injection quantity of 0.8 ml, the interval time for 30 minutes). The results show significant phase shifts increase as a function of injection volume. ICP and phase shift were directly proportional to the related, while HRV were stable around 200beats*min^{-1}. Our system has high sensitivity that even 0.8 ml can also be detected. In this study, the curves of inductive phase shift are significantly related to ICP. This observation suggests that the method could be valuable, in addition to continuous monitoring, also for early warning in emergency medicine and critical care units.

1. Introduction

In brain neurosurgery to date, deformation of the brain usually occurs due to the change of intracranial pressure, which can be caused by surgical intervention, medications received and acute cerebral hemorrhage [1]. The study of the cerebral hemorrhage is benefited from advances in medical imaging such as MRI and PET [2]. While advanced medical imaging techniques are extremely valuable in studying cerebral hemorrhage, the devices are expensive and not readily available either to patients or to researchers in areas with low economical resources [3, 4]. The goal of this study is to introduce a new inexpensive and valuable technology which could become a useful tool in research and clinical applications related to the study of brain injury in general and cerebral hemorrhage in particular.

The field of study relates to the electromagnetic properties of biological tissues. As shown by Tarjan and McFee [5], an electrode-less measurement of changing impedances in the human body can be achieved by measuring the effect of inducted eddy currents. It is established in the literature that different biological tissues have different electromagnetic properties and that it is possible to distinguish between different tissues on the basis of these properties [6]. The complex electrical properties of impaired tissue are substantially different from those of normal tissue, various non-invasive systems based on electro-magnetic measurements have been proposed for different medical applications [7, 8]. Magnetic induction tomography (MIT) has been studied as a potential tool for noninvasive detection of tissue oedema [7, 8, 9, and 10]. Also there are different approaches for detecting electromagnetic changes in tissue by non-invasive means [11]. Measuring changes in the electromagnetic properties of the tissue with two non-contact induction coils placed around the tissue could be used as an alternative technique for detection of the clinical changes in the tissue [12, 13]. While information is not in the form of an image it can produce temporal and even gross local information about changes in tissue [14], from non-invasive bulk measurements. In particular, detecting the change in the variable electromagnetic phase shift in time after the occurrence of a suspected clinical episode could serve as a first-order clinical warning to detect the presence and progression of change in tissue [11]. In this study, we show with experiments in the rabbit brain that the concept of non-invasive measurements of the electromagnetic properties of the brain can be used to detect the evolution of cerebral hemorrhage in the brain. The electromagnetic property we measure in this study is the change of electromagnetic phase shift in time in the brain. We describe here the fundamental principles, the device and show results obtained in the brain of cerebral hemorrhage rabbits.

2. Methods

2.1. Theoretical estimation

We have considered as a case study a cylinder of tissue of radius ***R*** and thickness ***t***, placed centrally and midway between an excitation coil and a detection coil spaced at a distance ***d***(t<<d). A sinusoidal current ***I***, of angular frequency ***ω***, flows in the excitation coil and produce a magnetic field **B.** The magnetic field matches in phase a reference voltage (**V**r) in the excitation coil. The non-magnetic cylinder has conductivity ***σ*** and relative permittivity ε_r. The skin depth ***δ*** is assumed to be much greater than ***t***, which means that the attenuation produced by the cylinder is small. We use spherical polar coordinates(***r***, ***θ***) as shown in Figure 1.

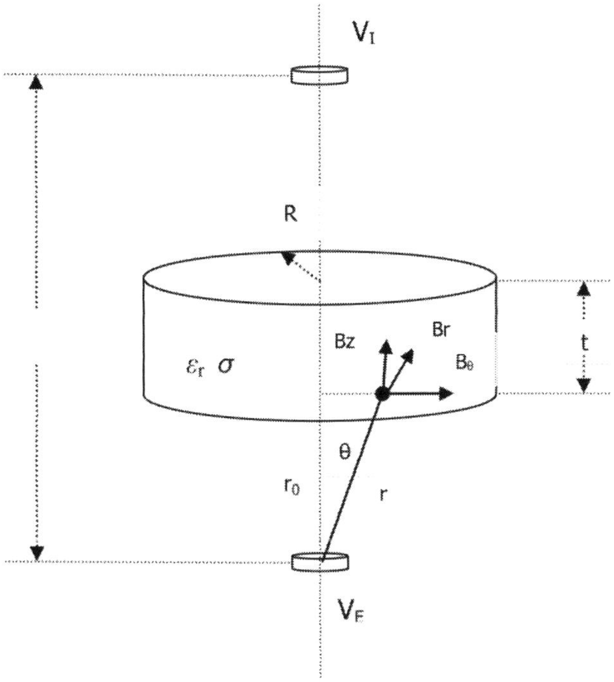

Figure 1.Magnetic field in a cylinder due to an excitation coil (shown in polar coordinates). The cylinder of radius ***R*** and thickness ***t*** is placed centrally and midway between a small excitation coil and a small sensing coil spaced a distance ***d***. The cylinder has conductivity σ_c, relative permittivity ε_c and is non-magnetic.

According to Griffiths et al [15], at a point ***P*** in the cylinder model, the magnetic vector potential

$$\frac{\Delta B}{B} = (\omega\varepsilon_0\varepsilon_r - j\sigma)\left(\frac{\omega\mu_0 t d^3}{16}\right)\left[\frac{1}{r_0{}^2} - \frac{r_0{}^2+2R^2}{(r_0{}^2+R^2)^2}\right] \quad (1)$$

The magnetic field and its perturbation can be obtained from the voltages induced in the detection coil, **V**i and △**V**i [16]. Hence, △**B/B** can be defined in terms of the induced voltage in the detection coil [16] by

$$\frac{\Delta B}{B} = \frac{\Delta V_i}{V_i} \quad (2)$$

We define a constant k

$$k = \left(\frac{\mu_0 t d^3}{16}\right)\left[\frac{1}{r_0^2} - \frac{r_0^2+2R^2}{(r_0^2+R^2)^2}\right] \tag{3}$$

Substituting (3) and (2) into (1), the phase of the total induced voltage θ_{Vi} in the detection coil with respect to the induced voltage by the primary magnetic field could be expressed as a function of frequency and electrical parameters in the cylinder between the coils [15]by

$$\theta_{Vi} = arctg\left(\frac{Im(\Delta V)}{V+Re(\Delta V)}\right) = arctg\left(\frac{-k\omega\sigma}{k\omega^2\varepsilon_0\varepsilon_r+1}\right) \tag{4}$$

The phase of the reference voltage θ_{Ve} in the presence of the cylinder sample can be estimated from the following expression [11]

$$\theta_{Ve} = arctg\left[\frac{\mathrm{Im}\left(\frac{Z_L+\Delta Z_L}{Z_{out}+Z_L+\Delta Z_L}\right)}{\mathrm{Re}\left(\frac{Z_L+\Delta Z_L}{Z_{out}+Z_L+\Delta Z_L}\right)}\right] \tag{5}$$

Where Z_L represents the impedance of the excitation coil. Z_{out} represents the output impedance of the excitation source. The total change in phase shift ($\Delta\theta$) between the reference and induced voltages in the excitation and sensing coil, respectively, is given by the expression:

$$\Delta\theta = \theta_{Vi} - \theta_{Ve} \tag{6}$$

2.2. Surgical procedure.

A rabbit (3.8kg) were anaesthetized with urethane (25% , 5ml/kg) via ear vein. We establish the model of ACH by means of stereotactic method. The coordinates for the brain were determined using a standard atlas [17]. A short midline skin incision was made and the point for a stereotactic approach through the skull determined. The head was leveled in such a way that the bared Bregma was 1.5mm higher than Lambda. With the position of Bregma as a reference point, the coronal zero planes (AP_0) passed through the coronal suture. A_1 referred to the coronary plane which was 1 mm before AP_0, while P_1 was 1mm after AP_0, The scope of the internal capsule was between A_5 and P_2. We chose coronary plane P_1. Puncture point is on the right of the coronal suture(R=6mm) and after the sagittal suture (P_1, AP=1mm). A 1mm drill bit was twisted by hand to perforate the skull. A needle (d=0.7mm) was then introduced to the proper depth (H=13mm) and autologous blood slowly injected. From 0.4 to 5.2 ml of autologous blood was injected. RM 6280B biological signal collecting and processing system was used to measure the changes of ICP and APP of paracele [18] and HRV. As shown in Figure2, rabbits were operated to monitoring ICP, APP and HRV.

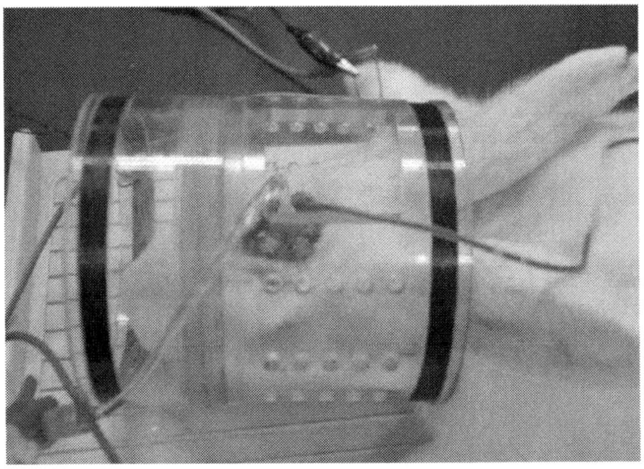

Figure2. Rabbit model

2.3. Experimental magnetic induction prototype

An experimental inductive setup was designed and constructed. The system consists of seven modules: signal generator, excitation coil, detection coil, phase detector, PC and biological signal collecting and processing part. A picture of the experimental set-up is shown in Figure 3.

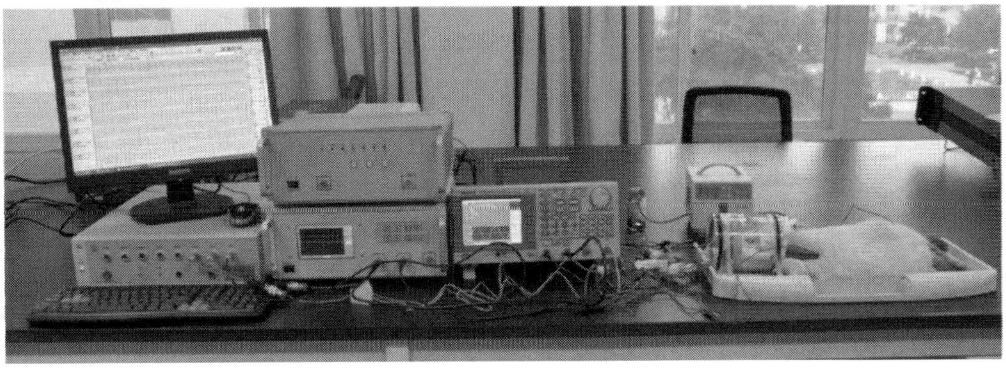

Figure3. Experimental setup for detection of cerebral hemorrhage in rabbits with magnetic induction

2.3.1 Signal generator

The function generator was implemented by a signal generator, which supplies a signal $I\cos(\omega t)$ of approximately 5v at 10.7MHz. The range of output power from the excitation source is 1.5~33dB. The frequency stability is the order of magnitude of 10^{-4}, the distortion reaches 10^{-2}~10^{-4} and the SNR is 30~60dB in the excitation source.

2.3.2 The coils

The excitation and detection coils were coaxially placed. Both coils were ten turns with radius $r = 5$ cm. The distance between coils was d=11 cm. To avoid inductive pickup the leads of the coils were twisted. The excitation coil induces a current in the detection coil by magnetic induction and an inductive phase shift as a function of the electrical properties of the rabbit brain.

2.3.3 Phase detector

The phase detector includes the band-pass filter, amplifier, AD, FDGA, DSP, flash memory and LCD. The phase detector can be calibrated and the temperature drift and the noise can be eliminated by calibration software. The parameters of our phase detector are as follows: the range of the phase measurement: 0~180^0, the phase precision: 0.05^0, the range of gain: -10~35dB, once measurement time: 3~7s. The 12h data measured can be saved in the phase detector.

2.3.4 Biological signal collecting and processing part

The data processing part (6280C, Chengdu Instrument Factory) was controlled by a personal computer.

2.4. Experimental design.

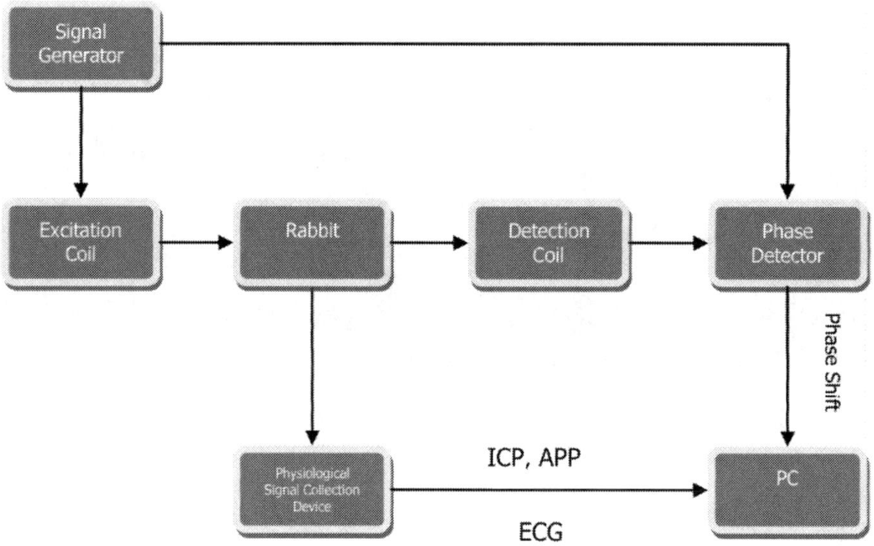

Figure 4 Work flow diagram

Figure4 shows a work flow diagram of the experimental prototype. ICP, APP, ECG and phase shift were synchronously collected for the comparison of measured results. In a typical procedure, baseline data were taken after the surgical procedure (t = 0). Signals were continuously monitored for 1.5h. Autologous blood was injected from 0.8 to 2.4 ml (Each injection quantity of 0.8 ml, the interval time for 30 minutes). When each 0.8ml blood was injected, the phase detector measured the phase shift for 30 minutes and got 500 data. We choose 10 stable and continuous measurements of phase shift before next injection. ICP, APP and HRV were processed according to the 10 measurements of phase shift. And then each data set was averaged.

The values of the phase shift changes were recorded with respect to baseline measurements taken after the procedure. Experimental excitation and detection coils are coaxial placed around the brain cavity of the rabbits. In all the experiments the coils were placed in such a way that the brain cavity was centered between the excitation and detection coils. Changes in phase shift as a function of the injection volume were recorded. The fundamental hypothesis in this study is that the changes in phase shift measured with the detection coil will be representative of the changes in electrical conductivity of the brain.

The Animal Experiments and Ethical Committee of the Institution approved all experimental protocols and the care of animals was carried out in accordance with the Declaration of Helsinki and IASP guidelines [19, 20].

3. Results and discussion

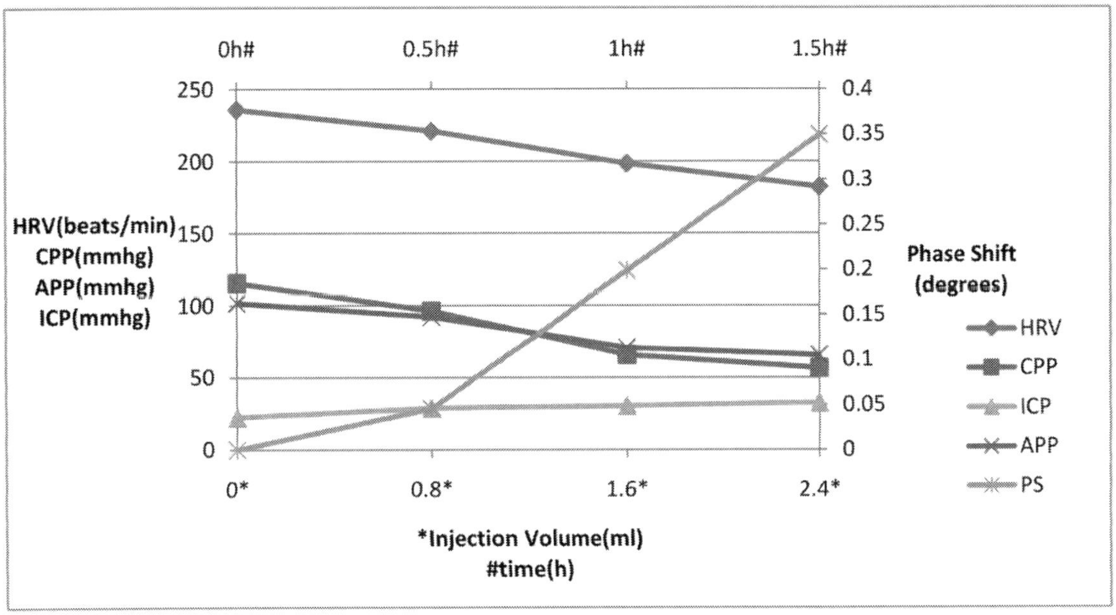

Figure 5.Relationship curves of intracranial pressure and inductive phase shift from rabbit brain hematoma model

Figure 5 showed relationship curves of ICP, APP, HRV and phase shift from rabbit brain hemorrhage model at exciting frequency 10.7MHz.

Along with the increase of injection quantity, the changes of phase shift also increased. The phase shift curve's slope was stable. Our study told us magnetic induction theory was reliable, and our system can detect a small amount of bleeding in the brain. ICP were directly proportional to phase shift, while HRV were stable around 200beats*min^{-1}. So the changes of phase shift reflected the changes of ICP.

Figure 5 also showed the change rate of ICP and phase shift were not identical. Reasons are as following: First, the collection of ICP and phase shift is not completely in sync. Second, manual injection brought error. Third, injection intervals were short (only for 30 minutes), so CBF (cerebral blood flow) and CSF (cerebrospinal fluid) regulatory function were not considered. As compared to this, CBF was reduced both around the hemorrhage and in the surrounding brain, and this change was strongly volume-dependent and was not accompanied by significant alterations in CPP [21].Fourth, the animal moved in the experimental process. In future we must optimize our system to ensure synchronous collection and use injection pump to avoid artificial error.

Also we may take long term monitoring up to 12h above. The coils inductance and electrical conductivity of rabbit brain were not calculated. Due to time restrictions, sample number was too small and had no control group. Further development aims at contrast research.

This was an early experiment and had many drawbacks. But this was the first study to reveal the relationship between the ICP and change of phase shift by animal experiment. A comparison of the analytical and experimental results with respect to others' previous studies [12, 13] suggests that various physiological conditions in the brain such as oedema and ischaemia can be distinguished through the analysis of the inductive phase shift change at specific frequencies. In this study, the curves of inductive phase shift are significantly related to ICP. This observation suggests that the method could be valuable for early warning in emergency medicine and critical care units in addition to continuous monitoring.

4. Conclusion

Measuring the magnetic induction phase shift changes as a function of time after brain injury has the potential for being a simple method for early and non-contact detection of the occurrence and progress of cerebral hemorrhage in the bulk of brain. This technique could be useful in both research and clinical applications in particular for low economical resources areas of the world. This is a preliminary study and much more research remains to be done to confirm these observations. Although the experimental evidence provided in this study has relevance to the clinical case, the cerebral hemorrhage formation in clinical practice is a more complex process than the experimental model evaluated in this study.

5. Acknowledgments

This work was supported by Foundation of Third Military Medical University and National Natural Science Foundation of China (61072254).

6. References

[1] Lighthall W J 1988 Controlled cortical impact: a new experimental brain injury model *J.*

Neurotrauma.5 1–15

[2] Chelsea S. Kidwell, Julio A. Chalela, Jeffrey L. Saver 2004 Comparison of MRI and CT for Detection of Acute Intracerebral Hemorrhage *JAMA*. 292(15) 1823-1830

[3] Otten D M, Onik G and Rubinsky B 2004 Distributed network imaging and electrical impedance tomography of minimally invasive surgery Technol. *Cancer Res. Treat.* 3 1–10

[4] Granot Y, Ivorra A and Rubinsky B 2008 A new concept for medical imaging centered on cellular phone technology *Plos.One*. 3 2075

[5] Tarjan P and McFee R 1968 Electrodeless measurements of the effective resistivity of the human torso and head by magnetic induction IEEE Trans. *Biomed. Eng.* 4 266–78

[6] Foster K R and Schwan H P 1989 Dielectric properties of tissues and biological materials: a critical review CRC Crit. Rev. *Biomed. Eng.* 17 25–104

[7] Al-Zeiback N S 1993 A feasability study of in vivo electromagnetic imaging Phys. *Med. Biol.* 38 151–60

[8] Korjenevsky A V and Cherepenin V A 1997 Magnetic induction tomography J. Commun. Technol. *Electron.* 42 469–74

[9] Korjenevsky A V and Cherepenin V A 1999 Progress in realization of magnetic induction tomography Ann. *N. Y.Acad. Sci.* 873 346–52

[10] Griffiths H 2001 Magnetic induction tomography *Meas. Sci. Technol.* 12 1126–31

[11] Gonzalez C A 2009 The detection of brain ischaemia in rats by inductive phase shift spectroscopy *Physiol. Meas.* 30 809–819

[12] Gonzalez C A and Rubinsky B 2006 A theoretical study on magnetic induction frequency dependence of phase shift in oedema and haematoma *Physiol. Meas.* 27 829–38

[13] Gonzalez C A and Rubinsky B 2006 Detection of brain oedema with frequency dependent phase shift electromagnetic induction *Physiol. Meas.* 27 539–52

[14] Rojas R, Rubinsky B and Gonz´alez C A 2008 The effect of brain hematoma location on volumetric inductive phase shift spectroscopy of the brain with circular and magnetron sensor coils: a numerical simulation study *Physiol. Meas.* 29 S255–S266

[15] Griffiths H, Steward W R and Gough W 1999 Magnetic induction tomography—a measuring system for biologicalmaterials Ann. *New York Acad. Sci.* 873 335–45

[16] Scharfetter H, Casa˜nas R and Rosell J 2003 Biological tissue characterization by magnetic induction spectroscopy(MIS): requirements and limitations *IEEE Trans. Biomed. Eng.* 50 870–80

[17] Sawyer CH, Everett JW, Green JD 1954 The rabbit diencephalon in stereotaxic coordinates. *J Comp Neurol.* 101 801-824

[18] Wl Jing 2007 Study on the model building of internal capsule hemorrhage and the changes of intracranial pressure and discharge of vagus nerve in rabbits shanxi medical journal 361

[19] Helsinki 1996 Declaration of Helsinki *Br. Med. J.* 313 1448–9

[20] ZimmermannM1983 Ethical guidelines for investigations of experimental pain in conscious animals *Pain.* 16 109–10

[21] Fredric P. NATH, F.R.C.S., et al 1986 Early hemodynamic changes in experimental intracerebral hemorrhage *J Neurosurg.* 65 697-703

First Latin-American Conference on Bioimpedance (CLABIO 2012) IOP Publishing
Journal of Physics: Conference Series **407** (2012) 012015 doi:10.1088/1742-6596/407/1/012015

Electrical Impedance Tomography Reconstruction Through Simulated Annealing using a New Outside-in Heuristic and GPU Parallelization

R. S. Tavares

E-mail: reseta@uol.com.br

T. C. Martins

E-mail: thiago@usp.br

M. S. G. Tsuzuki

Escola Politécnica da Universidade de São Paulo, São Paulo, Brazil.
Mechatronics and Mechanical Systems Engineering Department
Computational Geometry Laboratory

E-mail: mtsuzuki@usp.br

Abstract. Electrical Impedance Tomography (EIT) is an imaging technique that attempts to reconstruct the conductivity distribution inside an object from electrical currents and potentials applied and measured at its surface. The EIT reconstruction problem is approached as an optimization problem, where the difference between the simulated and measured distributions must be minimized. This optimization problem can be solved using Simulated Annealing (SA), but at a high computational cost. To reduce the computational load, it is possible to use an incomplete evaluation of the objective function. This algorithm showed to present an outside-in behavior, determining the impedance of the external elements first, similar to a layer striping algorithm. A new outside-in heuristic to make use of this property is proposed. It also presents the impact of using GPU for parallelizing matrix-vector multiplication and triangular solvers. Results with experimental data are presented. The outside-in heuristic showed to be faster when compared to the conventional SA algorithm.

1. Introduction

Electrical impedance tomography (EIT) is an imaging modality that estimates the electrical conductivity distribution within the body when a low amplitude current pattern is applied to a body surface and the potential at determined points of that surface is measured through electrodes or, alternatively, when a potential is applied and the current flowing through the surface is measured [1].

The two main forms of EIT are dynamic imaging and static imaging yielding differential and absolute images respectively. The images produced by differential imaging represent the conductivity changes of a region between two time intervals [2]. Imaging physiological function within the body largely relies on this technique. This work is focused on the reconstruction of static conductivity images, which requires more advanced numerical algorithms and, consequently, heavier computational load.

The spatial resolution of EIT is not comparable to other medical imaging techniques such as magnetic resonance, computerized tomography or ultrasonic imaging. It is also sensitive to electrode positioning and the boundary geometry. However, EIT presents some advantages over these techniques, such as being harmless to the patient, low cost and portable. Additionally, EIT has faster time-response characteristics, which enables it to monitor cyclic changes in the living tissues better than conventional imaging modalities.

EIT has a wide rage of medical applications, detection of acute cerebral stroke [3], breast cancer [4], monitor cardiac activity [5] and monitor lung aeration imposed by mechanical ventilation in critically ill patients [1, 6]. The last application is the main interest of our group, where EIT can be used to implement a protective lung strategy that can result in an improved lung function and survival [7].

In this work, it is shown that the reconstruction method proposed by Martins et al. [16] happens from the outside towards the interior of the domain. It is reasonable to presume that the objective function is more sensible to modifications on external cells, which can be explained by the fact that more information is present on the boundary. Additionally, in this work it is proposed an heuristic that uses this fact and accelerated the image reconstruction. The recent emergence of multiple-core architectures, mainly GPUs, as platforms that offer a large increase in computational power over traditional architectures motivates their use in scientific computation. Particularly, the high cost of reconstructing static EIT images through SA can be diminished by the use of parallel algorithms.

This work is structured as follows. Section 2 explains the problem formulation: direct problem, inverse problem and how to solve the inverse problem as an optimization problem. Section 3 explains how to apply SA to EIT, and briefly explains the SA with objective function incomplete evaluation [8]. The proposed outside-in heuristic is explained in section 4. Section 5 presents GPU parallelization algorithms for the sparse matrix-vector multiplication and triangular solvers. Section 6 shows some results and, finally, the conclusions are in section 7.

2. Formulation

The typical forward problem in EIT is, given the conductivity distribution σ and the current J injected through boundary electrodes, find the potential distribution ϕ within Ω and in particular the resulting potentials at the measurement electrodes ϕ_m. The frequencies used in EIT are low enough so that the quasi-static approximation hold, and thus we can ignore capacitive and inductive effects. Under such quasi-static conditions, the solution of the forward problem is rather simple as it only requires solving the Laplace equation

$$\nabla(\sigma\nabla\phi) = 0 \tag{1}$$

At the boundary, currents are injected through electrodes; thus the current density through the $l-th$ electrode surface J_l is given by

$$\sigma\frac{\partial\phi}{\partial\widehat{n}} = J_l \tag{2}$$

where $\widehat{n}$ is the external normal versor and zero elsewhere.

2.1. Current Patterns

Data is collected by injecting current with a single source and measuring voltage. Current is injected sequentially to the body using a pair of electrodes. There are several ways in which the pair of electrodes is switched and the voltage measurements are collected in the literature.

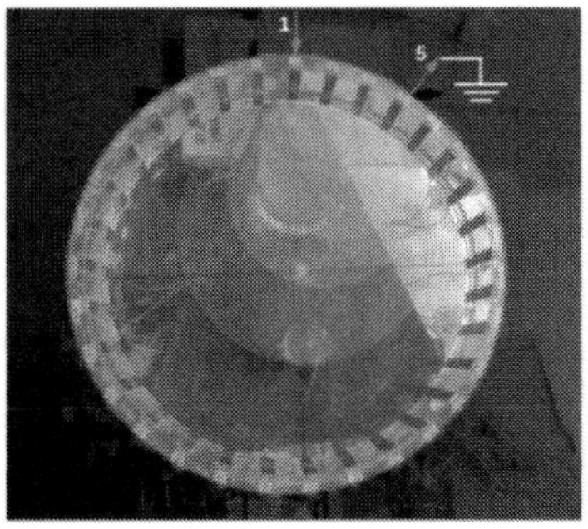

Figure 1: Current pattern with jump three.

Brown and Seagar [9] suggested a method whereby electrical currents are sequentially applied to the body using a pair of adjacent electrodes, and voltages between adjacent non current-carrying electrodes are measured. This procedure is repeated, applying current between each pair of adjacent electrodes to obtain a voltage data set. Hua et al. [10] suggested the cross method, where current is injected between a pair of electrodes which are separated from each other, obtaining a more uniform current distribution. In this method, two adjacent electrodes for current and voltage reference are chosen. Hua et al. [10] also introduced the opposite method whereby current is injected through two diametrically opposed electrodes and uses the electrode adjacent to the current injecting electrode as the voltage reference. Fig. 1 shows the current pattern used in this work.

2.2. *Finite Element Model*

The inverse problem is formulated as given the injected currents J and the potentials at measurement electrodes ϕ_m, find the electrical conductivity distribution σ within Ω. In practice, only a finite number of potential measurements is made through the electrodes, so the Dirichlet boundary condition is incomplete [11]. For an irregular domain and isotropic media, analytical solution to the Laplace equation (1) with boundary condition (2) are unknown; thus, the partial differential equations were approximated by the finite element method (FEM), the domain is discretized with triangular linear elements with constant conductivity and both problems, forward and inverse, are solved numerically. The virtual potential principle associated with the Laplace equation provides the local element matrices.

When the local element matrices are stated in terms of global coordinates of the mesh, the global conductivity matrix [1], which includes electrode contact impedance effects, is obtained; then the following relation holds

$$K \cdot \Phi = C \tag{3}$$

where $K(\sigma) \in \Re^{s \times s}$ is the conductivity matrix calculated at a given particular distribution σ_p, Φ is a matrix containing nodal potentials corresponding to each applied current pattern, and C represents p linearly independent current patterns.

It is known that for a symmetric positive definite matrix K, a factorization $K = L \cdot L^T$ exists and it is called Cholesky decomposition. The Cholesky decomposition is mainly used for the numerical solution of linear equations $K \cdot \Phi = C$. If K is symmetric and positive definite, then $K \cdot \Phi = C$ can be solved by first computing the Cholesky decomposition $K = L \cdot L^T$, then solving $L \cdot y = C$ for y, and finally solving $L^T \cdot \Phi = y$ for Φ [12, section 4.2.4]. Solving the forward problem, therefore, is done through Cholesky decomposition preconditioning and solving two triangular systems.

2.3. The Inverse Problem as an Optimization Problem

In the inverse problem, the objective is to find the conductivity distribution K given applied current patterns C and measured potentials Φ. Since there are known methods for efficiently solving the forward problem (such as FEM), one possible approach to the inverse problem is to look at it as an optimization problem, where the optimization variables are a parametrization of the conductivity inside the domain and the optimization function is some measure of how the solution of the forward problem applied to the conductivity distribution produced by the optimization variables matches the measured data.

One possible objective function $E(\sigma)$ is the Euclidean distance between the measured electric potentials and the calculated potentials for all the applied current patterns for a given conductivity distribution. Let ϕ^i_m be the measured potential vector *at the electrodes* and $\phi^i_c(\sigma)$ the corresponding calculated electrode potentials obtained from the solution of the forward problem for a conductivity distribution σ at the $i-th$ current pattern

$$E(\sigma) = \sqrt{\sum |\phi^i_m - \phi^i_c(\sigma)|_2} \tag{4}$$

An example of such approach is in [13], where the objective function (4) is minimized by Sequential Linear Programming yielding estimations of the conductivity distribution. Mello et al. [13] pointed that it is difficult to solve this problem by methods based on gradients of the objective function due to the fact that the problem is often ill-posed. Numerical errors in the calculation of the objective function are greatly amplified in its derivatives. That is why the interest on the Simulated Annealing (SA) applied to EIT is increasing, as it requires no evaluation of objective function derivatives.

3. Applying Simulated Annealing to EIT

As seen in section 2.3, the EIT inverse problem can be formulated as an optimization problem, and as such, can be approached with SA. Herrera et al. [14] minimized objective function (4) with SA and by doing so, managed to reconstruct very accurate conductivity distributions of the body, but at a very high computational cost. This is unsurprising, as each step of the SA involves the solution of a full FEM problem in order to evaluate the objective function. Martins et al. [8] proposed a SA with incomplete evaluation of the objective function that is briefly explained in the following.

3.1. Incomplete Evaluation of the Objective Function [8]

The evaluation of the objective function is responsible for the bulk of the SA computational cost. Therefore, methods to reduce its cost are of extreme interest, and a possible manner is to employ an incomplete evaluation of the objective function, as introduced by Lakshmanan and Derin [15]. Martins et al. [16] presented a study of how the SA process behaves in the presence of a partial evaluation of the objective function. The exact distribution of the objective function is presumed unknown, but boundaries for its values are estimated. As shown by Martins et al. [16], there is a relationship between those boundaries and the probability of deviation of the algorithm from an exact SA. Using those relationships and imposing upper limits for those probabilities, a stopping criteria for the interactive calculation of the objective function can be obtained.

It can be shown that, by imposing P_{err} as an upper limit on the probability of the process taking a "wrong" decision (rejecting a solution when it should accept it or accepting when it should reject), we can create conditions for the boundaries:

$$e^{-\frac{\Delta E_{max}}{kt}} \geq \begin{cases} 1 - P_{err} & \text{if } \Delta\widetilde{E} \leq 0, \\ e^{-\frac{\Delta\widetilde{E}}{kt}} & \text{if } \Delta\widetilde{E} > 0 \end{cases} \tag{5}$$

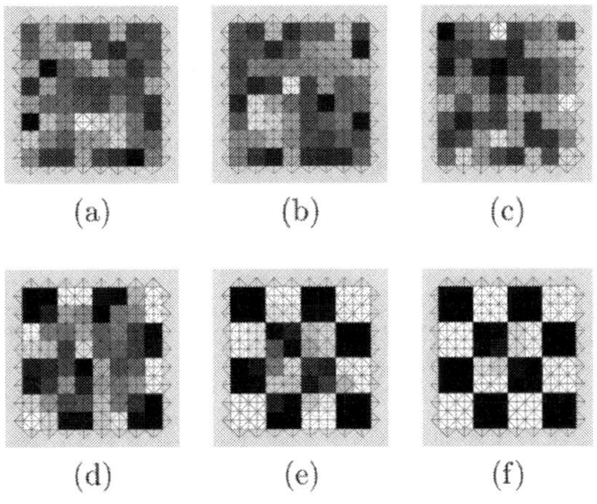

Figure 2: Convergence behavior for the "Checkerboard" problem [16].

$$e^{-\frac{\Delta E_{min}}{kt}} \leq min(1, P_{err} + e^{-\frac{\Delta \widetilde{E}}{kt}}) \qquad (6)$$

Where ΔE_{max} and ΔE_{min} are the partial solutions of two separated sets of FEM problems and $\Delta \widetilde{E}$ is variation of the partial evaluations of the objective function $\Delta \widetilde{E} = \widetilde{E}^j - \widetilde{E}^{j+1}$.

The conditions given by (5) and (6) do not translate directly into stopping criteria for the Conjugate Gradients (CG) algorithm. As such, every solution must have its objective function evaluation process stored so it may be possible to continue it from where it has stopped. In fact, it is often computationally more efficient to improve the boundaries on the evaluation of the *previous* SA solution than the current one. Since CG has an asymptotically geometric convergence and presuming that both the previous SA solution and the current one have similar convergence rates (a reasonable assumption considering that the K matrix does vary very little from one iteration to another and a pre-conditioner equalizes even more the convergence rates), a simple heuristic for picking which solution to evaluate further is to just pick the one with looser boundaries.

Figure 2 shows the convergence behavior of the SA with objective function incomplete evaluation. The impedance distribution is reconstructed from the outside towards the interior of the domain. It is reasonable to presume that the objective function is more sensible to modifications on external cells, which can be explained by the fact that more information is present on the boundary. The following section explains the proposed heuristic that makes use of this property.

4. Proposed Outside-In Heuristic

Considering the mesh presented in Fig. 3, all elements can be divided in circular layers according to each element's distance to the center of the mesh. Somersalo et al. [17] was the first to propose an algorithm based on the outside-in property. They assumed first that a current density with very rapid spatial variation is applied to the surface of the body. Since most of the corresponding current does not penetrate very deeply into the body but rather is affected mainly by the conductivity near the boundary, the corresponding voltage measurement can be used to estimate the conductivity in a thin layer at the boundary. Once the conductivity in this thin layer is approximately known, the outcome of the same kind of experiment is computed if the known boundary layer were stripped away. The conductivity on this next layer can be computed, strip it away, and so on through the body. In this way an estimate of the conductivity of the

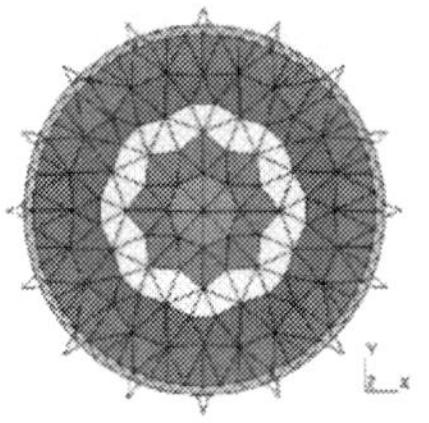

Figure 3: Mesh divided in circular layers.

body is found, layer by layer. However, as boundary elements are dependent on even innermost elements, it is necessary to start over from the external layers when changes were applied to inner elements. The heuristic proposed in this work uses a similar concept with the SA with objective function incomplete evaluation.

Martins et al. [8] used a SA with adaptive neighborhood proposed by Martins and Tsuzuki [18, 19]. Such a SA randomly selects one parameter and creates a modification based on the parameter crystallization factor. The crystallization factor modifies the probability distribution of the next candidate in way to increase the number of accepted solutions and keep the SA exploratory capacity.

In this work, the SA selects an element for modification in two steps. One must consider that the mesh elements are classified in layers. Initially, the SA selects one layer and, then, it selects one element from the previously selected layer. All elements from the selected layer have equal probability of being selected. However, the probability of each layer to be selected is different and controlled by the outside-in heuristic.

In its initial configuration, the SA selects with higher probability the external layer. As the temperature decreases, the external layer probability of being selected decreases and the internal layers probability increases. It is necessary that such process happens sequentially from external to internal layers, one after the other. Such heuristic is implemented as follows: if a candidate is accepted, its layer's probability is increased; if a candidate is rejected, its layer's probability is reduced and the immediately inner layer's probability is increased.

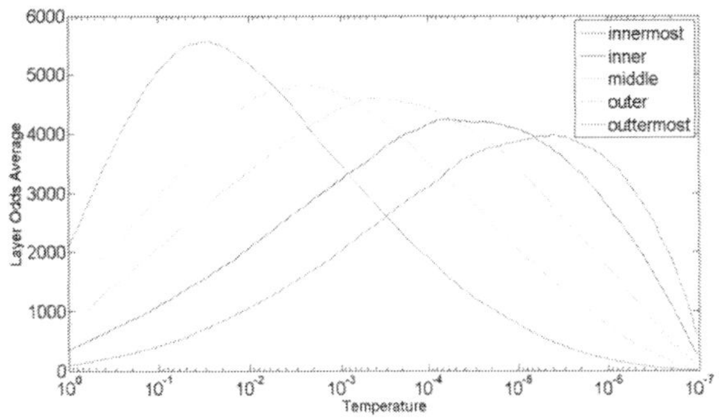

Figure 4: Each layer's average probability, per temperature.

In Fig. 4, it is shown a graph of each layer probability through the SA processing. Elements were divided in five layers. As it can be seen, the probability of each layer, representing its probability of having an element selected as a solution candidate, switches from the outermost layer to the innermost layer, passing through all intermediary layers. Therefore, along the cooling procedure, every layer becomes the most dominant one for a moment, and have the most elements selected.

5. GPU Parallelization

GPUs nowadays are specialized processing units with huge computational capability, and highly parallelized. Their role is mainly to stock big and complex sets of triangles in an ideal three dimensional environment, to assemble several of them as the CPU commands. In the last years GPUs grew capable of doing the most amazing graphic effects, like blur for foggy environment, transparencies for glasses, windows and liquids, reflections as well as partial reflections, for water effects. As the possibilities increased, instead of producing different hardware for each effect, the graphic units manufacturers started conceiving programmable graphic units, which could run arbitrary user defined programs in order to manipulate images in some of the key steps of the graphic rendering [20].

Current CPUs have between 4 and 8 available cores, against GPUs' hundreds of cores. Such capability led to researchers employing GPUs to solve problems with high computational costs, such as medical imaging, eletromagnetism and linear systems solvers. NVIDIA introduced an API, called CUDA, for general purpose computing on GPU (GPGPU), allowing researchers to use high level programming languages for implementing their applications to run on highly parallelized GPUs. The EIT reconstruction algorithm originally uses a serial, albeit efficient, library for matrix operations called Eigen. Such library will be compared to CUDA's in one of the most used matrix operations.

A graphic processing unit implements what is called the SIMT (Single Instruction Multiple Thread) architecture, that allows to have a conditional block based on a non constant value. This was impossible with the SIMD (Single Instruction Multiple Data) that is the architecture implemented by an ideal vector processor. Conditional (*if* statements) can greatly decrease performance inside an SM, as each branch conditional must be evaluated.

5.1. Sparse Matrix Data Structures and Sparse Matrix-Vector Multiplication

Typical sparse matrix formats incur storage and instruction overheads per non-zero element, since information is needed to keep track of which non-zero values have been stored. One aim in selecting a data structure is to minimize these overheads. A matrix is computationally represented by one dimensional array. When matrix $A_{m\times n}$ is dense, a common scheme stores element (i, j) at position $v[i+j\cdot m]$. This mapping allows random access to any matrix element.

Sparse matrix vector multiplication is one of the most important kernels in scientific and engineering applications. The challenge in developing high performance implementations of such kernels is choosing the data structure and code that best exploits the matrix structural properties.

It will be considered the sparse matrix vector multiplication $y \leftarrow y + Ax$, where A is an $m \times n$ sparse matrix with k non-zero elements, and x and y are dense vectors. The matrix vector multiplication can be defined as

$$\forall a_{i,j} \neq 0 : y_i \leftarrow y_i + a_{i,j} \cdot x_j$$

The conventional implementation of matrix vector multiplication is shown in Alg. 1.

Algorithm 1 Dense matrix vector multiplication.

```
for i = 0 to m - 1 do
    for j = 0 to n - 1 do
        y[i] ← y[i] + v[i + m · j] · x[j]
    end for
end for
```

In the case of sparse A, a sparse format is coordinate format, where A is stored using three arrays, v, ri and ci, each of length k. Array element $v[l]$ holds the value of the matrix element

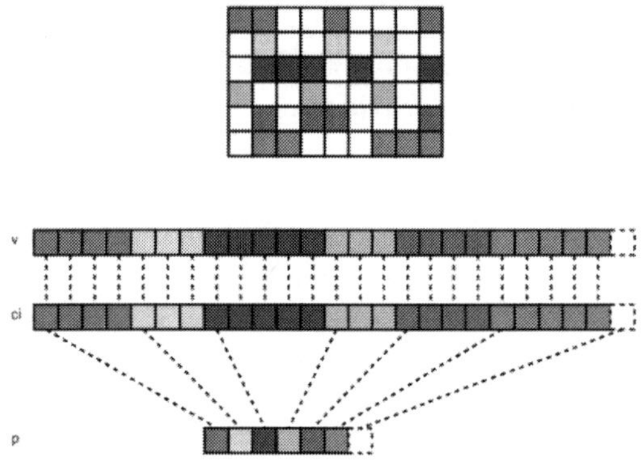

Figure 5: The elements of each row are shaded using the same color. Each row is stored as a sparse vector, and all rows (i.e., all sparse vectors) are stored contiguously in v and ci. The p array indicates where each sparse vector begins in v and ci.

at row $ri[l]$ and column $ci[l]$. Using this format, the matrix vector multiplication is shown in Alg. 2.

Algorithm 2 Matrix vector multiplication in coordinate format.

for l = 0 to k - 1 **do**
  $y[ri[l]] \leftarrow y[ri[l]] + v[l] \cdot x[ci[l]]$
end for

Another class of format is the compressed stripe storage that includes the compressed sparse row (CSR) format and compressed sparse column (CSC) format [21]. CSR is a collection of sparse vectors [1], allowing random access to entire rows and efficient enumeration of non-zeros within each row. CSR is illustrated in Fig. 5. The idea is to store each row (shown as elements having the same color) as a sparse vector. A single value array v stores all sparse row vector values in order, and a corresponding array of integers ci stores the column indexes. Each element p of a third array stores the offset within v and ci of row i. The array p has $m + 1$ elements, where the last element is equal to the number of non-zeros. This data structure allows random access to any row, and efficient enumeration of the elements of a given row. An implementation of sparse matrix-vector multiply using this format is shown in Fig. 3.

Algorithm 3 CSR matrix vector multiplication.

for i = 0 to m - 1 **do**
  for l = p[i] to p[i+1] - 1 **do**
    $y[i] \leftarrow y[i] + v[l] \cdot x[ci[l]]$
  end for
end for

In the matrix vector multiplication, the parallelism available across rows makes it a natural choice to distribute computations corresponding to a row to a thread block. The code shown in Alg. 4 allocates one thread to perform the computations corresponding to one row and a thread block to handle a set of rows. It corresponds to a mapping in which a one dimensional grid of thread blocks and a one dimensional block of threads are used to compute sparse matrix vector multiplication. This code is very similar to the one presented in Alg. 3.

[1] the non-zero elements of a sparse vector are stored contiguously in an array. An additional integer array stores the corresponding integer index for each non-zero value.

Algorithm 4 Sparse matrix vector multiplication implemented with CUDA.

```
int line = blockDim.x * blockIdx.x + threadIdx.x ;
if line < dim then
    double sum = y[line];
    sum = 0;
    int start = p[line];
    int end = p[line + 1];
    for i = start to end - 1 do
        sum += x[ci[i]] * v[i];
    end for
    y[line] = sum;
end if
```

5.2. Graph Colloring, Mesh Numeration and Triangular Solvers

In order to calculate the value for $x[r]$, all the values of $x[i]$, $0 \leq i < r$, above it must have been solved. This is called dependency for $x[r]$. Since L is a sparse matrix, $x[r]$ only depends on the columns in row r in which $A[i, r]$ is non zero. The performance of the triangular solver can be improved by expressing the data dependencies of the triangular solver for the lower triangular matrix L as a direct acyclic graph.

A dependence tree that represents the elements relationship is defined as follows: there is an edge from vertex i to vertex j is the result of $x[i]$ influences the value of $x[j]$. A simple way to check this condition is to look at row j in matrix L. For each column i with a non zero in row j, an edge from vertex i to vertex j is created [22].

Due to the parallel operation of the GPU, it is inevitable that there will be some contention when writing data. For example, if two elements share a node, the threads performing computation for each of these elements will both need to update the same location. In order to prevent data races, techniques that can be used include the use of atomic operations, or coloring schemes.

Coloring schemes may be used to remove the need for atomic operations, by ensuring that entities whose updates may be in conflict have different colors. The execution proceeds in parallel over each color, but each color is processed sequentially. This prevents any contention.

A graph is colored such that no two adjacent nodes share the same color. Such an algorithm guarantee that adjacent nodes in the mesh will have distinct colors. By reordering the elements, vertices with the same color are grouped and sequentially ordered, such that unknown values associated to these vertices can be calculated in parallel in the triangular system solver [23].

Consider a simple unidimensional FEM model with 5 nodes shown in Fig. 6. Its nodes are numbered sequentially, the original stiffness matrix represented by a sparse matrix is shown in the left, and the corresponding lower triangular portion is shown in the right. The dependence tree shows that no parallelization is allowed at all. Fig. 7 shows the same unidimensional FEM with renumbered nodes, the stiffness matrix is in the left, and the corresponding lower triangular portion is shown in the right. The dependence tree, in the renumbered FEM case, shows that the processing of the triangular solver needs just two passes with 3 threads.

It was implemented a recursive coloring algorithm for planar graphs with a maximum of five colors similar to the one proposed by [24]. The nodes associated with one specific color can be processed in parallel. The limitation is the available number of threads. The nodes associated to different colors are sequentially processed. The triangular linear system solver is implemented with code shown in Alg. 5. Three index vectors are used: two input index vectors (ii and di) and one output index vector (li).

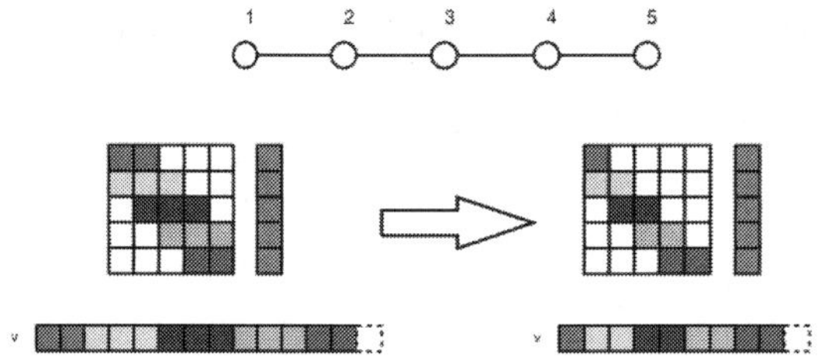

Figure 6: FEM model with 5 nodes numbered sequentially. In the left, the original sparse matrix structure is shown. In the right, the triangular sparse matrix after the incomplete Cholesky factorization is shown.

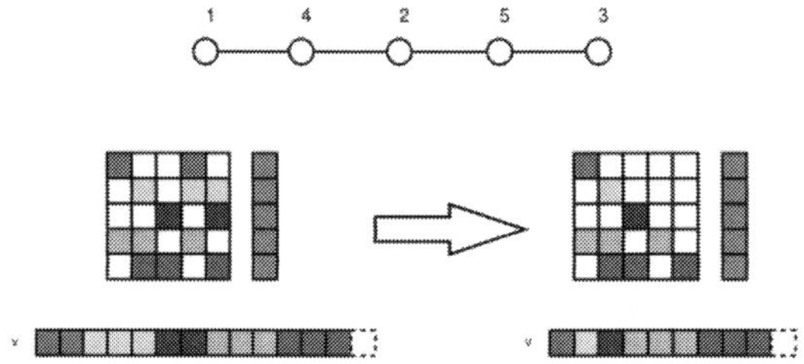

Figure 7: The same FEM with 5 nodes shown in Fig. 6 renumbered. In the left, the renumbered sparse matrix structure is shown. In the right, the triangular sparse matrix after the incomplete Cholesky factorization is shown.

Algorithm 5 Triangular solver implemented with CUDA.

```
int tid = blockDim.x * blockIdx.x + threadIdx.x;
if tid < cont[it] then
    if ii[tid] >= 0 then
        x[li[tid]] = x[li[tid]] - v[di[tid]] * x[ii[tid]];
    end if
end if
__syncthreads();
```

6. Results

For the evaluation of the proposed EIT reconstruction method, a simple model was built, composed of a cylindric container made of acrylic measuring 300 mm in diameter, with 32 electrodes equally spaced in the outer wall. The electrodes are prismatic, that is, their cross-sections are invariant. The container was filled with saline water up to a height of 25 mm. The current was applied to the model in "jump-three" patterns, that is, the current was applied on electrodes that were separated by three electrodes. Those patterns were applied to each of the 32 electrodes. The applied current was of about 10 mA (actual values varied at each application) AC at 125 Hz. In order to produce observable phantoms, three cucumber slices (see [25] for a study of cucumber as a material for EIT evaluation), slightly thicker than the solution height (but not enough to produce significant 3D effects) were inserted in a triangular pattern, as seen in Fig. 8.(a). Data was collected for both the empty saline solution and for the solution with cucumber slices.

For the simulated model, a FEM model was created with 750 triangular elements and 450 nodes (see Fig. 8.(b)). In a first step, the reconstruction process was executed with the

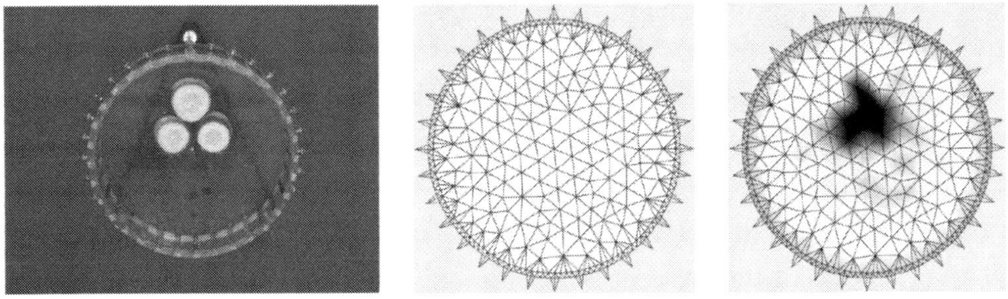

(a) (b)

Figure 8: (a) A picture of the experimental - the three cucumber slices can be seen. (b) Mesh used in the FEM model, containing 750 triangular elements and 450 nodes. (c) Reconstructed image from experimental data.

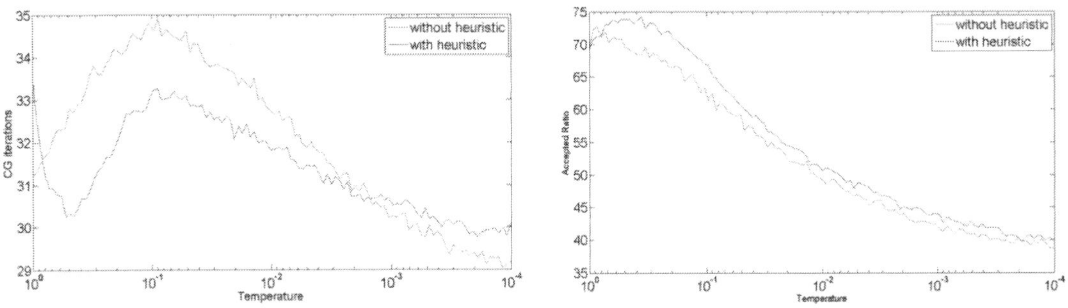

(a) (b)

Figure 9: (a) CG iterations per temperature for the reconstruction from experimental data - with and without the outside-in heuristics; (b) CG iterations per temperature for the reconstruction from experimental data - with and without the outside-in heuristics.

empty medium data, in order to obtain the impedance for each electrode. It was presumed uniform, yet unknown, conductivity inside the simulated model and independent conductivity impedance values for each electrode, yielding an optimization problem of $32 + 1$ unknowns. To reconstruct the cucumber phantoms, the conductivity distribution in the simulated model was parameterized using the same 1st-order interpolating functions used for the potential in the FEM (the parameters being the conductance value at each node). The conductance of the electrodes was fixed at the values obtained in the first step, and so was the conductance for the "outer ring" elements.

The FEM problems were solved using CG method with Incomplete Choleski decomposition preconditioning [26]. The neighborhood heuristic used by the SA was taken from [18], changing only a single conductivity parameter at each iteration and reducing the modifications on parameters that lead to rejected solutions. The divergence probability P_{err} was arbitrarily defined as $1/100$. The number of layers was set to 5.

The reconstructed phantom can be seen in Fig. 8 (c). Despite being heavily constrained by the mesh coarse discretisation, the reconstructed conductance distribution is a reasonable image of the physical phantom. The impact of the outside-in heuristic, and also the impact of using partial evaluation objective function, in the process performance can be seen in Fig. 9(a),

Table 1: Execution time for the Matrix Vector Product

Method	Iterations	Time (ms)
New	1	0.0196
	100	0.0792
CUSPARSE	1	0.0914
	100	7.3112
Eigen	1	1.1899
	100	177.56

showing the average number of CG iterations used by the process at each temperature, with and without the heuristic. Even without the outside-in heuristic, and considering that the system has over 450 nodes, it is interesting to note that the system is able to achieve those results while using in average less than 35 iterations of CG. The adoption of the proposed heuristic reduces the number of iterations even more, for most of the processing.

Finally, as it is shown in Fig. 9(b), there is also an increase in the acceptance rate of solution candidates. Considering that the number of accepted candidates defines the length of the internal loop, the increase in the acceptance rate leads to a decrease in computing time. Actually, on an intel $i7$ 2.8 GHz CPU, and using an initial temperature of 1.0 and final temperature of 0.0001, the image in Fig. 8 (c) was reconstructed in about 205 minutes when using the outside-heuristic. Without it, the reconstruction processing needed about 220 minutes, an increase of 7.3%.

This preliminary study shows that the outside-in heuristic can improve the speed of the SA with objective function incomplete evaluation. However, further research is necessary to verify if other implementations of the outside-in heuristic can have better performance.

The parallel sparse matrix-vector multiplication was tested using CUDA 4.1 on a NVIDIA GeForce GTX 480, containing 480 cores. The presented algorithm is compared with libraries available: CUSPARSE [2] [27] and Eigen [3]. The CUSPARSE libary represents sparse matrices in the CSR format, and it has parallelized versions for the matrix vector product and triangular solver. The Eigen library has serial implementation for matrix vector product and triangular solver.

6.1. Matrix Vector Multiplication

Table 1 shows the result for the matrix vector multiplication. It is possible to observe that the proposed representation and method for the sparse matrix $K(\sigma)$ showed to be much faster mainly when the number of iterations increases.

6.2. Triangular Solver

The proposed triangular solver was tested with the original and reordered matrices. The test was also executed with the CUSPARSE triangular solver as it is also implemented with two phases: dependence tree determination and triangular solver.

Figure 10(a) shows the original matrix and Fig. 10(b) shows the reordered matrix. The results are summarized in Table 2. Using the original matrix, 135 iterations were necessary to solve the triangular linear system. After reordering, with 455 nodes, the parallelization level increased and only 10 iterations were necessary to solve the same triangular system.

In the proposed method, it is possible to separate the two phases, and the first phase is executed just once. This fact reduces the impact of its computational cost. The performance

[2] `http://developer.download.nvidia.com/compute/DevZone/docs/html/CUDALibraries/doc/CUSPARSE_Library.pdf`
[3] `http://eigen.tuxfamily.org/`

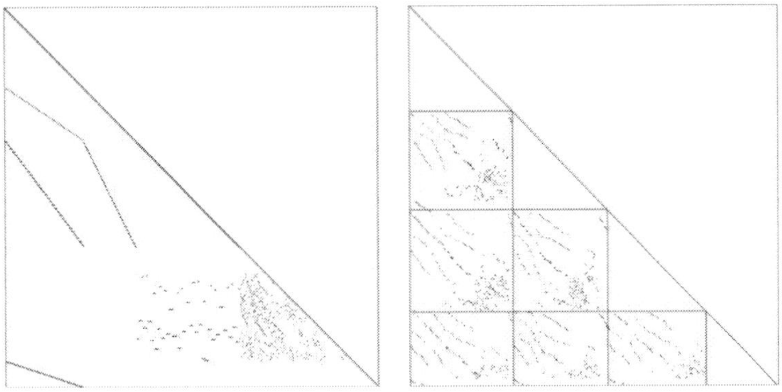

(a) (b)

Figure 10: (a) Original matrix. (b) Reordered matrix.

Table 2: Execution time for the Triangular Solver with CUSPARSE and the proposed method with original and reordered matrices. The pre-analysis creates the dependence tree explained in section 5.2

Method	Pre-Analysis (ms)	Solver (ms)
CUSPARSE	3.442	0.027
Original	3.727	0.638
Reordered	8.174	0.032

Table 3: Execution time (ms) and error for Conjugate Gradient Algorithm with and without the incomplete Cholesky preconditioner

Method		Iterations	Time	Error (10^{-11})
Eigen	with	348	471	4.7
	without	106	167	2.0
CUSPARSE	with	348	110	3.4
	without	106	50	2.4
New	with	348	35	3.3
	without	106	23	1.3

reached by the proposed method is similar to the CUSPARSE library. However, the CUSPARSE library requires that the pre-analysis to be executed every time with the triangular solver. This fact is a large drawback for the CUSPARSE library.

A conjugate gradient algorithm was implemented using the proposed parallel algorithms with and without the incomplete Cholesky preconditioner. A similar test was executed with the Eigen and CUSPARSE libraries. The results are in Table 3, and one can observe that the proposed method showed compatible convergence rates with smaller computational costs. The use of preconditioners in the conjugate gradient method improves the convergence rate by reducing the number of iterations to reach convergence. The execution time does not reduce in the same proportion because the preconditioner has a high computational cost.

7. Conclusions

The presented outside-in heuristic improved the computation time of the SA with objective function incomplete evaluation. The new heuristic selectively picks a layer with higher probability, such that the solution is found from external to internal layers, improving the efficiency of the SA algorithm without compromising its convergence, reducing the computing time by increasing the acceptance rate of candidates.

Also, it is shown how the use of GPU parallelization of matrix operations can effectively reduce the computational time. The use of the general purpose libraries might not be always the best approach. The proposed parallelization methods showed compatible convergence rates with smaller computational costs when compared to available libraries as CUSPARSE and Eigen.

Acknowledgements

This work was supported by FAPESP (Grant 2010/19380–0). R.S. Tavares was supported by FAPESP (Grant 2010/18658–4). M.S.G. Tsuzuki was partially supported by CNPq (Grant 309.570/2010–7).

References

[1] F. C. Trigo, R. G. Lima, M. B. P. Amato, Electrical impedance tomography using the extended Kalman filter, IEEE Transactions on Biomedical Engineering 51 (2004) 72–81.

[2] D. C. Barber, B. H. Brown, Applied potencial tomography, Journal of Physics E: Scientific Instruments 17 (1984) 723–733.

[3] M. T. Clay, T. C. Free, Weighted regularization in electrical impedance tomography with applications to accute cerebral stroke, IEEE Transactions on Medical Imaging 21 (2002) 629–637.

[4] T. J. Kao, D. Isaacson, J. C. Newell, G. J. Saulnier, A 3D reconstruction algorithm for EIT using a handheld probe for breast cancer detection, Physiological Measurements 27 (2006) S1–S11.

[5] B. M. Eyüboglu, B. H. Brown, D. C. Barber, In vivo imaging of cardiac related impedance changes, IEEE Engineering in Medicine and Biology Magazine 8 (1989) 39–45.

[6] P. Hua, E. J. Woo, J. G. Webster, W. J. Tompkins, Finite element modeling of electrode-sking contact impedance in electrical impedance tomography, IEEE Transactions on Biomedical Engineering 40 (1993) 335–343.

[7] M. B. P. Amato, C. S. V. Barbas, D. M. Medeiros, R. B. Magaldi, G. P. Schettino, G. Lorenzi-Filho, R. A. Kairalla, D. Deheinzelin, C. Munoz, R. Oliveira, T. Y. Takagaki, C. R. R. Carvalho, Effects of a protective ventilation strategy on mortality in the acute respiratory distress syndrome, The New England Journal of Medicine 338 (1998) 347–354.

[8] T. C. Martins, E. D. L. B. Camargo, R. G. Lima, M. B. P. Amato, M. S. G. Tsuzuki, Image reconstruction using interval simulated annealing in electrical impedance tomography, IEEE Transactions on Biomedical Engineering 59 (2012) 1861–1870.

[9] B. H. Brown, A. Seagar, The Sheffield data collection system, Clinical Physics and Physiological Measurements 8 (1987) A91–A97.

[10] P. Hua, J. G. Webster, W. J. Tompkins, A regularised electrical impedance tomography reconstruction algorithm, Clinical Physics and Physiological Measurements 9 (1988) 137–141.

[11] F. S. Moura, K. C. C. Aya, A. T. Fleury, M. B. P. Amato, R. G. Lima, Dynamic imaging in electrical impedance tomography of the human chest with online transition matrix identification, IEEE Transactions on Biomedical Engineering 57 (2010) 422–431.

[12] G. H. Golub, C. F. van Loan, Matrix Computations, 3rd Edition, John Hopkins, 1996.

[13] L. M. Mello, C. R. Lima, M. B. P. Amato, R. G. Lima, E. C. N. Silva, Three-dimensional electrical impedance tomography: a topology optimization approach, IEEE Transactions on Biomedical Engineering 55 (2008) 531–540.

[14] C. N. L. Herrera, M. F. M. Vallejo, F. S. Moura, J. C. C. Aya, R. G. Lima, Electrical impedance tomography algorithm using simulated annealing search method, in: Proceedings of International Congress of Mechanical Engineering, ABCM, Brasília, 2007.

[15] H. D. S. Lakshmanan, Simultaneous parameter estimation and segmentation of Gibbs random fields using simulated annealing, IEEE Transaction on Pattern Analysis and Machine Intelligence 11 (1989) 799–813.

[16] T. C. Martins, M. S. G. Tsuzuki, Simulated annealing with partial evaluation of objective function applied to electrical impedance tomography, in: Proceedings of the 18th IFAC World Congress, Milano, Italy, 2011, pp. 4989–4994.

[17] E. Somersalo, M. Cheney, D. Isaacson, E. Isaacson, Layer stripping: a direct numerical method for impedance imaging, Inverse Problems 7 (1991) 899–926.
[18] T. C. Martins, M. S. G. Tsuzuki, Placement over containers with fixed dimensions solved with adaptive neighborhood simulated annealing, Bulletin of the Polish Academy of Sciences Technical Sciences 57 (2009) 273–280.
[19] R. S. Tavares, T. C. Martins, M. S. G. Tsuzuki, Simulated annealing with adaptive neighborhood: a case study in off-line robot path planning, Expert Systems with Applications 38 (2011) 2951–2965.
[20] R. Nath, S. Tomov, T. T. Dong, J. Dongarra, Optimizing symmetric dense matrix-vector multiplication on GPUs, in: Proceedings of 2011 International Conference for High Performance Computing, Networking, Storage and Analysis, SC '11, ACM, New York, NY, USA, 2011, pp. 6:1–6:10.
[21] C.-Y. Lin, Y.-C. Chung, J.-S. Liu, Efficient data compression methods for multidimensional sparse array operations based on the EKMR scheme, IEEE Transactions on Computers 52 (12) (2003) 1640 – 1646.
[22] P. Kumar, M. Kumar, A. Basu, Parallel algorithms for sparse triangular system solution, Parallel Computing 19 (2) (1993) 187 – 196.
[23] P. Berger, P. Brouaye, J. C. Syre, A mesh coloring method for efficient mimd processing in finite element problems, in: Proc. of the International Conference on Parallel Processing, IEEE Computer Society, 1982, pp. 41–46.
[24] G. N. Frederickson, On linear-time algorithms for five-coloring planar graphs, Information Processing Letters (1984) 219–224.
[25] A. R. D. S. Holder, Y. Hanquan, Some practical biological phantoms for calibrating multifrequency electrical impedance tomography, Physiological Measurement 17 (4A) (1996) A167–77.
[26] S. Kershaw, The incomplete Cholesky-conjugate solution of systems, Journal of Computational Physics 65 (1978) 43–65.
[27] NVIDIA, CUDA CUSPARSE library, Tech. rep., NVIDIA (Aug. 2010).

Journal of Physics: Conference Series **407** (2012) 012016 doi:10.1088/1742-6596/407/1/012016

Variable impedance cardiography waveforms: how to evaluate the preejection period more accurately

VV Ermishkin[1], VA Kolesnikov[1], EV Lukoshkova[1], VP Mokh[1], RS Sonina[1], NV Dupik[2], and SA Boitsov[2]

[1] Russian Cardiology Research Center, Moscow, Russia
[2] Russian Research Center of Prophylactic Medicine, Moscow, Russia

E-mail: v.v.erm@mail.ru, elena.lukoshkova@gmail.com

Abstract. Impedance method has been successfully applied for left ventricular function assessment during functional tests. The preejection period (PEP), the interval between Q peak in ECG and a specific mark on impedance cardiogram (ICG) which corresponds to aortic valve opening, is an important indicator of the contractility state and its neurogenic control. Accurate identification of ejection onset by ICG is often problematic, especially in the cardiologic patients, due to peculiar waveforms. An essential obstacle is variability of the shape of the ICG waveform during the exercise and subsequent recovery. A promissing solution can be introduction of an additional pulse sensor placed in the nearby region. We tested this idea in 28 healthy subjects and 6 cardiologic patients using a dual-channel impedance cardiograph for simultaneous recording from the aortic and neck regions, and an earlobe photoplethysmograph. Our findings suggest that incidence of abnormal complicated ICG waveforms increases with age. The combination of standard ICG with ear photoplethysmography and/or additional impedance channel significantly improves the efficacy and accuracy of PEP estimation.

1. Introduction

Electromechanical systole includes two functionally important systolic time intervals (STI): the preejection period (PEP) and left ventricular ejection time (LVET). Evaluation of PEP and LVET, as well as the ratio PEP/LVET, remains to be a subject of particular physiological and clinical importance for noninvasive assessment of cardiac function [1]. In impedance cardiography, LVET is usually used for stroke volume evaluation [2], while PEP proved to be an important indicator of the state of contractility and its neurogenic control. Assessment of PEP and LVET intervals require accurate identification of ejection onset, which is often problematic due to individual and physiological variability of the B-point [3-5]. In part, it can be achieved using other 'events' in impedance cardiogram (ICG) also related to the beginning of ejection [4-6]. Another side of this problem concerns the peculiar shapes of the ICG waveforms encountered in patients [7-9]. Some authors [7, 9] pointed out that the abnormalities in the ICG waveform *per se* can be a sign of cardiovascular pathology, but the underlying mechanisms of such 'pathologic' changes remain unclear. Unfortunately, few studies are focused at the peculiarities of 'pathologic ICG waveforms'. In any case, the significant corruption of dZ/dt waveform leads to ambiguity in detecting the onset of ejection even at rest. The shape of Z-signals can be drastically distorted not only by motion artifacts, but also due to physiological changes in central hemodynamics triggered by exercise. All these factors make evaluation of PEP by ICG method rather uncertain and inaccurate. Therefore, the development of novel and simple methods improving reliability of PEP measurements during active motion conditions is urgent. To meet this end, we introduced: 1) an additional channel of Z-signal led from the neck region and 2) the earlobe photoplethysmography (PPG).

2. Methods

Twenty eight healthy volunteers, aged 19 to 64, and 6 patients with coronary heart disease (CHD; aged 57 to 76) were studied at rest (standing or sitting) and during treadmill or bicycle exercise test. ICG signals from the aortic and neck regions and ECG in CM5 lead were recorded simultaneously with a dual channel impedance electrocardiograph RPKA-2-01 (*MEDACC*, Moscow). A tetrapolar ICG-electrode configuration was used in both impedance channels, which worked independently at the frequentcies of 30 and 50 kHz. The band-type current electrodes fixed to the forehead and a leg were common for both tracts. A pair of voltage spot electrodes was located at the upper thorax along the ascending aorta projection [5]. Two other spot electrodes used for the second channel were placed in the lower and upper part of the neck. As a supplementary method of tracing PEP an earlobe PPG was used, according to [10]. The signals were continuously sampled at the rate of 1 kHz with a 16-bit analogue-to-digital converter (*National Instruments*, Texas) and stored on a hard disk. The first and the second derivatives of the impedance signal were calculated point-by-point using a 3-rd order polynomial fitting over 61 points (Savitzky-Golay method [11]). The time coordinates of $d^2Z/dt^2{}_{max}$ were automatically determined, and the time interval between the points of the ECG R wave peak and $d^2Z_{Ao}/dt^2{}_{max}$, was calculated in each cardiac cycle. These values were used for PEP estimation from impedance aortogram [5, 12-13]. Similarly, PEP_{Neck} and PEP_{Ear} intervals were determined from other both signals, as specified above, by $d^2Z_{Neck}/dt^2{}_{max}$ and $d^2PPG/dt^2{}_{max}$, respectively. A method of synchronous averaging of pulse waves [14] for 5-20 cardiac cycles was used to eliminate the motion artefacts. The changes in ICG and PPG waveforms, as well as the trends of RR interval, PEP_{Ao}, PEP_{Neck}, and PEP_{Ear} during functional tests were analysed. In order to show the related blood volume changes, ΔZ curves and their derivatives are shown inverted on the plots, i.e., the positive deflections of the curves correspond to the decrease in impedance.

3. Results and Discussion

3.1. Abnormal ICG waveforms in elderly people and in cardiologic patients

All healthy subjects at the age below 40 years demonstrated canonical ICG waveform (figure 1, a and b) depicted in the frequently cited publications [2, 3]. Most of the healthy subjects under 60 years also had near canonical ICG waveform (figure 1, c). In virtually all subjects above 60, the characteristic abnormalities in the shape of both ΔZ and dZ/dt ICG waves were clearly seen (figure 1, d). Similar alterations were also typical of patients with CHD (figure 1, e-f). The percentage of abnormal Z-waves observed at rest in the groups of various age is shown in table 1. The characteristic feature of such age- and pathology-dependent changes is a remarkable stepwise preejection wave superposing on the ejection wave front (figure 1, d-f, middle traces). By our hypothesis, there are two major processes contributing to formation of the impedance pulse wave: the changes in the heart geometry and surrounding vessels produced by ventricular contraction, which precede the ejection onset, and an increase in the volume of aorta and other major arteries during ejection. The former process initiates the preejection wave and the latter triggers the ejection wave.

Table 1. Occurrence of abnormal ICG waveforms in respect to age and cardiac pathology.

Groups, age	Number of subjects	% of abnormal ICG
Healthy:		
< 28	15	0 %
37-46	5	20 %
51-55	4	25 %
58-64	4	75 %
CHD patients		
57-73	6	100 %

When the preejection changes become especially pronounced, the superposition of the two waves gives rise to abnormal ΔZ waveforms. As a result, both dZ/dt and d^2Z/dt^2 curves become double-humped (d-

Journal of Physics: Conference Series **407** (2012) 012016 doi:10.1088/1742-6596/407/1/012016

f, bottom traces), and the first peak in both of these derivatives, associated with the preejection wave, becomes nearly equal or even greater than the second one, corresponding to the ejection wave. No doubt, such waveforms are extremely complicated for establishing the exact time of PEP termination.

Note that identification of B-point, which is commonly used as a mark of PEP end, hits many problems even in simpler ICG waveforms [3-5, 13]. The alternative approaches based on seeking for maximum of either dZ/dt [6] or d^2Z/dt^2 [5, 12, 13] also appeared to be problematic in such cases because it can not be determined definitely which peak in dZ/dt or d^2Z/dt^2 corresponds to the beginning of ejection. In that case an independent method of central pulse recording is urgent.

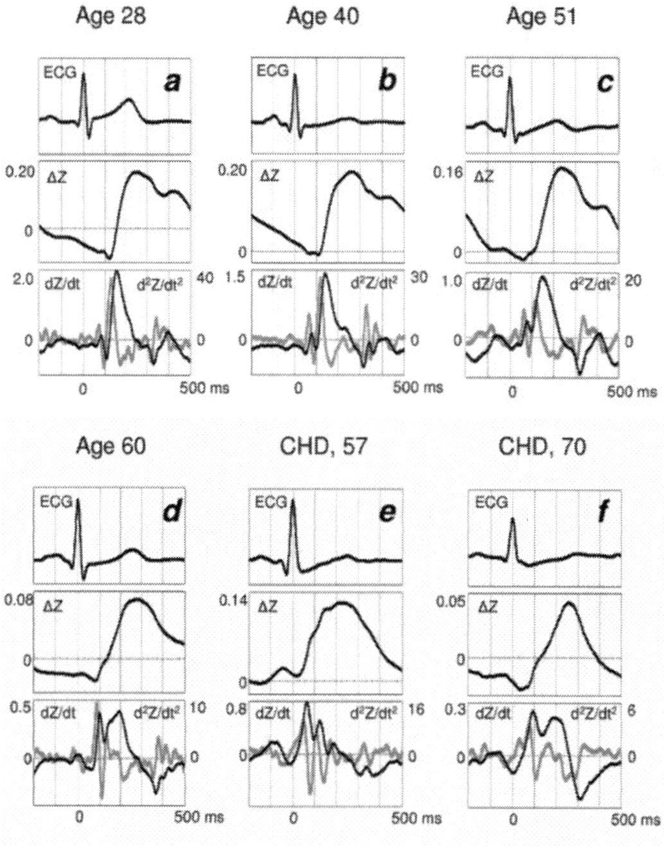

Figure 1. The shape of ICG waves in healthy subjects of various age (a, b, c and d) and in patients with CHD (e, f). *From top to bottom*: ECG, ΔZ (Ohm), dZ/dt (black, left scale, Ohm/s) and d^2Z/dt^2 (grey, right scale, Ohm/s^2).

Manifestations of the preejection processes were most evident in the elderly subjects and in patients, however, some distortions of dZ/dt waveform can also be discerned in the young subjects (figure 1, b and c). We have not found any indications on the age-dependent changes of ICG waveforms in literature. The factors which could affect the shape of preejection waves in our study were the locality of impedance recording from the aortal area and a vertical posture. Previously, the alterations in the ICG shape with change in posture were observed in patients with heart disease [15].

3.2. Corruption of ICG waveform during exercise

In some healthy subjects, initially normal ICG waveforms were drastically corrupted during physical exercise, Valsalva maneuver or other tests. An example of such phenomenon is presented in figure 2.

A healthy man (aged 52) performed a prolonged (30 s) expiration accompanied with abdominal muscles contraction. The trend of PEP estimated by d^2Z/dt^2 maximum (the thick red line on the upper trace) displayed a sudden and significant decrease in its value during this maneuver, which could be attributed to an abrupt rise in the left ventricular contractility. In fact, it was an artifact related to the enormously raised preejection wave which masked the ejection upstroke. As a result, the second peak of d^2Z/dt^2 which normally indicates the onset of ejection disappeared and PEP was erroneously estimated by the first d^2Z/dt^2 peak, corresponding to the beginning of the preejection wave (the left panel). This was confirmed by the absence of changes in PEP trend assessed with PPG signal recorded simultaneously from the earlobe. The PPG trend (the thin blue line on the upper trace) is shown taking into account the time correction for the pulse wave propagation from aorta to the ear. On the right panel showing the corresponding PPG waveforms and their derivatives, it is clearly seen that the time of the major d^2PPG/dt^2 maximum remained unchanged during the maneuver.

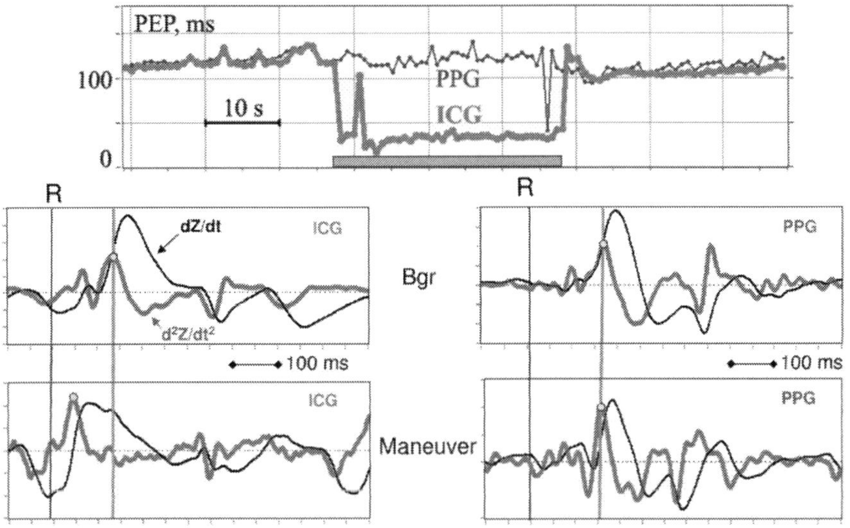

Figure 2. Corruption of ICG waveforms during an expiratory maneuver resulted in erroneous PEP estimation while the shape of PPG signals remained normal. *Upper panel* shows trends for PEP_{Ao} and corrected PEP_{Ear}. The waveforms of dZ/dt (black) and d^2Z/dt^2 (red) for ICG recorded before and during the expiration are displayed on the *left panels below*. The corresponding waveforms of dPPG/dt (black) and d^2PPG/dt^2 (violet) are given on the *right panels*. R shows the time of ECG R peak; the circles indicate the second derivative maxima used for PEP determination.

In another subject (age 55), pronounced alterations in the shape of ICG waveforms were observed during cycling at a constant load of 60 W (figure 3). As a rule, PEP abruptly shortened at the beginning of cycling due to effect of Starling mechanism [16]. It was followed by a few minute period of slow PEP changes coming to a prolong plateau that persisted up to the end of cycling. Similar course of PEP changes could be observed for the intervals determined from ear PPG or additional impedance recording from the neck region (upper trends in figure 3). By contrast, PEP trend for impedance aortogram continued to decline throughout the exercise reaching significantly lower values to its end (the lowest, black trend). The difference between the trends can be attributed to transformation of the ICG waveforms recorded from the aortic region: the main peak of $d^2Z_{Ao}/dt^2{}_{max}$ became double-humped and the B-point disappeared completely (bottom panels in figure 3), which results in erroneously shorter PEP estimates due to premature detection of the ejection onset by any algorithm. The shapes of the signals recorded from the neck and the ear (not shown) did not changed noticeably. It is worth to note that the transformations of the aortic ICG waves depicted in figures 2 and 3 were of hemodynamic origin since the noisy component in both cases was negligible.

3.3. The usage of Z-neck or PPG-ear channels in treadmill or cycling tests

Analysis of more than 30 records of dual-channel ICG and PPG signals performed during treadmill and cycling tests demonstrated pronounced motion artifacts in the neck Z-signals and to a lesser extent in aortic ICG. The most noise-immune was the earlobe PPG. This implies that PPG can be a good assistant to aortic ICG when PEP is evaluated during exercise. Previously, Spodick [10] demonstrated efficacy of the earlobe PPG method for STI evaluation. His method assumed a relative constancy of the time shift between the pulse waves arriving to the carotid artery and to the ear. In fact, it is not exactly so. Our findings indicated that, in subjects over 50 years old, a delay between the aortic impedance and ear PPG waves increased (by about 30 % in average) in the early stage of treadmill test, and it decreased (by 10 %) in the recovery period. The latency variations were less pronounced during the bicycle test (< 10 % of its baseline value).

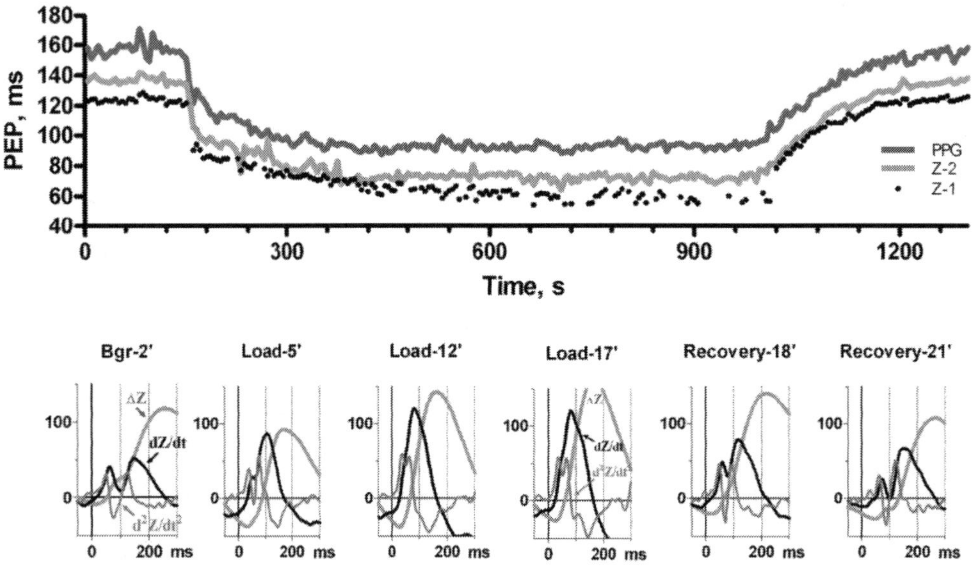

Figure 3. Changes in aortic ICG waveforms during cycling (in a healthy man). *Upper traces:* PEP_{Ao} (Z-1), PEP_{Neck} (Z-2) and PEP_{Ear} (PPG). *Below:* the aortic waveforms of ΔZ, dZ/dt and d^2Z/dt^2 at rest (Bgr-2'), during load (on 5, 12 and 17 min from the start of the record), and during recovery (on 18 and 21 min).

Thus, PPG method can not be simply substituted for the original ICG method because the corresponding pulse transition time is not stable during the exercise. However, the PPG signals can improve PEP assessment when the strong motion artifacts in aortic or neck ICG pulse waves cannot be eliminated by synchronous averaging. After evaluation of the time lags between ICG and PPG signals at rest and during the recovery period, PPG method can provide a satisfactory approximation to PEP changes during the tests. Also, PPG signals can help to avoid ambiguity in tracing the time course of PEP responses when the detection of the end of PEP with a computer algorithm is confused due to abnormal or corrupted ICG waveforms. If the proper 'mark' in the ICG signal pointing to the aortic valve opening could be identified using somewhat other technique (e.g., echocardiography) just before the start of the exercise, then the PPG signals would help in tracing PEP changes during motion. Another way to approximate the start point of ejection is to record the neck impedance. In contrast to aortic ICG, the neck impedance has a simple pulse waveform even in elderly humans and patients. Also, it has a much smaller and stable latency, compared to the ear PPG. Together with aortic ICG, the neck signal yields the most accurate evaluation of PEP in the cases of complicated ICG waveforms, though it can be often useless in moving subjects due to strong artifacts. So, we consider that this channel is useful mainly for appropriate identification of PEP termination 'mark' in complicated ICG waveforms at rest, in the recovery period, and during other stages with relatively small motion artifacts in the neck impedance.

4. Conclusion

Thus, a combination of standard ICG with earlobe PPG and/or additional impedance channel can significantly improve the efficacy and accuracy of PEP estimation, especially in cardiologic patients and elderly people or in the cases with pronounced motion artifacts. PPG is most useful during treadmill or bicycle exercise tests, because it is far less sensible to motion-related disturbances. The time shift between PPG and ICG signals is not stable. The method requires time-to-time corrections for these changes. The neck impedance is more preferable to use under the rest conditions or when a maneuver does not require a strong physical activity or neck muscles tension. The advantage of the latter is a small latency between the neck and the aortic pulses.

Supported by Russian Foundation for Basic Research, Grant No 10-04-01765-a

References

[1] Oh J K and Tajik J 2003 *JACC* **42** 1471–4
[2] Kubiček W G, Kottke J, Ramos M U, Patterson R P, Witsoe D A, Labree J W, Remole W, Layman T E, Schoening H and Garamela J T 1974 *Biomed. Eng.* **9** 410–6
[3] Sherwood A, Allen M T, Fahrenberg J, Kelsey R M, Lovallo W R and van Doornen L J P 1990 *Psychophysiology* **27** 1–23
[4] Lozano D L, Norman G, Knox D, Wood B L, Miller B D, Emery C F and Berntson G G 2007 *Psychophysiology* **44** 113–9
[5] Ermishkin V V, Lukoshkova E V, Bersenev E Y, Saidova M A, Shitov V N, Vinogradova O L and Khayutin V M 2007 *IFMBE Proc.* **17** 655–8
[6] Meijer J H, Boesveldt S, Elbertse E and Berendse H W 2007 *IFMBE Proc.* **17** 596–9
[7] Mohapatra S N 1981 *Noninvasive Cardiovascular Monitoring by Electrical Impedance Technique* (London: Pitman Medical) p 286
[8] Deshpande A K et al 1990 *J. Postgrad. Med.* **36** 207–12
[9] Bour J and Kellett J 2008 *Eur. J. Intern. Med.* **19** 399–405
[10] Spodick D H and Lance V Q 1976 *Circulation* **53** 673–6
[11] Savitzky A and Golay M J E 1964 *Analytical Chemistry* **36** 1627–39
[12] Khaiutin V M, Lukoshkova E V, Ermishkin V V and Sonina R S 2009 *Usp. Fiziol. Nauk* **40** 3–20
[13] Khaiutin V M, Lukoshkova E V, Ermishkin V V and Sonina R S 2009 *Kardiologiia* **50** 38–44
[14] Muzi M, Ebert T J, Tristani F E, Jeutter D C, Barney J A and Smith J J 1985 *J. Appl. Physiol.* **58** 200–5
[15] Richards N T and McBrien D J 1988 *Int. J. Cardiol.* **20** 365–72
[16] Horwitz L D, Atkins J M and Leshin S J 1972 *Circ. Res.* **31** 868–75

First Latin-American Conference on Bioimpedance (CLABIO 2012) IOP Publishing
Journal of Physics: Conference Series **407** (2012) 012017 doi:10.1088/1742-6596/407/1/012017

Behaviour of the electrical impedance myography in isometric contraction of biceps brachii at different elbow joint angles

A B B Coutinho, B Jotta, A V Pino and M N Souza[1]
Production Editor, *Journal of Physics: Conference Series*,
Programa de Engenharia Biomédica,COPPE, UFRJ, Rio de Janeiro, Brazil
E-mail: brand@peb.ufrj.br, bjotta@yahoo.com.br, pino@peb.ufrj.br, souza@peb.ufrj.br

Abstract. Electrical impedance myography (EIM) can be understood as an experimental technique applied to evaluate bioelectrical impedance associated to the muscular activity. With the development of technique, some studies are trying to associate the EIM parameters with the morphological and physiological changes that occur in the muscle during contraction. In this context this work sought to associate EIM parameters observed during isometric contractions of the biceps brachii muscle at different elbow joint angles with the correspondent muscular force. Differently from previous works that did not observe significant correlation between those data, our findings point to high correlations between the some EIM resistive parameters and the muscle force. Despite the need of further investigation, our results indicated that EIM technique can be used to estimate muscle force in a noninvasive way.

1. Introduction

Electrical impedance myography (EIM) can be understood as an experimental technique applied to evaluate bioelectrical impedance associated to the muscular activity [1].

EIM has been used in the assessment of neuromuscular diseases [2] and also to investigate some aspects of muscle contraction [3,4,5]. In the last context, changes in EIM have been attributed to morphological and biochemical changes observed in the muscle during contraction. Despite those works, a better understanding of EIM behavior under isometric or dynamic muscle contractions is still a challenge.

Some studies have reported low correlation between EIM data and muscular contraction associated to joint angles. For example, Zagar [5] compared EIM parameters during rest and isometric contraction of biceps brachii muscle in order to investigate if it is possible to distinguish bioimpedance components related to biochemical and structural changes of the muscle. Although the results indicated the possibility of distinguishing these two components from EIM data, the observed low correlation might be associated to a poor estimate of the muscle strength associated to the studied joint angles.

Some authors [2,3] have also reported a low sensitivity of EIM technique in detecting changes associated to the muscular contraction using the classical tetrapolar bioimpedance electrodes arrangement (with sinusoidal constant current of 800μA, and frequency of 50 kHz) to measure segmental bioimpedance on a muscle. However, alternative methods have been reported to perform bioimpedance spectroscopy (BIS). For example, Neves and Souza [6] proposed a bioimpedance

[1] To whom any correspondence should be addressed.

spectroscopy method based on the current response to a voltage step excitation (BIS-STEP) that has been applied in many bioimpedance applications [7,8,9].

Aiming to contribute to a better understanding of EIM, this study investigated the relationship between the force produced by the biceps brachii muscle in isometric contraction in different elbow joints and EIM parameter achieved by BIS-STEP technique.

2. Methods

2.1. Sample

Twelve subjects (9 males and 3 females), aging 27.3 ± 4.9 years, height 1.73 ± 0.07 m, and weight 80.3 ± 23.1 kg, participated in the study. All subjects were right-handed without previous registration of neuromuscular injuries of upper and lower limbs. Six of them were physically active and six sedentary.

2.2. Instrumentation

As mentioned before, EIM data were achieved using a BIS-STEP prototype developed in the Biomedical Instrumentation Laboratory at Federal University of Rio de Janeiro - Brazil. Such equipment supplies data somehow equivalent to a multifrequency bioimpedance analyzer ranging 1-500 kHz [3].

Because the BIS-STEP is a bipolar technique that considers the effect of electrode impedances, two $12cm^2$ metallic plate electrodes with electrolyte gel were used to minimize the electrode-tissue impedance during the data acquisition of the muscle segmental bioimpedance. Each electrode was placed 2 cm apart the line defining the belly of the biceps brachii muscle. Removal of the stratum corneum layer by rubbing with cotton and saline solution was accomplished before electrodes placement.

2.3. Experimental protocol

The experimental protocol consisted of measuring EIM in volunteers' biceps brachii muscle at different elbow joint angles, while the subjects were holding a 2.2 kg load in their hands. The studied angles were 45, 60, 90, 120 and 135 degrees, and all of them were quantified by a metallic universal goniometer. EIM measurements for each elbow joint angle were performed in a random order, adopting a rest interval of two minutes between successive measurements.

2.4. Analysis and data processing

The current response associated to the BIS-STEP method was interpreted using the electric model depicted in Figure 1. A gradient descent method was used to optimize the electric model parameters that best fit experimental data, in particular the bioimpedance ones (extracellular resistance – *Re*, intracellular resistance – *Ri*, membrane capacitance – *Cm*). The parameter *Ce* models the electrode capacitance and was also estimated by the gradient descent method. The resistive part of electrode impedance was fixed and calculated from geometric and dimensional features.

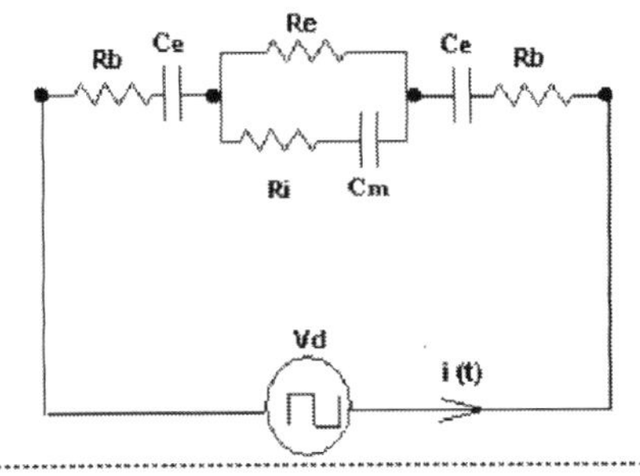

Figure 1. Electric model used to analyze the BIS-STEP current response. The extracellular resistance – *Re*, intracellular resistance – *Ri*, and membrane capacitance – *Cm*, stand for the segmental bioimpedance model. The capacitances – *Ce* and resistances – *Rb* are associated to the impedances of the electrode-tissue interfaces.

The behavior of EIM parameters at different elbow joint angles was compared to the behavior of an estimate of the biceps brachii force obtained using a biomechanical model (Figure 2) proposed by Tozeren [10].

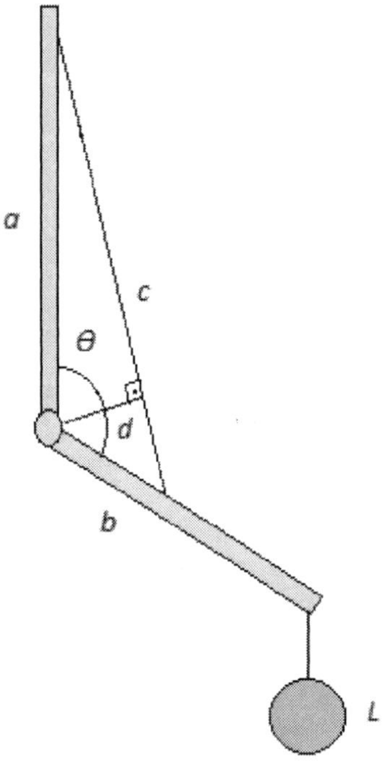

Figure 2. Biomechanical model used to estimate the force of the biceps brachii muscle. *a* – length of the arm, *b* – length of the forearm + half length of the hand, *c* – length of the biceps brachii, *d* – distance between the muscle biceps and the elbow, *L* – load mass *M*

Anthropometric variables used in the biomechanical model were estimated from percentages of subjects' heights according [11], as shown in Table 1. Mean values for the length of the arm, forearm and hand were used to estimate the general behavior of the biceps brachii force at the studied elbow joint angles.

Table 1. Anthropometric variables of the studied sample

Height	Length of humerus	Length of forearm	Length of hand	Half length of hand	Hand and forearm
163,0	28,19	26,08	09,64	4,82	30,90
174,5	30,18	27,92	10,33	5,16	33,08
172,0	29,75	27,52	10,18	5,09	32,61
165,0	28,54	26,40	09,76	4,88	31,28
173,5	30,01	27,76	10,27	5,13	32,89
185,0	32,00	29,60	10,95	5,47	35,07
174,0	30,10	27,84	10,30	5,15	32,99
174,0	30,10	27,84	10,30	5,15	32,99
180,0	31,14	28,80	10,65	5,32	34,12
185,0	32,00	29,60	10,95	5,47	35,07
179,0	30,96	28,64	10,59	5,29	33,93
163,5	28,28	26,16	09,67	4,83	30,99
Mean	30,10	27,84	10,30	5,15	32,99

Journal of Physics: Conference Series **407** (2012) 012017 doi:10.1088/1742-6596/407/1/012017

Considering isometric contraction, and that the elbow flexion is produced just the biceps brachii muscle, equation (1) holds.

$$M_b = m\,g\,\frac{b}{2}\sin\theta + M\,g\,b\sin\theta \tag{1}$$

here M_b is the torque produced by the biceps brachii muscle, m is the estimate mass for the forearm, b is the length of the forearm, θ is the elbow joint angle, and M is load mass.

The biceps torque must the equal to the product of the muscle force by the distance d (Figure 2) from the force direction (direction of the muscle) to the rotation point of the elbow, i.e.:

$$M_b = F_m\, d \tag{2}$$

where it can be demonstrated that

$$d = \frac{\sqrt{4\,a^2c^2-(-b^2+a^2+c^2)^2}}{2\,c} \tag{3}$$

where a is length of the arm, b is length of the forearm + half length of the hand, c is length of the biceps brachii.

From equations (1), (2), and (3) can estimate the biceps brachii force F_m as a function of the elbow joint θ. Those estimates were compared with the corresponding values of resistive parameters of EIM. Besides the comparison with values of *Re* and *Ri*, values of *Rinf*, defined as the parallel association of *Re* and *Ri*, were also studied.

3. Results

The behavior of the estimated biceps brachii force F_m as a function of the elbow joint θ is depicted in Figure 3. The F_m estimates, computed considering equations (1), (2) and (3), used mean values of the anthropometric variables for the studied sample.

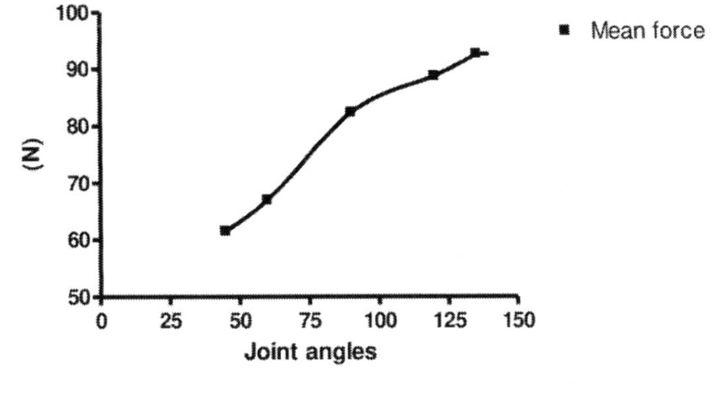

Figure 3. Behavior of the estimated force of the biceps brachii muscle at different elbow joint angles

The behaviors (mean and standard deviation) of EIM resistive parameters are depicted in Figure 4.

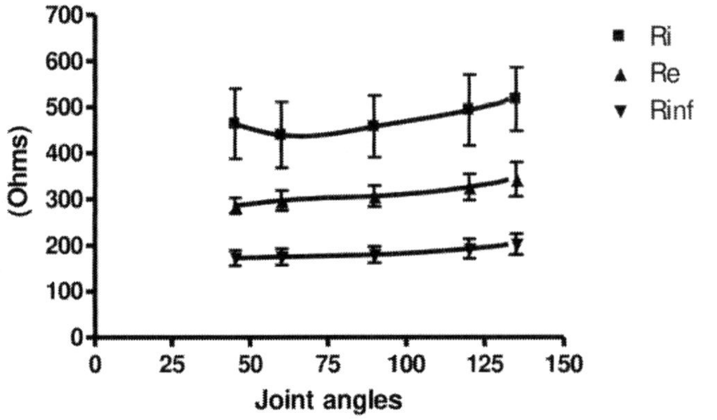

Figura 4. Behavior of the EIM resistive parameters for the studied elbow joint angle.

The linear correlation between the series of the estimated force of the biceps brachii muscle and the resistive EIM parameter can be seen in Table 2.

Table 2. Linear Person correlation (*r*) between the estimated force of the biceps brachii muscle and the resistive EIM parameter

	Ri	*Re*	*Rinf*
Average Force	0,891	0,984	0,971
Angle	0,919	0,999	0,991

4. Discussion

Our findings agree with some previous studied that observed a high correlation between EIM parameters and both joint angle and associated muscle force [2,12]. Some previous works, have been consider three aspects for explanation changes of bioimpedance parameters during muscle contraction in different joint angles: change in distance between electrodes, muscle contraction physiology and morphologic changes during variations of muscle joint studied [3,12].

It has been reported that changes in inter-electrodes distance at different joint angles is a factor that can influence bioimpedance parameters [2,3,6]. Our estimates using the biomechanical model indicate that the length of biceps brachii muscle changes from 24,57 cm (at 45°) to 37,02 cm (at 135°), representing a change of about 33%. Thus, it is seems reasonable that this change the muscle length can lead to an associated change in the effective distance between the electrodes that explains part of the behavior observed in EIM resistive parameters.

Modifications in the transverse section of muscle at different joint angles can be another factor that can change EIM resistive parameters. However, the change in the transverse section can be considered correlated to the above-mentioned influence related to the muscle length, i.e., as the muscle contracts it length decreases and its transverse section increases. Thus, since the resistive parameters should be somehow proportional to the muscle length and inverse proportional to the transverse section, both effects would contribute to change the resistive parameters in the same direction. Kim [12] also reported high correlation between joint angles variations and bioimpedance variables and attributed such findings to morphological modifications of the muscle.

Physiological components could have also influenced our findings. Even holding a low weight load (2.2kg), especially the sedentary volunteers (6 of 12) could have experienced changes in their physiological muscle activity at the different joint angles studied. This physiological activity is basically changed in concentrations of ions in intracellular and extracellular liquids. For example, as muscle contracts potassium (K^+) concentration decreases in intracellular medium and increases in extracellular medium. Others ions as chlorine (Cl^-), sodium (Na^+) and calcium (Ca_2^+) have a opposite drift comparing with potassium during muscle contraction [13]. Shiffman [3] reported an increase of EIM values even before the muscle force generation. Those results were only associated to the physiological factors. However, the way each ion concentration affects EIM parameters is still a challenge.

5. Conclusion

We conclude that the BIS-STEP technique seems to be sensitive enough to EIM application and that it capable to supply parameter that are high correlated with muscle contraction. Part of the changes observed for those parameters can be explained by morphological changes of the muscle and another part associated to physiological changes in ionic concentrations of the extracellular and intracellular mediums. Further investigation must be performed in order to better distinguish these two components.

6. References

[1] Rutkove SB, Fogerson PM, Garmirian LP, Tarulli AW. Reference values for 50-kHZ electrical impedance myography. *Muscle Nerve*. 2008 set;38(3):1128–32.

[2] Rutkove SB. Electrical impedance myography: Background, current state, and future directions. Muscle Nerve. 2009 dez;40(6):936–46.

[3] Shiffman CA, Aaron R, Rutkove SB. Electrical impedance of muscle during isometric contraction. *Physiol Meas*. 2003 fev;24(1):213–34.

[4] Liao TJ, Nishikawa H. The variation of action potential and impedance in human skeletal muscle during voluntary contraction. *Tohoku J. Exp. Med.* 1994 jul;173(3):303–9.

[5] Zagar T, Krizaj D. Multivariate analysis of electrical impedance spectra for relaxed and contracted skeletal muscle. *Physiol Meas*. 2008 jun;29(6):S365–372.

[6] Neves CE, Souza MN. A method for bio-electrical impedance analysis based on a step-voltage response. *Physiol Meas*. 2000 ago;21(3):395–408.

[7] Neves EB, Pino AV, Almeida RR, Souza MN. Objective assessment of knee osteroarthritis in parachuters by bioimpedance spectroscopy. *Conf Proc IEEE Eng Med Biol Soc.* 2008;2008:5620–3.

[8] Alvarenga RL, Souza MN. Estimation of the lactate threshold using bioelectrical impedance spectroscopy: a new noninvasive method. *Conf Proc IEEE Eng Med Biol Soc.* 2007;2007:3052–5.

[9] D. O. Mendonca MNS. Application of step response impedance spectroscopy for detection of skin irritation. 2003;3215– 3217 Vol.4.

[10] Tozeren A. *Human Body Dynamics: Classical Mechanics and Human Movements*. Washington: Springer; 2000.

[11] Plagenhoff S, Evans F, Abdelnour T. Anatomical data for analyzing human motion. *Res Q Exerc Sport*. 1983:(54):169–78.

[12] Kim SC, Nam KC, Kim DW, Ryu CY, Kim YH, Kim JC. Optimum electrode configuration for detection of arm movement using bio-impedance. *Med Biol Eng Comput*. 2003 mar;41(2):141–5.

[13] Cairns SP, Lindinger MI. Do multiple ionic interactions contribute to skeletal muscle fatigue? *J. Physiol. (Lond.)*. 2008 set 1;586(Pt 17):4039–54.

Bioelectrical impedance analysis (BIA): a proposal for standardization of the classical method in adults

C H González-Correa[1], **J C Caicedo-Eraso**[1,2]

[1] Research Group on Electrical Bioimpedance, University of Caldas, Manizales, Colombia, South America

[2] Department of Systems and Informatics, University of Caldas, Manizales, Colombia, South America

E-mail: clara.gonzalez@ucaldas.edu.co

Keywords: bioelectrical impedance analysis, standardization, body composition, sources of variation, BIA protocol, Bioimpedanciometry.

Abstract: The accuracy of BIA measurements is limited by different sources of error such as physical model, cross sectional area, ethnicity, body hydration, age and level of body fat among other variables. Equation for each population is required as they can produce overestimation when manufacturer's equations are used. The classical measurements hand to foot has shown better correlation against hydrodensitometry than foot to foot or hand to hand. However there is a lack for an accepted standard of BIA procedures. This is compounded when there is not a good report of the BIA study's methodology; hence the comparability between the results is poor and reduces the reliability of the method. Perhaps, standardization of methods would be the first step for BIA studies to move forward and subsequently improve its accuracy. Standardized procedures could also minimize the impact of these variables on studies results. The aim of this study was to propose a protocol as a checklist to standardize BIA procedures and produce comparable results from future studies performed with the classic hand-foot configuration in adults.

1. Introduction
Bioelectrical impedance analysis (BIA) method for body composition evaluation has been used in many clinical situations [1] and new medical applications are appearing [2, 3]. This is due to its multiple advantages: ease of use, noninvasive, minor technician error, relatively cheap, good degree of reproducibility, inter-observer reproducibility, and safety of operation [4]. However, there are variable and contradictory results obtained by BIA [5] which calls into question its accuracy and precision.

Evaluation of performance and clinical usefulness of BIA measurements requires studies conducted under standard methods [5]. Because BIA results are influenced by at least 30 variables or factors, a lack of standardization in procedures may contribute to conflicting findings [6]. BIA studies are also limited by the poor quality of reporting of research methodology which prevents readers evaluate the validity and applicability of the study's results [7].

If BIA is to gain reliability in clinical assessment and monitoring treatments, further efforts will be needed to develop, implement and report universal protocols [5, 8]. Therefore, this study aimed to propose a protocol as a checklist for future users of classical hand to foot BIA.

2. Methods
We collected information available from the scientific literature where variables affecting BIA results were reported. Afterwards, these variables were categorized according to their source, and a list in keeping with their impact on BIA results was made. Finally a checklist for standardized BIA procedures was developed which would be useful for future studies.

PUBMED, PROQUEST, SCIENCE DIRECT AND HINARI databases were consulted. The search strategy included the following terms related to BIA in conjunction with factors affecting its results: Resistance OR Bioelectric$ OR Bio-electric impedance OR Bioimpedanciometry OR Bioelectrical impedance spectroscopy; dual frequency bioelectric bioimpedance OR multi multiple frequency impedance; AND menstrual cycle, temperature, etc. Additionally, a hand searching of review articles on BIA was developed.

3. Data acquisition

Methods for recording the percentage change in the results obtained by modifying the specific factor in the studies were twofold:

1. Direct record of the percentage reported in a study.
2. Calculation of percentage of change (equation 1) in results of direct BIA parameters (resistance, impedance) or biological estimations of total body water, fat mass, fat free mass obtained from BIA measurements at the beginning and end of a experiment.

$$\% \ change = [(Final \ value - initial \ value) / initial \ value] * 100\% \qquad (1)$$

4. Results

Table 1 shows the possible variables influencing BIA results, the percentage of change on BIA results and a classification according subjects characteristics or external sources of error. Table 2 lists variables mentioned in the literature but from which there is not enough information. Table 3 is a summary of variables according to their influence on the results obtained by BIA. Table 4 displays the same variables as in Table 1 but ordered according to their maximum impact on BIA results. Table 5 is a useful checklist to help researchers to include all variables that affect BIA results.

Table 1. Variables and their percentage of error in BIA measurements.

Item	Variable	%	Conditions	Ref
OPERATOR	Expiration	0.1	At the end of normal expiration	9
	Jewellery removed	0.2	Using or not using metallic watch, necklace and ring	10
	Non-conductive surface	0.2	A metallic hospital bed or a plastic bed	10
	Skin preparation	0.4	Cleaning or not with alcohol	10
	Inaccurate body weight	0.5	Inaccurate 1.0 kg	11
	Number of trials	1.0	2 trials	12
	Electrode position	2.1	moving the electrode 1.0 cm beyond the standard positions	10
	Inaccurate body height	2.6	Inaccurate 2.5 cm	11
	Inspiration	4.9	At end of normal inspiration	9
	Movement	7.0	Arm contracted	9

	Skin and core temperature	2.0	Changing temperature in 2.8 °C	10
		8.0	Mean skin temperature at the hand and foot raised and lowered to 33.4 and 24.1 °C	13
	Limb position	2.7	Arms separated from the trunk 30 or 90 degrees	10
		18.0	Contact of crossed legs to the waist, errors occur with skin-to-skin contact conditions	13
		43.0	Contact of hands to waist, errors occur with skin-to-skin contact conditions	13
MEDICAL CONDITION & TREATMENTS	Electrolyte concentrations / Electrolyte infusions	0.8	Iso-osmolality, variations less than 0.6 mmol/kg	14
		0.8	Hypo-osmolality: decreasing 13 mmol/kg	14
		6.3	Hyper-osmolality: increasing 11 mmol/kg	14
	Diseases and medications affecting hydration status	1.3	Use of Tibolone in postmenopausal women	15
		15.0	Surgical and cancer patients with malnutrition (without ascites or edema)	16
		16.3	Growth hormone replacement therapy in hypopituitarism	17
		23.4	Cardiac patients	18
		24.5	Lymph edema	19
		25.0	Preoperative treatment with growth hormone	20
		26.1	Edema (cancer)	21
		34.5	Pulmonary patients	18
REPRODUCIBILITY	Reproducibility between observers	0.5-0.9	Between observers	22
		0.7-1.4	Within observers	22
	Reproducibility within day	0.9	Within day, SF BIA, foot-foot	23
		1.0-2.0	Within day	13
	Reproducibility between day	2.1	Between day, SF BIA, foot-foot	23
		2.0-3.5	Day to day	13
	Reproducibility within subjects	1.5-2.4	Within subjects	22
		9.1-22.9	Between subjects variation	22

SUBJECT CHARACTERISTICS	Bladder voided	1.0	Voiding o not the bladder	10
	Phase or menstrual cycle	1.2	Different phases of ovarian cycle	10
	Menopause	2.1	1 year and maximum 15 years after natural menopause	15
	Dominance	2.3	From right to left	10
	Food or beverages or alcohol consumption	2.9	Two hrs after a meal	10
	Ethnicity	2.5	White vs. Black, men & women USA	24
	Fat distribution, abdominal obesity	2.7	When %BF was >33.0% in women	25
		4.3	When %BF was >25.0% in men	25
	Hydration status (dehydration)	6.4	Loosing 2.0 kg with diuretics	10
	Physical Exercise	2.1	After 1 hr of exercise	10
		3.0	R change immediately post exercise	13
		8.0	Xc change immediately post exercise	13
	Body position	3.0	Standing to lying over the first 10'	13
		3.3	From standing to supine position	10
		4.7	Supine position 12 hr	26
		2.0-5.0	Extracellular volume	27
		1.8-8.0	Intracellular volume	27
	Age	8.4	TBW by BIA 20-29 vs. 70-79 years old men	28
	Weight loss	1.3	3 days diet very low calories	29
		7.8	Diet + exercise for 16 wk	30
		7.8	Diet + exercise for 10 wk	31
		8.6	Diet for16 wk in obese women	30
		11.8	Extreme obesity vs. massive weight loss (gastric bypass surgery 14 ± 2 months)	32
	Gender	21.3	White vs. Black, men & women USA	24

EQUIPMENT	Electrode size	0.2	Electrodes area from 14.5 to 4.0 cm^2	10
	Electrode contact vs. gel	2.9	SF BIA, foot-foot, Pressure contact vs. gel electrode	23
	Electrode type	5.2	ECG electrodes	33
	Analyzer MF-BIA	1.1	Inter 3 foot-to-foot MF-BIA	34
		2.5	Inter 3 MF-BIA analyzers	35
		3.0	Inter 2 MF-BIA analyzers	36
	Analyzer SF-BIA vs. MF-BIA	6.7	50kHz vs. 100kHz	37
		10.4	FFM at 50kHz vs. all freq	38
ENVIRONMENT	Environment temperature	5.2	From 15 to 35 °C	39

Ref: Reference

Table 2. Variables without enough information

	Variables
1	Environment humidity
2	Environment electrical interferences
3	Cables
4	Body shape
5	Body density
6	Differences in proportional limb lengths
7	Muscle mass
8	Hypothyroidism
9	Orthopedic prosthesis/implants
10	Pacemakers, defibrillators

Table 3. Categorized variables according the maximum error in BIA results.

	Source of error	%
1	Operator	0.1 - 43.0
2	Medical condition & treatments	0.8 - 34.5
3	Reproducibility	0.5 - 22.9
4	Subject characteristics	1.0 - 21.3
5	Equipment	0.2 - 10.4
6	Environment	5.2

Table 4. Variables classification by influence on BIA results

	Variable influence	%
1	Limb position	2.7 - 43.0
2	Diseases and medications affecting hydration status	1.34 - 34.5
3	Reproducibility within subjects	1.5 - 22.9
4	Gender	21.3
5	Weight loss	1.25 - 11.8
6	Analyzer SF-BIA vs. MF-BIA	6.7 - 10.4
7	Age	8.4
8	Physical Exercise	2.1 - 8.0
9	Skin and core temperature	2.0 - 8.0
10	Body position	1.8 - 8.0
11	Movement	7.0
12	Hydration status (dehydration)	6.4
13	Electrolyte concentrations / electrolyte infusions	0.8 - 6.3
14	Electrode type	5.2
15	Inspiration	4.9
16	Environment temperature	5.2
17	Fat distribution, abdominal obesity	2.7 - 4.3
18	Reproducibility between day	2.1 - 3.5
19	Ethnicity	2.2 - 3.3
20	Analyzer MF-BIA	1.1 - 3.0
21	Electrode contact vs. gel	2.9
22	Food or beverages and alcohol consumption	2.9
23	Inaccurate body height	2.6
24	Dominance	2.3
25	Menopause	2.1
26	Electrode position	2.1
27	Reproducibility within day	0.9 - 2.0
28	Reproducibility between observers	0.5 -1.4
29	Phase or menstrual cycle	1.2
30	Bladder voided	1.0
31	Number of trials	1.0
32	Inaccurate body weight	0.5
33	Skin preparation	0.4
34	Electrode size	0.2
35	Jewellery removed	0.2
36	Non-conductive surface	0.2
37	Expiration	0.1

Journal of Physics: Conference Series **407** (2012) 012018 doi:10.1088/1742-6596/407/1/012018

Table 5. Checklist for standardized BIA measurements

Stage	Variable	Ideal BIA measurement
Before the patient Appointment	Reproducibility	Report laboratory reproducibility concerning: variation within subject, between subjects, within day, between days, within observers, between observers.
	Previous questionnaire	Fill a previous questionnaire: age, sex, race, phase menstrual, prosthesis, pregnancy, pacemaker, stent, amputation, co-morbidities, treatments and medication (steroids, growth hormone, diuretics, contraceptives and others).
		Instruct patient about previous preparation
Before the patient arrives	Analyzer	Calibrate with the manufacturer's equivalent electronic circuit testing
	Cables	Verify integrity, appropriate length for subject size, avoid cross linked cables
	Electrode type	Use the manufacturer's recommendation or validated electrodes
	Electrode size	Use manufacturer's recommendation electrodes, > 4 cm^2 or validated size electrodes
	Environmental interference	Neutral environment, isolation from electrocardiography devices and electrical interferences
	Scale	Minimal accuracy 0.1 kg and calibrate
	Statiometer	Minimal accuracy 0.5 cm and calibrate
	Examination table/surface	Use a non-conductive surface, isolated from ground and suitable subject's size
Check patient preparation	Hydration status (dehydration)	Verify hydration status
	Food or beverages: coffee, tea, chocolate, and alcohol consumption	Defined fast for 8 or 12 hr
	Physical Exercise	Avoid moderate or strenuous exercise 8 hr before test
	Phase or menstrual cycle	Avoid in premenstrual and menstrual phase
	Medical condition impacting fluid and electrolyte balance	Ask for diarrhea or fever
	Treatments and medication	Previous questionnaire

Journal of Physics: Conference Series **407** (2012) 012018 doi:10.1088/1742-6596/407/1/012018

Table 5 (continued)

Stage	Variable	Ideal BIA measurement
Check Just before the test	Environment temperature	Use a electric heater and record and control room temperature: 22 ± 1.5 °C
	Environment humidity	Use a dehumidifier and record and control room humidity: Manufacturer's recommendation or less than 70% RH at °C 40 (non condensing)
	Bladder voided	30 minutes before the test
	Jewellery removed	Remove body ornaments
	Accurate body height	Standard protocol
	Accurate body weight	Standard protocol
	Dominance	Verify dominance
	Differences in proportional limb lengths	Verify
	Body position	Lying down
	Limb position	Arms abducted 45^0 from the trunk and legs separated
	Avoiding fluid shift	Lying down for 5 minutes
	Skin Landmarks	Midpoint between anatomical references
	Skin preparation	Clean with alcohol and dry with a piece of paper
	Electrode position	Over skin landmark on standard position
	Skin and core temperature	Using a standard hospital gown wearing underwear clothes
During the measurements	Movement	Avoid movement while measurements
	Respiration	Make measurements at the end of an exhalation
	Number of trials	Average a set of 3 measurements

Table 5 (continued)

Stage	Variable	Ideal BIA measurement
Analysis with caution	Fat distribution, abdominal obesity	Use positioning isolating restrains
	Body density	
	Body shape	
	Gender and Age	Use ethnic, gender and age specific equation if available
	Ethnicity	Try to determine ethnicity and record
	Muscle mass	
	Menopause	
	Weight loss	
If future measurement	Timing during different days	Take following measurements at the same time of day and by the same operator
	Timing within day	Take following measurements with the same protocol and by the same operator
	Timing within subject	Be sure measurements are made with subject in the same basic conditions

5. Discussion

The validity, sensitivity, and specificity of biological parameters estimated with BIA need to be established [5]. For this purpose authors have claimed for standard BIA procedures and reports to obtain reliable estimations of body composition in diverse clinical situations [7, 40]. It has been widely recognized that ethnic and clinical conditions produce considerable variability in BIA studies but we need to control also a number of other factors that can influence BIA results.

There have been some efforts to draw attention to all factors affecting BIA measurements and the need for standard measurements [41, 42]. The best attempt to show the variables to be considered for a possible standardization was made by Kyle et al, [43] however, there are still sources of error to think over and control to provide information on the usefulness of BIA data.

It has been said that BIA technique does not require many skills of the technician who only causes minor errors. Surprisingly, the results showed that if the operator is not careful with the variables related to the recommendations before the test, electrode placement and patient positioning, He or She can be the major source of errors.

Although BIA has been validated against reference techniques such as dual-energy X-ray absorptiometry [44], and other body composition methods, [45, 46, 47] its usefulness is still doubtful and concerns about the number of variables affecting the estimations of body composition and their relative contribution remain. We know the level of errors that cause the variables that have a greater effect as temperature, degree of hydration and the position of the subject but there are a number of variables with less effect of which we do not know if add or not to the total errors produced by the technique.

However, BIA features such as low-cost, low invasiveness and best performing when compared to body mass index, or skin folds, continues to do so as an attractive method that everybody wants to standardize and use with confidence.

In this paper an effort was made for performing a comprehensive literature search for articles reporting BIA factors affecting its results and put them into a consideration to obtain a standard protocol. However, there was not enough information on ten that were only mentioned in this paper. It is thus required to define the specific influence of these variables.

This study may have limitations. We are aware of the possibility of missing some studies. In addition, the categorization may not have been adequate in some cases or some factors may belong to more than one category or have been improperly classified. Calculating the percentage of change in a factor should be using raw data but in many cases they were not reported. Finally, some factors or variables may be lacking.

We know that the proposal is not complete and other variables may appear in the light of new treatments or new requirements for the estimation of body composition (breast, buttocks or muscle implants, bariatric surgery and mobile phones or wireless interference), but the list is an attempt to standardize the technique and put back the debate on this topic. We suggest that researchers try to use the checklist and make their contributions to improve it and finally, report all controlled variables.

6. References

[1] Jaffrin MY 2009 Body composition determination by bioimpedance: an update Curr Opin *Clin Nutr Metab Care* **12**(5):482-6.
[2] Barbosa-Silva MC, Barros AJ 2005 Bioelectrical impedance analysis in clinical practice: a new perspective on its use beyond body composition equations *Curr Opin Clin Nutr Metab Care.* **8** (3):311-7.

[3] Van Puyenbroeck K, Roelandts L, Van Deun T, Van Royen P, Verhoeven V 2012 The Additional Value of Bioelectrical Impedance Analysis-Derived Muscle Mass as a Screening Tool in Geriatric Assessment for Fall Prevention *Gerontology*. Mar 27. [Epub ahead of print]

[4] Anja Walter-Kroker, Axel Kroker, Muriel Mattiucci-Guehlke3 and Thomas Glaab 2011 A practical guide to bioelectrical impedance analysis using the example of chronic obstructive pulmonary disease *Nutr J.*; **10**: 35.

[5] Lukaski HC 1999 Requirements for clinical use of bioelectrical impedance analysis (BIA) *Ann N Y Acad Sc.* **20**; 873:72-6.

[6] Buchholz AC, Bartok C, Schoeller DA 2004 The validity of bioelectrical impedance models in clinical populations *Nutr Clin Pract.* **19**(5):433-46.

[7] Barrera-Ariza L , CH. González-Correa and CA González-Correa 2009 Quality of reporting of bioelectrical impedance analysis (BIA) studies evaluating body fluid volumes: the need for standardization *IFMBE Proceedings,* **25**(7) 244-246.

[8] Gonzalez CH, Evans JA, Smye SW, Holland P 2002Total body water measurement using bioelectrical impedance analysis, isotope dilution and total body potassium: a scoring system to facilitate intercomparison *Eur J Clin Nutr.* **56**(4):326-37.

[9] Kaldoudi E 1989 Body Composition Studies Using Bioelectric Impedance Measurements Physics Department University of Surrey *Thesis* (England)

[10] Gonzalez CH, Evans JA, Smye SW and Holland P 1999 Variables affecting bioimpedance analysis measurements of body water *Med Biol Eng Comput* **37**: 106-107

[11] Kushner RF, Schoeller DA, Fjeld CR, Danford L 1992 Is the impedance index (ht2/R) significant in predicting total body water? *Am J Clin Nutr* **56** (5) pp 835-839

[12] Buchholz AC, Bartok C, Schoeller DA 2004 The validity of bioelectrical impedance models in clinical populations *Nutr Clin Pract* **19**(5):433-46

[13] Kushner RF, Gudivaka R, Schoeller DA 1996 Clinical characteristics influencing bioelectrical impedance analysis measurements *Am J Clin Nutr* **64** (3 Suppl) pp 423S-427S

[14] Berneis K, Keller U 2000 Bioelectrical impedance analysis during acute changes of extracellular osmolality in man *Clin Nutr* **19**(5) pp 361-366

[15] Meeuwsen IB, Samson MM, Duursma SA, Verhaar HJ 2001 The effect of tibolone on fat mass, fat-free mass, and total body water in postmenopausal women *Endocrinology* Nov **142**(11) 4813-4817

[16] Cohn SH 1985 How valid are bioelectric impedance measurements in body composition studies? *Am J Clin Nutr* **42**(5) pp 889-890

[17] Chrisoulidou A, Beshyah SA, Rutherford O, Spinks TJ, Mayet J, Kyd P, Anyaoku V, Haida A, Ariff B, Murphy M, Thomas E, Robinson S, Foale R, Johnston DG 2000 Effects of 7 years of growth hormone replacement therapy in hypopituitary adults *J Clin Endocrinol Metab* **85** (10) 3762-3769

[18] Katch FI, Solomon RT, Shayevitz M, Shayevitz B 1986 Validity of bioelectrical impedance to estimate body composition in cardiac and pulmonary patients *Am J Clin Nutr* **43**(6) pp 972-973

[19] Cornish BH, Ward LC, Thomas BJ, Bunce IH 1998 Quantification of lymphoedema using multi-frequency bioimpedance *Appl Radiat Isot* **49**(5-6) pp 651-652

[20] Plank LD, Hill GL 2000 Use of bioimpedance spectroscopy to assess effects of perioperative treatment with growth hormone on fluid changes in patients undergoing major surgery *Ann NY Acad Sci* **904**: 190-192

[21] Itobi EO 2007 The impact of post operative oedema on clinical recovery and its potential causes University of Southampton Developmental Origins of Health and Disease Doctoral *Thesis* 149pp

[22] Hansen S, Cold S, Petersen PH, Rose C 1997 Estimates of the sources of variation (variance components) of bioelectric impedance and anthropometric measurements in an epidemiological case-control study of breast cancer *Eur J Clin Nutr* **51**(11) pp 764-770

[23] Nuñez C, Gallagher D, Visser M, Pi-Sunyer FX, Wang Z, Heymsfield SB 1997 Bioimpedance analysis: evaluation of leg-to-leg system based on pressure contact footpad electrodes *Med Sci Sports Exerc* **29**(4) pp 524-531

[24] Sun SS, Chumlea WC, Heymsfield SB, Lukaski HC, Schoeller D, Friedl K, Kuczmarski RJ, Flegal KM, Johnson CL, Hubbard VS 2003 Development of bioelectrical impedance analysis prediction equations for body composition with the use of a multicomponent model for use in epidemiologic surveys *Am J Clin Nutr* 77(2) pp 331-340

[25] Sun G, French CR, Martin GR, Younghusband B, Green RC, Xie YG 2005 Comparison of multifrequency bioelectrical impedance analysis with dual-energy X-ray absorptiometry for assessment of percentage body fat in a large, healthy population *Am J Clin Nutr* **81**(1) 74–78

[26] Slinde F, Bark A, Jansson J, Rossander-Hulthén L 2003 Bioelectrical impedance variation in healthy subjects during 12 h in the supine position *Clin Nutr* **22**(2) 153-157

[27] Scharfetter H, Monif M, László Z, Lambauer T, Hutten H, Hinghofer-Szalkay H 1997 Effect of postural changes on the reliability of volume estimations from bioimpedance spectroscopy data *Kidney Int* **51**(4) 1078-1087

[28] Ellis KJ 2000 Human body composition: in vivo methods *Physiol Rev* **80**(2) 649-680

[29] Deurenberg P, Weststrate JA, Van Der Kooy K 1989 Body composition changes assessed by bioelectrical impedance measurements *Am J Clin Nutr* **49**(3) pp 401-403

[30] Evans EM, Saunders MJ, Spano MA, Arngrimsson SA, Lewis RD, Cureton KJ 1999 Body-composition changes with diet and exercise in obese women: a comparison of estimates from clinical methods and a 4-component model *Am J Clin Nutr* **70**(1) 5-12

[31] Ross R, Léger L, Martin P, Roy R 1989 Sensitivity of bioelectrical impedance to detect changes in human body composition *J Appl Physiol* **67**(4) 1643-1648

[32] Das SK, Roberts SB, Kehayias JJ, Wang J, Hsu LK, Shikora SA, Saltzman E, McCrory MA 2003 Body composition assessment in extreme obesity and after massive weight loss induced by gastric bypass surgery *Am J Physiol Endocrinol Metab* **284**(6) 1080-1088

[33] Caicedo-Eraso JC, Gonzalez-Correa CH, Gonzalez-Correa CA Use of electrocardiogram (ECG) electrodes for Bioelectrical Impedance Analysis (BIA) *J. Phys.: Conf. Ser.* (In press 2012)

[34] Macfarlane DJ 2007 Can bioelectric impedance monitors be used to accurately estimate body fat in Chinese adults? *Asia Pac J Clin Nutr* **16**(1) 66-73

[35] Ward LC 2009 Inter-instrument comparison of bioimpedance spectroscopic analysers *Open Med Dev J* **1** 3-10

[36] Stroud DB, Cornish BH, Thomas BJ, Ward LC 1998 A comparison of two multi-frequency bioimpedance analysers *Appl Radiat Isot* **49**(5-6) 479-480

[37] Khaled MA, Kabir I, Goran MI, Mahalanabis D 1997 Bioelectrical impedance measurements at various frequencies to estimate human body compositions *Indian J Exp Biol* **35**(2) pp 159-161

[38] Thomson R, Brinkworth GD, Buckley JD, Noakes M, Clifton PM 2007 Good agreement between bioelectrical impedance and dual-energy X-ray absorptiometry for estimating changes in body composition during weight loss in overweight young women *Clin Nutr* **26**(6) pp 771-777

[39] Michael J Buono et al 2004 The effect of ambient air temperature on whole-body bioelectrical impedance *Physiol Meas* **25**: 119

[40] Kyle Ursula G, Ingvar Bosaeus, Antonio D. De Lorenzo, Paul Deurenberg, Marinos Elia, Jose Manuel Gomez, et al 2004 Composition of the ESPEN Working Group. Bioelectrical impedance analysis part I: review of principles and methods *Clin Nutr* **23**, 1226–1243

[41] Barillas MC, Vettorazzi C, Molina S, Pineda O 1987 Experience with bioelectrical impedance analysis in young children: sources of variability. In: Ellis KJ, Yasumura S, Morgan WD. Eds. *In vivo body composition study*. London: IPSM, 3: 87-90

[42] Gonzalez CH, Oldroyd B, Evans JA, Smye SW, Holland P 2000 Standardized limb abduction for bioimpedance measurements using position restraints *Physiol Meas* **21**(2):263-70.

[43] Kyle Ursula G, Ingvar Bosaeus, Antonio D. De Lorenzo, Paul Deurenberg, Marinos Elia, Jose Manuel Gomez, et al 2004 Bioelectrical impedance analysis—part II: utilization in clinical practice *Clin Nutr* **23**, 1430–1453
[44] Rebecca Thomsona,b, Grant D. Brinkwortha,_, Jonathan D. Buckleyb, Manny Noakesa, Peter M. Cliftona 2007 Good agreement between bioelectrical impedance and dual-energy X-ray absorptiometry for estimating changes in body composition during weight loss in overweight young women *Clin Nutr* **26**, 771–777
[45] Goran MI, Khaled MA 1995 Cross-validation of fat-free mass estimated from body density against bioelectrical resistance: effects of obesity and gender *Obes Res* **3**(6):531-9.
[46] Bosy-Westphal A, Mast M, Eichhorn C, Becker C, Kutzner D, Heller M, Müller MJ 2003 Validation of air-displacement plethysmography for estimation of body fat mass in healthy elderly subjects *Eur J Nutr* **42**(4):207-16.
[47] Bhat DS, Yajnik CS, Sayyad MG, Raut KN, Lubree HG, Rege SS, Chougule SD, Shetty PS, Yudkin JS, Kurpad AV 2005 Body fat measurement in Indian men: comparison of three methods based on a two-compartment model *Int J Obes* **29**(7):842-8.

7. Conflict of interest
The authors declare that there is no conflict of interest for this work.

Interfacing the AD5933 for bio-impedance measurements with front ends providing galvanostatic or potentiostatic excitation

Uwe Pliquett, Andreas Barthel
Institut für Bioprozess- und Analysenmeßtechnik, Heilbad Heiligenstadt, Germany
Email: uwe.pliquett@iba-heiligenstadt.de

Abstract The AD5933 [1], a specialized single chip impedance analyzer, made by Analog Devices, is basically not intended for use with four electrode interface. Due to electrochemical phenomena at the electrodes connecting the material under test (MUT), especially in the low frequency region below 100 kHz, a two electrode interface generates considerable errors during the measurement. Thus, for most application in bio-impedance measurement only a four electrode interface can guarantee reliable results. Here we show how a four electrode interface with galvanostatic excitation but also for potentiostatic excitation can be realized by just a few external components.

1. Introduction

Bio-impedance measurement is a simple an innocuous way for electrical characterization of cells and tissue. If the important information is expected to be the conductivity of a suspension medium or the integrity of the cell membranes, low frequency measurements between some Hz up to MHz are advisable. The frequency range of the AD 5933 without external components - from 5 kHz up to 100 kHz - is therefore well suited for many applications in bio-impedance measurements. Using an external divider, one can synthesize lower frequencies. This however, is behind the scope of this paper.

A great advantage of the AD 5933 is the simple programming via I^2C-bus and a fully functioning device with just a few components. However for bio-impedance measurements some shortcomings are essential and need to be addressed:

- The dynamic range for the impedance of the material under test (MUT) is only two orders of magnitude. Although numerous bio-impedance applications for MUT with only small variation exist, the impedance of biological MUT ranges often over several orders of magnitude.
- In low frequency applications below 100 kHz, the contact between the MUT and the electronics is usually mediated by electrodes. Therefore, the electrode polarization will add considerable serial impedance (polarization impedance) to the MUT which can be on the order of the MUT-impedance or even bigger, yielding quite wrong results. A widely used approach to compensate the polarization impedance is the four electrode interface where a current is applied between two electrodes and the voltage monitored across a second pair of electrodes. The AD5933 is not intended for use with a four electrode interface.
- Due to the trans-impedance at the input of the AD5933, a measurement with respect to ground is only possible with external components.
- Most bio-impedance measurements require offset free excitation which is only possible with external signal conditioning.

Several applications of the AD5933 for measuring bio-impedances using a four electrode interface are already shown [2-3]. The principle of the analog interface is a symmetric current source, controlled by the output of the AD5933 and a voltage-current converter feeding the voltage dropping at the voltage monitor electrodes into the input. In [2] a comprehensive analysis of the precision achieved with this circuit connected to the analog interface in comparison to a commercial impedance meter showed useful performance of this circuit.

Our activities concentrate on the use of miniaturized electrodes with either bare metal surface but also passivated electrodes for online measurements on fluids, cell suspensions and other organic material. Therefore, we adapted the AD 5933 with a four electrode interface for galvanostatic (similar to [2] and [3]) but also potentiostatic excitation of the material under test.

2. Principle of the AD5933

The AD5933 is made by Analog Devices (Norwood, MA, USA) as single chip impedance converter. It contains everything from frequency synthesis, signal conditioning and –processing up to the I^2C-interface. Both, the excitation signal (intern, no user access) and the answer signal are sampled (1024 sample points) and Fourier transformed in order to get the real and imaginary part of the ratio between signals. In case of voltage excitation and current monitored, it is directly the electrical impedance. Due to the low frequency, the group delay within external components is not important. Thus, interfacing with additional operational amplifiers will not seriously compromise the performance of the device. The internal signal conditioning with programmable gain amplifier (1x or 5x) allows a limited dynamic range but requires for most applications additional conditioning using the analog front end.

3. Galvanostatic excitation

For most applications of bio-impedance measurement using a four electrode interface, galvanostatic excitation is advantageous [4]. The basic principle of the symmetric current source, controlled by the output of the AD5933 is a Howland-circuit using an operational amplifier with symmetric output (AD8132) together with a reference resistor (Fig.1). The signal after the R_{REF} is feed back to the non-inverting input via buffer amplifier. The resistors R5/R6 are used for attenuation (1:10) of the 1.98 V – output of the AD5933. A second R_{REF} with the same value is used at the negative side in order to ensure symmetry of the current flow with respect to ground. The offset of the AD5933 was removed using a feedback loop as shown in detail in [4]. The voltage dropping across the inner electrodes (voltage monitor electrodes) is amplified using a difference amplifier (AD8130) which is optimized for a gain of 1. If a higher gain is necessary, this circuit should be exchanged for the AD 8129. Although not shown here, we always use input buffers for the AD8130 since the input impedance is rather low and would cause a mismatch with the high impedance miniaturized electrodes reaching sometime up to many kΩ. Since the AD5933 – input is not designed for offset free operation, we added the offset using the reference input of the AD 8130/8129.

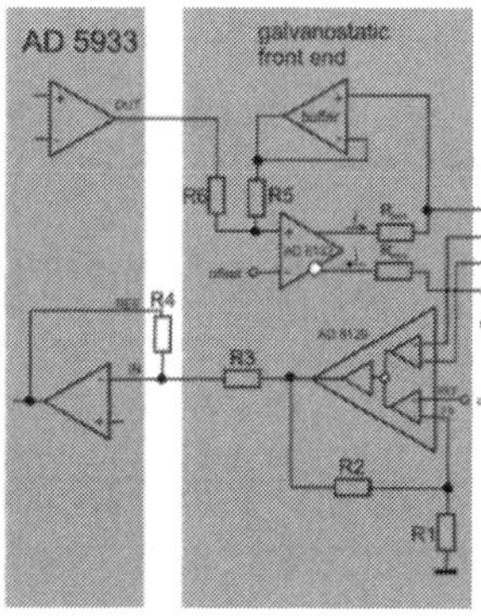

Fig.1 principle of AD5933 coupled to a four electrode interface with galvanostatic excitation

The reference resistor is between 100 Ω and 10 MΩ which allows setting the excitation current between 200 nA and 2 mA. A lower current is possible but does not make sense for the electrode devices we used because of considerable parasitic capacitances resulting in a very low signal between the inner electrodes.
It should be noted, that the output is the current divided by the voltage and is therefore the reciprocal of the impedance.

4. Potentiostatic excitation

If a passivized electrode is used with current excitation at low frequency, the voltage across the electrode reaches high values up to the supply voltage. Although this is usually not a problem, it becomes critical if the distance of the electrodes is low.

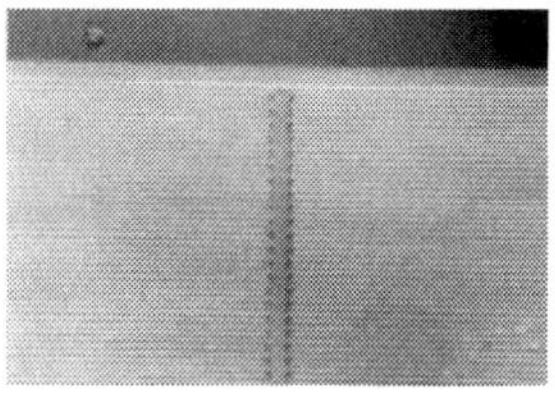

Fig.2 Miniaturized interdigitated electrode (3 μm structure repetition) with surface passivation (SiN) showing damage (bottom right) due to material migration in a high electric field (5 V applied across electrodes) .

With distances on the order of 1 μm, the field strength may reach up to 5 MV/m with a supply voltage of 5 V. This is sufficient for causing material migration and finally a failure of the electrode (Fig.2.)

In order to prevent electrode damage but also for measurements where the electric field strength plays an important role, we constructed a front end where the voltage between the voltage monitor electrodes is kept constant (Fig.3). Therefore a difference amplifier (AD8130) was used where the difference voltage between the inner electrodes was feed back to the inverting input. Moreover, for the offset compensation we used the inverting input.

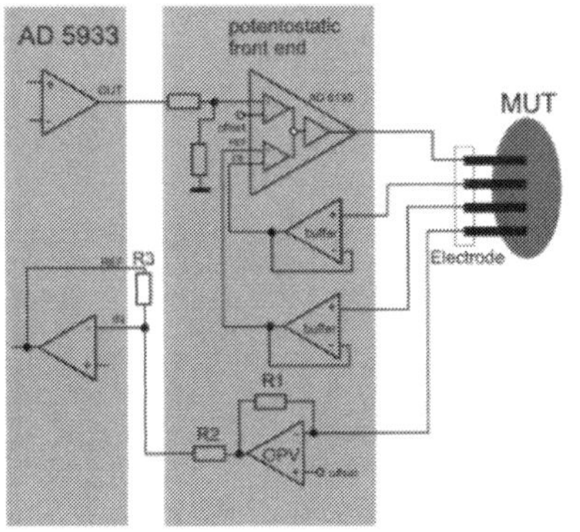

Fig.3 AD5933 coupled to a four electrode interface with potentiostatic excitation

The current through the MUT could be feed directly into the input of the AD5933. But during our tests we realized that the dynamic range of the current is not sufficient for our electrodes used. Therefore an external trans-impedance with adjustable gain (R1) was used. R2 and R3 were adjusted to meet the input requirements of the AD5933. Although it is possible to add the required offset simply by adding a voltage to the non-inverting input of the OPV used for the trans-impedance, a more sophisticated circuitry allows an offset voltage independent of the feedback resistor which is behind the scope here. Moreover, in case of small impedances, i.e. by using large electrodes and highly conductive objects like potato or meat, a very simple external circuitry consisting of only the AD8230 is possible since the input impedance of the difference input is high compared to the electrode impedance of the voltage monitor electrodes. Moreover, if an offset of VDD/2 is used, the trans-impedance input of the AD5933 can be connected directly because the non-inverting input is internally connected to VDD/2. This simple front end however, works only under the mentioned circumstances.

5. Sample results

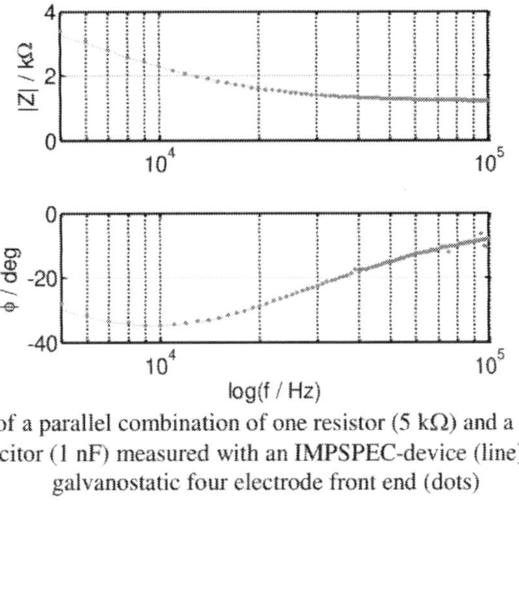

Fig.4 impedance spectrum of a parallel combination of one resistor (5 kΩ) and a serial combination of another resistor (1 kΩ) and a capacitor (1 nF) measured with an IMPSPEC-device (line) and with an AD5933 with galvanostatic four electrode front end (dots)

Simple comparison between a reference system (IMPSPEC, MEODAT Germany) and the AD5933 with both, galvanostatic and potentiostatic front end shows good coincidence. A result with a typical test dummy (5 kΩ in parallel to a series of 1 kΩ and 1 nF) is shown in Fig.4.

Using a real object, in this case a potato (Fig.5) interfaced by four electrodes (needle electrodes, 10 mm long, 2 mm diameter and 5 mm apart, shows the suitability of this circuit as well. Although not shown here, potentiostatic excitation prevented the damage of miniaturized electrodes and is as precise as the galvanostatic front end within the range of the tested impedances from 100 W up to 1 MW. A more sophisticated analysis of the precision of the analog front ends is behind the scope of this contribution and will be presented elsewhere.

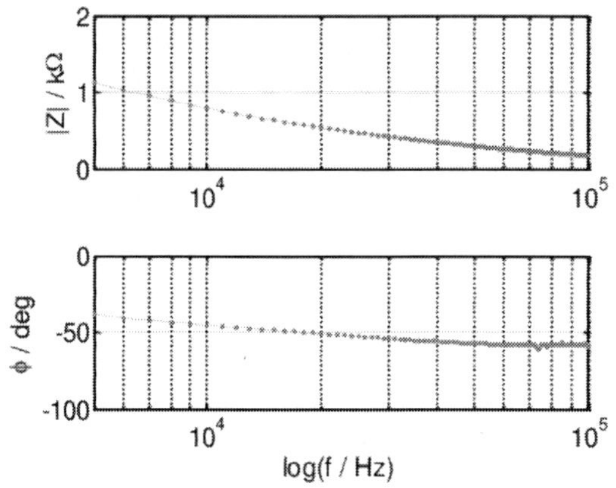

Fig.5 impedance spectrum of potato measured with an IMPSPEC-device (line) and with anAD5933 with galvanostatic four electrode front end (dots)

6. Conclusion

Although the impedance converter AD5933 is not designed for four electrode operation, a relatively simple analog front end can circumvent this shortcoming making this IC an interesting alternative for many bio-impedance applications. Potentiostatic and galvanostatic excitation can be chosen depending on the requirement of the application.

7. References

1. Analog Devices *AD5933*: http://www.analog.com/en/AD5933/productsearch.html.
2. Seoane, F., Ferreira, J., Sanchez, J.J., Bragos, R., (2008), An analog front-end enables electrical impedance spectroscopy system on-chip for biomedical applications, Physiol. Meas.,29, pp.S267-S278
3. Ferreira, J., Seoane, F., Ansede,A., Bragos, R., (2010), AD5933-based Spectrometer for Electrical Bioimpedance Applications, (2010), International Conference on Electrical Bioimpedance, Journal of Physics: Conference Series **224,** 012011
4. Pliquett, U., Schönfeldt, M., Barthel. A., Frense, D., Nacke, T., Beckmann, D., (2011), Front end with offset-free symmetrical current source optimized for time domain impedance spectroscopy, Physiol.Meas., 32, pp.927-944
5. Nacke, T., Barthel, A., Friedrich, J., Helbig, M., Sachs, J., Schäfer, M., Peyerl, P., Pliquett, U., (2007), ICEBI 2007, IFMBE Proceedings 17, pp. 194–197

Bio-impedance detector for Staphylococcus aureus exposed to magnetic fields

Haval Younis Yacoob Aldosky[1], **Waleed Jameel Omar Barwari**[2] **and Janan M Salih Al-mlaly** [2]

[1]Physics Department, Faculty of Science, University of Duhok. Kurdistan Region, Iraq
[2]School of Medicine, Faculty of Medicine, University of Duhok. Kurdistan Region, Iraq

E-mail: yacoobaldosky@uod.ac

Abstract. Rapid detection of viability and growth of pathogenic microorganisms is very important in many applications such as food and drug production, health care, and national defense. Measurements on the electrical characteristics of cells have been used successfully in the past to detect many different physiological events. The effect of electromagnetic fields on the growth of bacteria (Staphylococcus aureus) was studied with the bio-impedance technique. The growth situations of bacteria in the absence and presence of different intensities of static and alternative magnetic fields were examined and analyzed. The results show that the impedance of bacteria fell in the presence of DC magnetic fields. In contrast the impedance increased when the bacteria were exposed to AC magnetic fields. Based on these results the bacterial growth indicated by the change in the impedance is inhibited under DC magnetic fields and enhanced under AC fields.

1. Introduction

Many biological parameters and processes can be sensed and monitored using their impedance as marker with the advantage that it is a non-invasive and a relatively cheap technique [1 and 2]. Cell growth, cell activity, changes in cell composition, shapes and cell location are only some examples of processes which can be detected by the impedance method. The electrical impedance of a biological sample reflects actual physical properties of the tissue. Electrical bio-impedance can be used to assess the properties of biological materials involved in processes such as cancer development [3]. Because the cells of healthy tissues and cancer are different in shape, size and orientation, abnormal cells can be detected using their impedance as a marker. Impedance Z can be represented by the ratio between the voltage and current. Thus, any alteration in the components of the current will alter the voltage current relationship and result in an impedance change. The impedance measured depends on a number of variables, such as adhesion tightness, cell type, and surface area of the electrode, frequency, and confluence of cells [4]. The exposure of living tissue to various types of magnetic fields is a commonly encountered event: extremely low frequency from power lines, high frequency electromagnetic fields (EMF) from cellular phones, and computers [5]. Since this is a task of medical and technological importance, a number of attempts have been given to clarify the effects of magnetic fields on biological cells. [6]. The resurgence of interest in the interactions between electromagnetic fields and biological systems has mainly focused on AC (time-varying) fields. However, there have been studies showing that DC (static) magnetic fields can also interact with living systems at various levels. Nosocomial infections remain a serious problem in medical institutions. Methicillin-resistant Staphylococcus aureus (MRSA) is a major nosocomial pathogen and continues to trigger outbreaks of infection. Efforts to control such outbreaks are of crucial importance not only for the sake of protecting patients but also in preventing the tremendous costs that arise within the context of an epidemic [7]. Experimental investigations on how static magnetic fields and low-frequency magnetic fields can affect the growth dynamics of bacteria Staphylococcus aureus have been done by many other scientists [8 and 9]. The objective of this study is to test the effect of DC and AC magnetic fields on the Staphylococcus aureus growth recorded by the bio-impedance technique.

2. Theoretical aspects

Impedance can be defined as the resistance to the flow of an alternating current as the current passes through a conducting material. When two metal electrodes are immersed in a conductive medium the test system behaves as a resistor and a capacitor in series. When the system is treated as a series combination of the resistance and the capacitance an applied sinusoidal potential will produce a resultant current which depends on the impedance of the system [10]. The impedance in turn is a function of the resistance, capacitance and applied frequency. Microbial metabolism usually results in an increase in both conductance and capacitance causing a decrease in impedance and a consequent increase in admittance. The change in conductance occur in culture medium due to metabolism of uncharged or weakly charged substrates which are converted to highly charged end products, e.g. proteins to amino acids. Capacitance is related to the behavior of ions at the surface of the electrodes and can be monitored separately from conductance [11].

3. Materials and methods

3.1. Impedance System Fabrication

The impedance system was fabricated in a plastic tube of a diameter of 0.5 cm and a length of 4 cm with four silver electrodes (Ag 92.5gm to Cu 7.5gm) fixed inside the tube. These electrodes had a diameter of 0.5 cm and a thickness of 0.2 cm and they were arranged at equal distances of 0.5 cm apart as shown in figure (1).

3.2. *Bacteria Culturing*

Specimens were collected from patients in Azadi hospital in Duhok city-Kurdistan region, Northern Iraq. The isolates were recovered from nasal swabs of patients. For isolating the bacteria (Staphylococcus aureus), samples were cultured on the blood agar (HI-MEDIA) in an aerobic atmosphere. After incubation at 37°C for 24 hours, colonies of Staphylococcus aureus on blood agar appeared yellowish and usually surrounded by a zone of haemolysis. Then the sample was sub cultured on mannitol salt agar (HI-MEDIA) and incubated at 37°C for 24 hours. Colonies characterization was generally yellow surrounded by a yellow zone caused by the fermentation of mannitol. Standard microbiological methods for the identification of Staphylococcus aureus were followed including Gram staining, biochemical tests (catalase, coagulase and DNase) and the bacteria were identified further by API staph method (Biomerieux- France).

3.3. *Setup of Magnetic Fields*

A pair of conducting circular coils, (Helmholtz Coils with a radius of 14 cm) each coil having 320 turns, carrying different values of current I and with the coils separated by a distance equal to half of the radius (7 cm) of the circular loops, produces a homogeneous magnetic field B in the mid-plane between the two circular coils. The strength of the magnetic field was measured at a point in space called the field point. This point was located in the mid-plane between the two coils. Direct (DC) and alternative (AC) currents (50 Hz) were applied separately in simple series circuits as shown in figure (1).

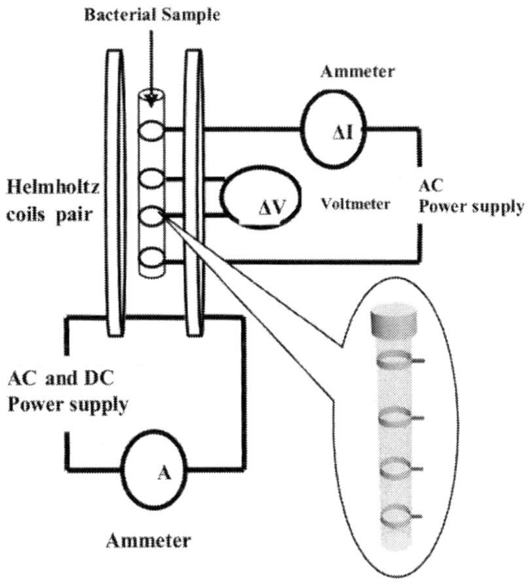

Figure (1) circuit diagram for impedance measuring.

3.4. Impedance Measurement

The microbial impedance changes were measured with a pair of electrodes placed in a growth medium or reacting solution. This measurement was taken over a frequency range of 100 Hz to 20 KHz from the tube filled with bacterial broth. The tube was incubated at 37°C for the duration of the experiment.

The voltage of the applied sinusoidal signal was about 1V*rms*. The impedance of the bacterial samples was determined by the ratio of the measured potential drop ΔV between the two inner electrodes and the electrical current ΔI which depends on the measurement frequency and is supplied at the two outer electrodes. In addition to the control samples (not exposed to magnetic field), the impedance of dead bacteria was determined similarly by heating up a test sample to 75 ℃ and exposing the sample to different magnetic fields. Another test was carried out to measure the impedance of the control sample as a function of time

3.5. Statistical Analysis

The statistical analysis *ANOVA single factor* was used to perform a simple analysis of the variance of the data. The values described with ($p < 0.05$) were considered as significant (Excel 2003).

4. Results

4.1. Impedance Measurements

The impedance changes associated with the growth of Staphylococcus aureus were calculated as a function of frequency between a pair of Helmholtz Coils. At the beginning two preliminary experiments were done. In the first experiment the impedance of the heated dead Staphylococcus aureus was acquired as a function of the frequency. The result is shown in figure (2).

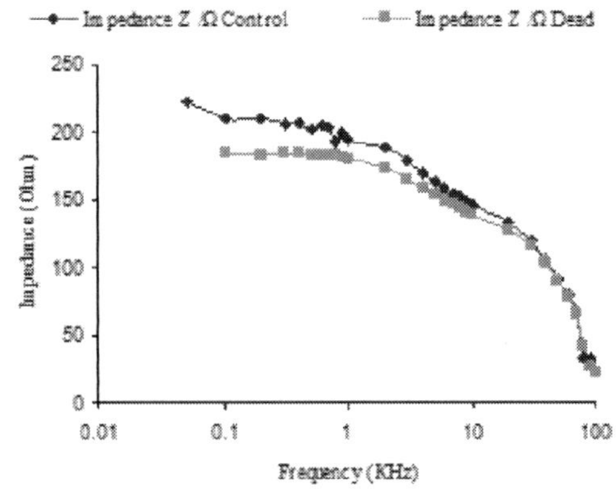

Figure 2. frequency responses for the impedance of control and dead Staphylococcus aureus .

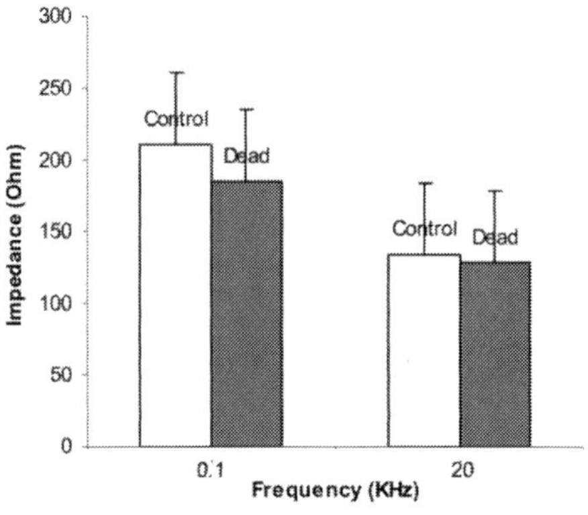

Figure 3. the significant changes of impedance values as a function of frequency for the control and dead Staphylococcus aureus .

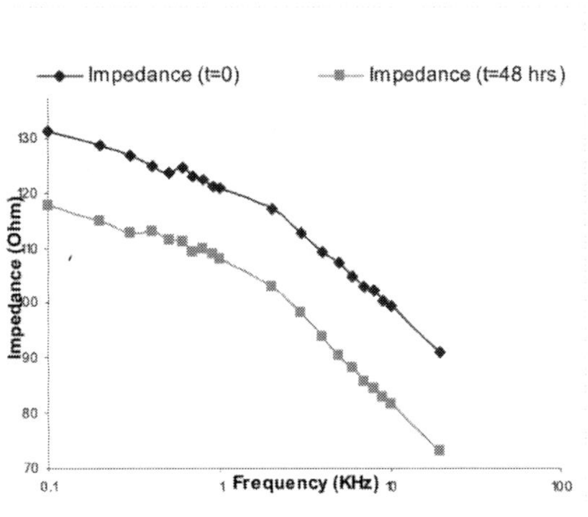

Figure 4. the significant changes of impedance values as a function of frequency for the control and two days growth of Staphylococcus aureus.

In the second experiment the impedance of the Staphylococcus aureus grown for two days was measured as a function of the frequency. The results are shown in figure (4) and indicate that variations of impedance caused by the bacterial growth could be detected by the bio-impedance system. The changes ($P < 0.05$) were significant in comparison to the control sample which clearly appears in figure (3). These results were in good agreement with data reported in [12, 13 and 14].

4.2. DC Magnetic Fields

To test bacterial growth exposed to static magnetic fields using the bio-impedance system, the impedance values of magnetized (test) and un-magnetized (control) samples of bacteria were measured (table 1). The test samples were exposed to the magnetic field densities of 0.1, 0.2, and 0.3

mT at different frequencies ranging from 100 Hz to 20 KHz. The effect of DC magnetic fields on impedance variations is shown in figure (5). The impedance variation was significant (P< 0.05) only at low frequencies (50-100 Hz) as shown in figure (6).

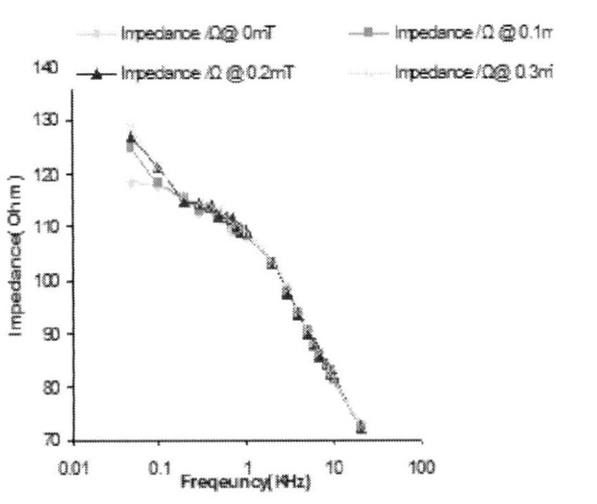

Figure 5. frequency responses for the impedance of Staphylococcus aureus exposed to different DC magnetic fields.

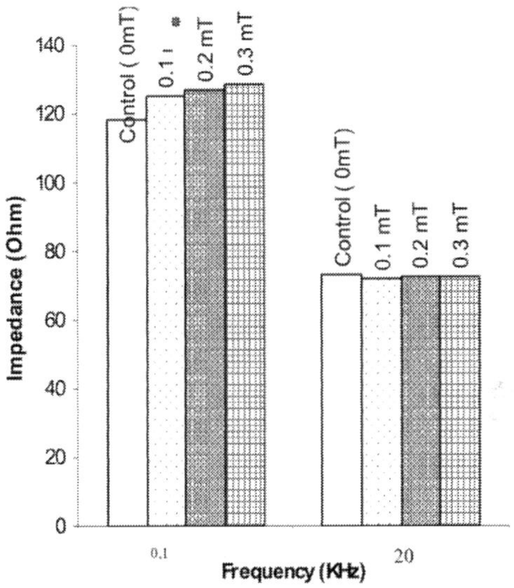

Figure 6. the significant changes of impedance values as a function of frequency (100Hz and 20 KHz) for Staphylococcus aureus exposed to different DC magnetic fields.

4.3. AC Magnetic Fields

To test bacterial growth exposed to alternative magnetic fields using the bio-impedance system, the impedance values of magnetized (test) and un-magnetized (control) samples of bacteria were measured (table 2). The test samples were exposed to the magnetic field densities of 0.11, 0.21, 0.5, and 0.7 mT at different frequencies ranging from 100 Hz to 20 KHz. The effect of AC magnetic fields on impedance variations is shown in figure (7). The significant changes of the control and magnetized samples are also summarized in fig. 8. The most important contribution to the impedance change at low frequencies is related to the capacitance of the medium, which is dependent on the frequency.

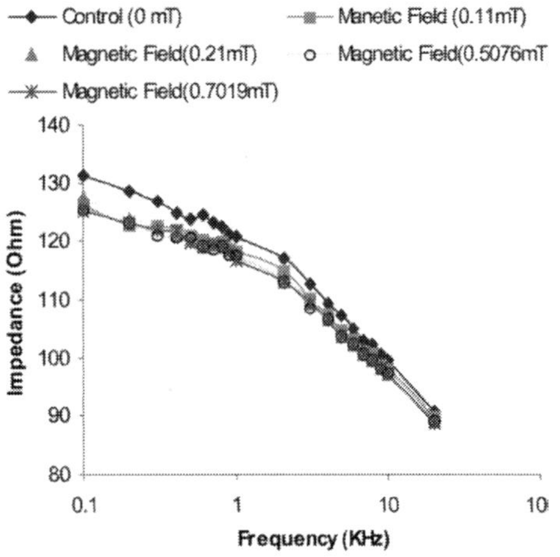

Figure 7. frequency response for the impedance of Staphylococcus aureus exposed to different AC magnetic fields

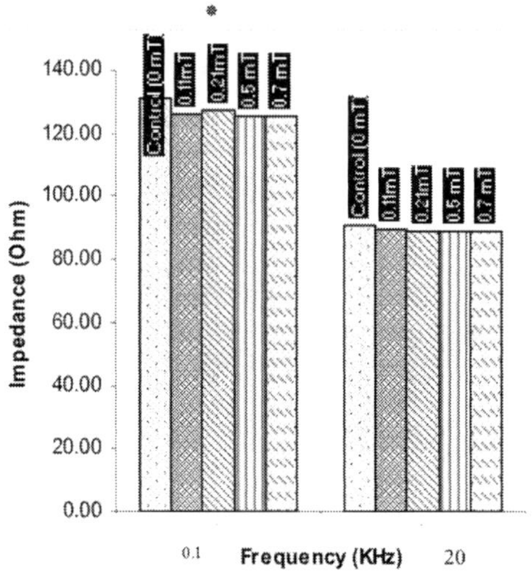

Figure 8. the significant changes of impedance values as a function of frequency (100 Hz and 20 KHz) for Staphylococcus aureus exposed to different AC magnetic fields.

Table 1 impedance response vs. frequency for Staphylococcus aureus exposed to different DC magnetic fields.

Frequency /KHz	Impedance (Ohm)			
	Control 0 mT	DC Magnetic Field 0.1 mT	DC Magnetic Field 0.2 mT	DC Magnetic Field 0.3 mT
0.05	118.42	125.00	126.76	128.57
0.10	117.58	118.18	121.18	121.43
0.40	112.90	113.33	113.56	113.56
0.50	111.45	111.90	112.10	113.01
0.60	111.03	111.45	111.63	111.72
0.80	109.72	109.29	110.22	110.29
0.90	108.84	109.09	109.22	109.29
1.00	108.00	108.22	109.09	109.15
2.00	103.05	103.13	103.16	103.18
3.00	98.25	97.62	97.59	98.18
4.00	93.71	93.60	93.53	94.08
5.00	90.40	90.29	90.17	90.70
6.00	88.20	87.57	88.00	87.93
7.00	85.56	85.39	85.80	86.29
8.00	84.44	83.80	84.18	84.18
9.00	82.87	82.68	82.58	82.58
10.00	81.77	81.11	82.02	81.46
20.00	73.03	71.91	72.32	72.32

Table 2 impedance response vs. frequency for staphylococcus aureus exposed to different AC magnetic fields

Frequency /KHz	Impedance (Ohm)				
	Control 0 mT	AC Magnetic Field 0.11 mT	AC Magnetic Field 0.21 mT	AC Magnetic Field 0.5 mT	AC Magnetic Field 0.7 mT
0.10	131.33	126.19	127.38	125.58	125.29
0.40	125.00	122.03	121.01	120.66	121.67
0.50	123.77	120.97	120.00	120.63	119.84
0.60	124.60	120.16	119.23	119.08	119.23
0.80	122.39	119.85	119.12	118.98	119.12
0.90	121.17	118.71	118.71	117.86	117.99
1.00	120.86	118.31	117.61	117.61	116.90
2.00	117.11	114.94	113.64	112.99	112.99
3.00	112.58	109.94	109.32	108.70	109.38
4.00	109.26	107.32	106.71	106.75	106.75
5.00	107.36	104.82	104.22	103.61	103.61
6.00	104.85	102.99	102.40	102.40	102.40
7.00	103.01	101.19	101.19	100.60	100.60
8.00	102.41	100.60	99.41	99.41	99.41
9.00	100.60	98.82	98.22	98.22	98.22
10.00	99.40	98.22	97.06	97.63	97.06
20.00	90.91	89.76	89.22	89.22	88.69

5. Discussion

A simple and rapid impedance method was proposed to detect bacterial growth exposed to DC and AC magnetic fields by making use of the properties of bacterial cell suspensions and by using a four electrodes system. The impedance values of the bacterial growth were measured to find their dependency on the magnetic fields applied to control the bacterial growth.

The significant variations of impedance shown in figures (3 and 5) indicate that the change in impedance figures (2 and 4) is suitable as a parameter indicating bacterial growth. Absolute values of the changes in impedance could be obtained as a consequence of bacterial growth. These changes in time are explained by the metabolic products created during the growth of microorganisms. The metabolic products modify the composition of the medium, thus changing the ionic content, which in turn produces a change in the conductivity of the culture media [15, 12, 16 and 17].

The results reinforce the concepts that the change in impedance is the consequence of bacterial growth. This significant relationship between growth of bacteria and its electrical properties (Impedance change) has been harnessed to clarify the effects of magnetic fields on biological cells. These effects on the bacteria have been recorded through impedance measuring techniques.

In case of applying DC magnetic fields, the impedance values of the exposed bacterial sample increase which are significantly higher than that of the control sample as shown in figures (5 and 6). This indicates that the bacterial growth was inhibited when exposed to DC fields. In other words the growth of these bacteria was negatively affected by increasing the static magnetic field which aligns rather well with results reported in [8, 18, 19, 20, and 21]. Invariably the growth of the culture was associated with a topping up in impedance of about 5.2, 6.6 and 7.9 % for 0.1, 0.2, and 0.3 mT magnetic fields respectively comparing to the impedance of the control cells. This agrees with the observation that the bacterial growth decreases with increasing DC magnetic fields intensity [8, 22].

In contrast, the drop in the impedance values of bacteria exposed to AC magnetic fields figures (7 and 8) indicates that bacterial growth was enhanced. This means that when the bacteria were exposed to an alternating field, a positive effect was observed on the rate of growth of the colonies [8, 19, 23, 24, and 25]. Table 2 shows that the growth of the bacteria was associated with a drop in impedance of about 3.9, 3, 4.3 and 4.6 %, for 0.11, 0.21, 0.5, and 0.7 mT AC fields respectively in comparison to the impedance of the control cells.

Opposite to the observations when DC fields were applied, these values suggest that there is no substantial contribution of the increasing AC fields to the impedance change. This means that the bacterial growth is independent of AC field intensities since approximately a similar relative impedance drop was produced for all field intensities examined [26]. It is clear that there is no linear decrease of impedance values depending on increase of magnetic fields. The major contribution to the impedance change for the bacteria cells can be viewed at low fields.

6. Conclusion

It can be concluded that the impedance variation is a good indicator of bacterial growth in colonies of bacteria exposed to magnetic fields. The changes in the medium during bacterial growth subject to magnetic fields can be detected by measuring impedance only at low frequencies. In DC fields the impedance values increase which reflects a decline in bacterial growth. On the other hand, impedance values drop in AC fields which reflect enhancing bacterial growth.

7. Reference

[1] Alberto Yúfera, Daniel Cañete and Paula Daza 2011 [A Microelectrode- Cell Sensor Model for Real Time Monitoring] *SENSORDEVICES: The Second International Conference on Sensor Device Technologies and Applications*

[2] Meas. Sci. Technol, Sanchez B, Vandersteen G, Bragos R and Schoukens J 2011[Optimal multisine excitation design for broadband electrical impedance spectroscopy]. *Meas. Sci. Technol* **22** 115601

[3] Rohit Dua, Daryl Beetner G, William Stoecker V, and Donald Wunsch C 2044 [Detection of Basal Cell Carcinoma Using Electrical Impedance and Neural Networks] *IEEE TRANSACTIONS ON BIOMEDICAL ENGINEERING*, **51** 1

[4] Wiertz R.W. F., Rutten W. L. C. , and Marani E 2010 [Impedance Sensing for Monitoring Neuronal Coverage and Comparison With Microscopy] *IEEE TRANSACTIONS ON BIOMEDICAL ENGINEERING*, **57** 10

[5] Ikehata, M., Koana T, Suzuki Y, Shimizu H and Nakagawa M, 1999 [Mutagenicity and co-mutagenicity of static magnetic fields detected by bacterial mutation assay]. *Mutation Research* **427**, 147-156

[6] Nakasono S. and Saiki H 2000 [Effect of ELF magnetic fields on protein synthesis in Escherichia coli K12]. *Radiation Research* **154** 208-216

[7] Wichlhaus, T, Hunfeld, K, Boddinghaus, B, Kraiczy P, Schafer V and Brade V 2001 [Rapid molecular typing of MRSA by PCR-RFLP] *Infection Control and Hospital Epidemiology* **22** 294-298

[8] Pérez Medina 2010 [EFFECT OF WEAK MAGNETIC FIELDS ON BACTERIUM STAPHYLOCOCCUS] *AUREUSInicio* **1** 1

[9] Masahiro Kohno, Muneyo Yamazaki, Isao Kimura and Moriyasu Wada 2000 [Effect of static magnetic fields on bacteria: Streptococcus mutans, Staphylococcus aureus, and Escherichia coli] *Pathophysiology* **7** 2 143-148

[10] Yang L, Ruan C and Li Y 2003 [Detection of viable Salmonella typhimurium by impedance measurement of electrode capacitance and medium resistance] *Biosens Bioelectron* **19** 5 495-502

[11] Bolton F J In 1991[Conductance and Impedance Methods for detecting Pathogens] *Rapid Methods and Automation in Microbiology and Immunology* 176-181

[12] Nardo Ramírez, Angel Regueiro, Olimpia Arias and Rolando Contreras 2009 [Electrochemical impedance spectroscopy: An effective tool for a fast microbiological diagnosis] . *Biotecnología Aplicada* **26** 72-78

[13] STRATEN, VAN R and SONDERKAMP H J 1978 [Impedance as a parameter for bacterial growth in medical microbiology] *Antonie van Leeuwenhoek* **44**

[14] Tao Zhu, Zhenhua Pei, Jianyong Huang,ac Chunyang Xiong, Shenggen Shib and Jing Fanga 2010 [Detection of bacterial cells by impedance spectra via fluidic electrodes in a microfluidic device] *Lab Chip* **10**, 1557–1560

[15] Owicki J and Parce J 1992 [Biosensors based on the energy metabolism of living cells: the physical chemistry and cell biology of extracellular acidification]. *Biosens Bioelectron* **7** 257-72.

[16] James B Allison, John A Anderson and William H Cole 1938 [The Method of Electrical Conductivity in Studies onBacterial Metabolism] *J. Bacteriol* **36** 6 571

[17] Lasik M and Nowak J 2010 [Electrical impedance for bacterial metabolic activity screening-evaluation of single and mixed bacterial consortia for wastewater biodegradation] *International Food Research Journal* **17** 591-599

[18] Wenjin Ji , Huimin Huang, Aihua Deng and Chunyang Pan 2009 [Effects of static magnetic fields on Escherichia coli] *Micron* **40** 8 894-898

[19] Piatti E, Albertini MC, Baffone W, Fraternale D, Citterio B, Piacentini MP, Dacha M, Vetrano F and Accorsi A 2002 [Antibacterial effect of a magnetic field on Serratia marcescens and related virulence to Hordeum vulgare and Rubus fruticosus callus cells]. *Mol Biol* **132** 2 359-65

[20] Zhang S, ei W, Zhang J, Mao Y and Liu S 2002 [Effect of static magnetic field on growth of Escherichia coli and relative response model of series piezoelectric quartz crystal] *Analyst. Mar* **127** 3 373-7

[21] Kohno M, Yamazaki M, Kimura I and Wada M. 2000 [Effect of static magnetic fields on bacteria: Streptococcus mutans, Staphylococcus aureus, and Escherichia coli] *Pathophysiology* **7** 2 143-148

[22] Molouk Mohammed Khazan Alkhazan1 and Amna Ali Nasser Saddiq 2010 [The effect of magnetic field on the physical, chemical and microbiological properties of the lake water in Saudi Arabia] *Journal of Evolutionary Biology Research* **2** 1 7-14

[23] Luiz F.C. NASCIMENTO, Galdenoro BOTURA Jr. and Rogério P. MOTA 2003 [GLUCOSE CONSUME AND GROWTH OF E. coli UNDER ELECTROMAGNETIC FIELD] *Rev. Inst. Med. trop. S. Paulo* **45** 2 65-67

[24] Andrea Amaroli, Francesca Trielli,1 Bruno Bianco, Stefano Giordano,2Elsa Moggia, and Maria Umberta Delmonte Corrado1 2006 [Effects of a 50 Hz Magnetic Field on Dictyosteliumdiscoideum (Protista)] *Bioelectromagnetics* **27** 528-534

[25] Bersani, B Del Re · F, Agostini C, Mesirca P and Giorgi G 2004 [Various effects on transposition activity and survival of Escherichia coli cells due to different ELF-MF signals] *Radiat Environ Biophys* 43:265–270

[26] Luigina Cellini,1 Rossella Grande,1 Emanuela Di Campli, Soraya Di Bartolomeo,1Mara Di Giulio,1 Iole Robuffo, Oriana Trubiani, and Maria A. Mariggio 2008 [Bacterial Response to the Exposure of 50 Hz Electromagnetic Fields] *Bioelectromagnetics* **29** 302-311

MINIMIZING THE PARTIAL VOLUME EFFECT USING THE RESPIRATORY SIGNAL TIMING IN THE ART OF CAPTURING IMAGES OF THE PET CT

Welder Souza Goulão[a] & Araken dos Santos Werneck Rodrigues [b]

[a] *Faculdade UnB Gama (FGA) – Área Especial Quadra I, Setor Central- Gama – DF - Brazil*
[b] *Faculdade UnB Ceilândia (FCE) – QNN 14 Área Especial, Guariroba, Ceilândia Sul - DF - Brazil*

***Resumo**. This work describes the development of an algorithm capable of filtering a signal captured by a breath of the NTC thermistor and then synchronizes this signal to a pulse generator. The measurement is essential for biomedical signals to obtain physiological data for diagnostic or research purposes. The information was obtained from an analysis of signals available in the database of phisionet. The greater relevance of this work is the possibility of contributing to medical research to identify the cancer cells present in small volumes of organs located in the thoracic region of patients undergoing clinical evaluation.*
KEY WORDS: signal conditioning, Sync, respiratory cycle, Nuclear Medicine, Lung cancer, the partial volume error.

1. INTRODUCTION

In the thorax and abdomen, there may be considerable movement of internal organs and structure, in response to respiratory and cardiac motion. A tumor present in this region and identified by a radioactive tracer, widely used in diagnostic imaging examinations, will also move, especially during the breathing cycle. The time required for one image volume of the lungs in a modern CT (Computational Thomography) scanner is of the order of seconds and it is relatively smaller than the breathing cycle. The image volume thus provides an indication of the position of internal organs at a point in time and does not provide information about the potential movement of abnormal tissue volumes during the respiratory cycle.

In comparison, data are acquired from a PET (Positron Tomography-Emition) very differently in the time scale and the interpretation of the image obtained becomes problematic. An image obtained PET is usually over a period of 3-5 minutes, and therefore, many images of breathing cycles. The image data thus depict a distribution that is time-averaged and based on the residence time distribution of the radioactive tracer in each point of the image thus the tumor seen in CT gives only a position where it moves during breathing, while the image seen by PET shows the entire area of this movement. The PET remains a lower resolution imaging technique to CT and this compromises the detection and quantification of activity in small volumes. Thus, the tumor seen on CT has only one position where it moves during breathing while the image seen by PET shows the entire area of this movement, Figure 1 illustrates this difference.

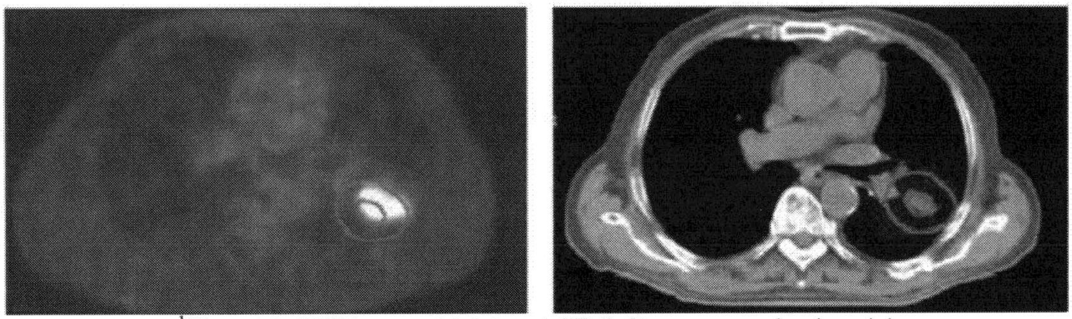

[1]**Figure 1.** PET images (right) and CT (left) reconstruction in axial

The implications of using these devices against the movement of the entire thoracic structure are basically two. First, the difference in time required for image acquisition between PET and CT can lead to disturbances in the feature image and a quantification incorrect attenuation correction is applied to the image reconstruction. Second, the movement of radioactive tracer during the acquisition will result in an image reconstruction with distribution activity outside the real target, and staining the lesion showing a greater volume than actually exists, that phenomenon is presented as a partial volume error.

Thus, the principal effect of the respiratory motion is a spread of the tumor detected predominantly in the chest and, to a lesser extent, in areas below the diaphragm. However, since the first clinical application of PET in RTP (Radiotherapy) is in lung cancer, the issue of respiratory motion has a predominance of challenges to identify accurately determining the volume and concentration of functional activity of the markers. In particular, the case of PETCT scanners (Positron Tomography-Emition Computational Thomography) presents additional challenges, since the presence of respiratory movements produces inaccuracy in image reconstruction, not only as a result of the injury spot, but also as a result of registry error between PET and CT acquisitions. With the use of hybrid scanners, the CT maps are also used to correct the effects of attenuation on data from PET, an extra inaccuracy can be introduced when the data set obtained by CT and PET are not perfectly aligned due to error partial volume.

This can compromise efforts to delineate treatment volumes or to quantify the absorption of the lesion. These issues are being addressed by the introduction of monitoring respiratory movements that allow segmentation of images through the acquisition of timing between the movement of the marker and the acquisition, which promotes significant improvement in attenuation correction coefficients accurately and reconstruction of volumes of activity and concentrations of target cells.

2. METHODOLOGY

To display this work was performed in a research database of Physionet in order to locate signs of breathing that used sensors built based on the use of NTC (Negative Temperature Coefficient) thermistors. The database MIT-BIH Polysomnographic Database (slpdb) has basic breathing signals captured by sensors thermal and free from any kind of filtering, which represented a significant gain time to perform the initial tests of efficiency and robustness of the proposed filter this project.

We selected a short duration signal to better visualize the nuances of the curves of the respiratory cycle, and the signal was sampled lasting 10 seconds at a rate of 250 Hz, as shown in Figure 2 below.

[1] SERGIO L. Faria *et al.*, 2007 pg. 347

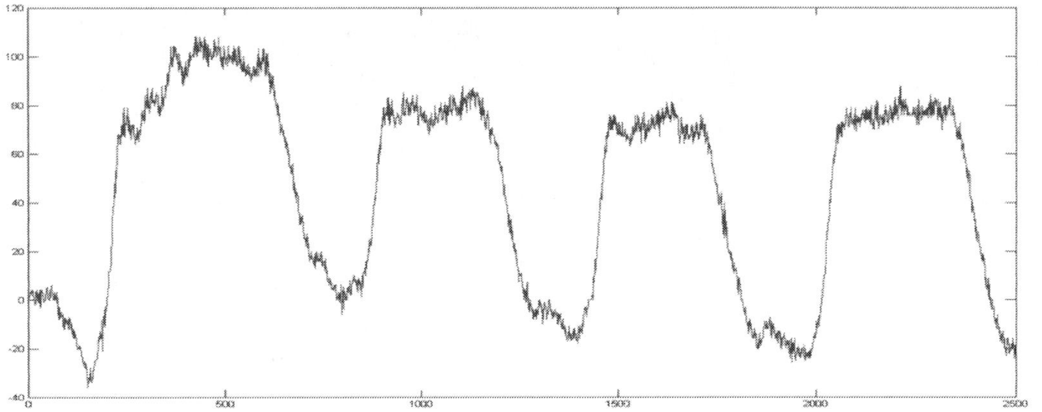

Figure 2. Free respiration signal filter (MATLAB).

3. RESULTS

This topic will discuss the technical aspects related to the algorithm used in MATLAB to achieve the filtering stages, taking as starting point the analysis of images generated by the program and the presentation of the characteristics necessary for the development of synchronized signals at selected points in the curve breathing.

3.1. IMAGES GENERATED BY ALGORITHM

After reading the selected signal in the database of phisionet using the MATLAB program has been possible to manipulate the variable responsible for storing the values of the curve assossiados respiration signal.

Initially the technique used for windowing convolution signal with a filter of the type FIR (Finite Impulse Response) frequency of interest in order to obtain a significant attenuation at frequencies assossiadas the noise shown in Figure 2. With only one filtering stage, ultilizando pass filter ranges of length N = 25 obtained a regular sinusoid, as indicated in figure 3.

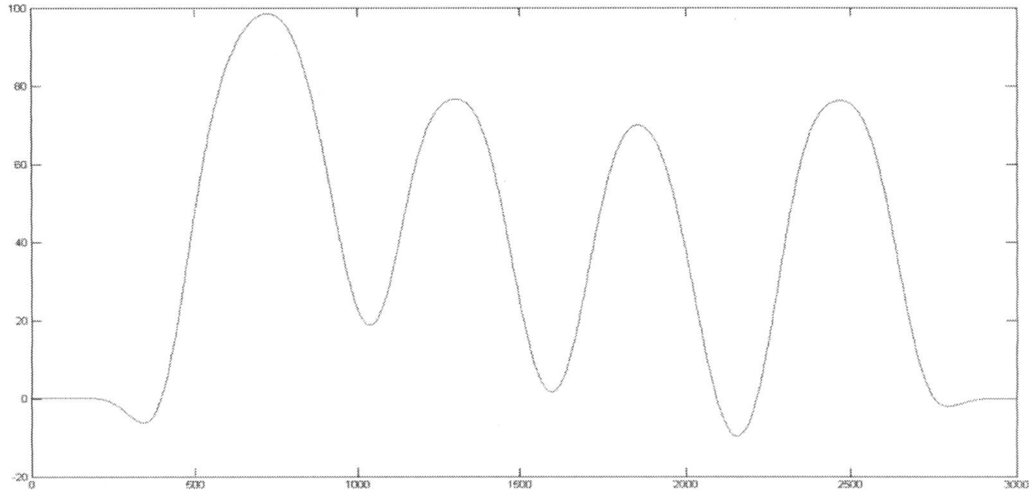

Figure 3. Respiration signal after the first stage of FIR filter (MATLAB).

3.2. PULSE OF TIMING

To synchronize a pulse generator with the respiration signal, it was necessary to implement an algorithm capable of identifying the peaks and valleys of the original signal and starting this identification locate points of interest on the curve, which represent the stages of inspiration and expiration breathing cycle.

Figure 5 is a sync pulse generator with the curve of breath using as reference the beginning of the expiration phase, ie at the end of a peak.

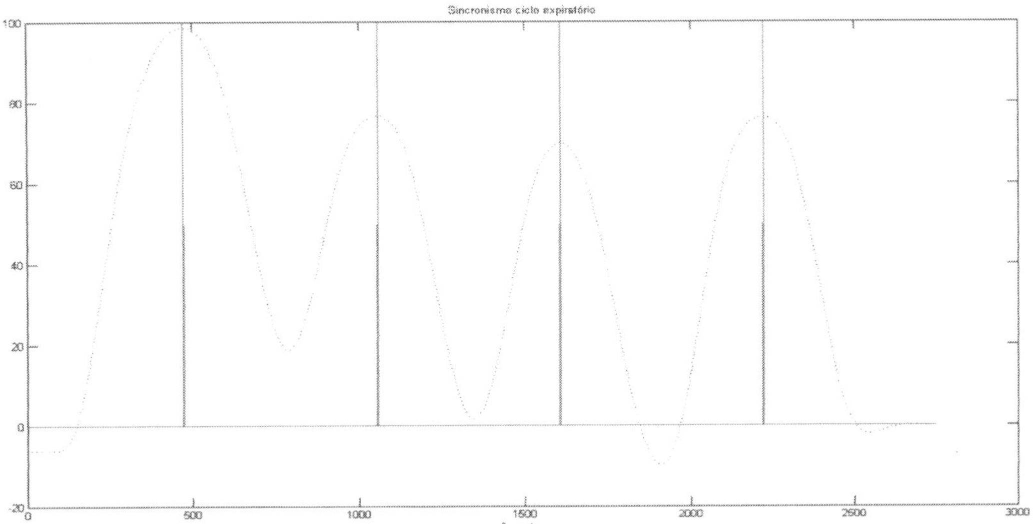

Figure 5. Timing of the respiratory cycle 0% (MATLAB).

In Figure 6 the signal is synchronized with the pulse generator in 50% of the descent stage of the cycle.

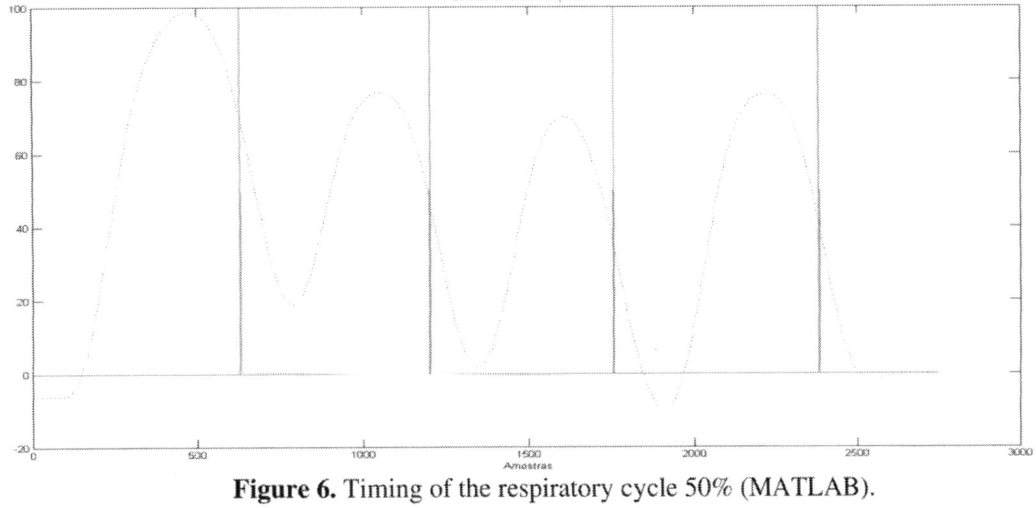

Figure 6. Timing of the respiratory cycle 50% (MATLAB).

3.3. PROCESSING

The respiration signal can be quantified in various ways. One can, for example, measuring their maximum amplitude, which is however a rough measure of the degree of muscular activity. Other methods consist in counting the number of peaks, or by the zero crossing number, or the number of times that a certain amplitude value is exceeded in a given time interval.

The method was used to filter the signal (band pass filter), identify the peaks and valleys and measure the distances between these points, thereby building the information base needed to sync with the pulse generator. This methodology enabled ensure that the filter after the interaction with the temporal signal was possible to maintain a linear relationship with the pulse generator.

4. CONCLUSIONS

Despite the demonstration of robustness and efficiency of the filter, this activity was an sample collected by third parties with no details about the initial conditions of data collection, or even the characteristics of the patients, thus an important variable for confirmation consistency of the algorithm.

For the commercial development of prototypes designer should consider implementing IIR (Infinite Impulse Response) type filters because of the small length and the reduction of multiplication stages, and therefore the lower computational cost and possibly faster response time series for analysis of real time data.

4.1. DISCUSSIONS

The use of FIR filter proved efficient for implementation of pulse generator synchronized to the respiration signal, however due to the need for greater attenuation of unwanted frequencies, it was necessary to increase the size of the filter, resulting in a significant delay in pulse generator. Despite the adjustment made by the small devices MATLAB, we also observed that in practice for real-time signals the best option will be the type IIR filter, which can in turn benefit the project with a smaller size and the length of the filter stages feedback, which ensure greater robustness to the filter before signals in real time, since the technical details of the filter specification are met efficiently.

5. BIBLIOGRAPHY

CARVALHO, Luis Carlos. Apostila de Instrumentação Biomédica – Capítulo IV: Eletromiografia e eletrodiagnóstico. Universidade Federal da Paraíba. UFPB.

BUTTON, Vera Lúcia da Silveira Nantes . Eletromiógrafo, Eletrocardiógrafo e Monitor Cardíaco. In: Saide Jorge Calil. (Org.). Equipamentos Médico-Hospitalares e o Gerenciamento da Manutenção. 1 ed. Brasília: Editora Ministério da Saúde, 2002, v. , p. 331-363.

BRONZINO, J. D., 2000, "The Biomedical Engineering HandBook", Second Edition, Boca Raton: CRC Press LLC.

SERGIO L. Faria; Robert Lisbona; Jerry Stern; Slobodan Devic; Luis Souhami; R Carolyn. Use of FDG-PET/CT scan in the planning of radiation therapy for nonsmall cell lung cancer. Radiologia Brasil. v. 40, p. 345–348, 2007

Analysis and Imaging in Magnetic Induction Tomography using the Impedance Method

Julia Grasiela B. Wolff[1], **Heric Dênis Farias**[2], **Airton Ramos**[3] **and Antônio Flavio Licarião Nogueira**[4], IEEE *member*

[1] Master of Appplied Sciences and Electrical Engineering Department's Professor, Udesc Joinville, SC.
[2] Electrical Engineering's Student, Udesc Joinville, SC.
[3] PhD. and Electrical Engineering Department's Professor, Udesc Joinville, SC.
[4] Doctor and Electrical Engineering Department's Professor, Udesc Joinville, SC.

E-mail: wolff1980@gmail.com, dee2ar@joinville.udesc.br and antonioflavio@ieee.org

Abstract. This article discusses the utilization of the impedance method in computation of the forward problem in magnetic induction tomography (MIT). The algorithms for the inverse problem were also developed. The new approach for solving the resulting ill-conditioned linear system of the inverse problem is proposed and the quality of images obtained is discussed based on a quality index proposed in the literature. The results show the prevalence of TAS in relation to linear correlation between the real image and obtained image. With respect to contrast TRT prevalece in relation TAS. The indices of average luminance presents similarity for the both methods. TAS prevails for smaller objects and TRT for larger objects, showing the greater robustness of TAS.

Key-words – Eddy currents, tomography, impedance methods, numerical modelling.

1. Introduction

Magnetic Induction Tomography (MIT) is a non-invasive technique for obtaining low-resolution images for the visualization of electromagnetic properties of an object using eddy currents. MIT is used for medical applications, such as checking the condition of biological tissues [1], and in industrial applications such as a non-destructive testing in conductive materials [2].

MIT is a system consisting of several sources and magnetic field sensors, formed by coils distributed around the object of interest. When the object is excited with an alternating magnetic field, eddy currents circulate inside and produce a secondary magnetic field. The magnitude and direction of the eddy currents and the field produced by them depend on the electromagnetic properties of the material medium of which the object of interest is formed. MIT is based on the possibility of obtaining images of the distribution of one or more electromagnetic properties within the object, from the measurement of the resultant magnetic field which is distributed around the object.

This study uses the impedance method [3],[4],[5] for computing eddy currents and the Biot-Savart law for calculating magnetic fields. These methods use quasi-static approximations for Maxwell equations, which are valid if the distances involved are small compared to the wavelength of the field.

In these imaging techniques, the process for the computation of eddy currents and the resulting magnetic field is known as the forward problem. And the process for achievement of electrical conductivity distribution inside the object is named inverse problem.

2. Tomograph Model

The presentation of numeric methods used in MIT forward and inverse problems will be done in Figure 1. Here, there is the schematic representation of a tomograph with 16 coils. In this illustration, the shaded circles are radially distributed in relation to the geometric center of the tomograph and represents, for example, 12 sources and 4 field sensors.

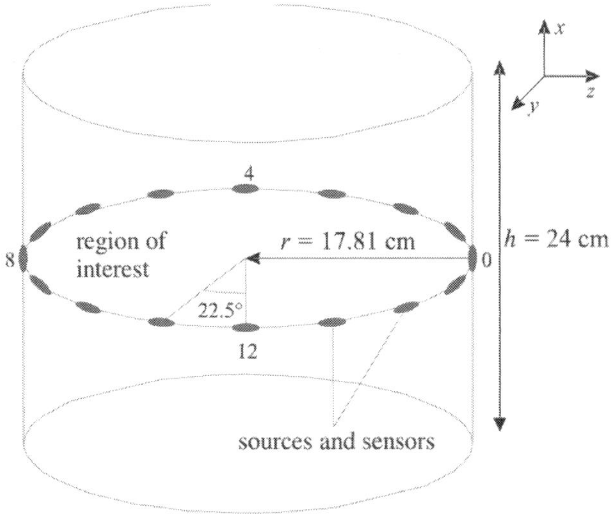

Figure 1 – Geometry of the tomograph used in the numerical model.

3. Calculation of Magnetic Inductions

Biot-Savart's Law is used for computing primary distributions of magnetic induction $\boldsymbol{B}$ also known as primary fields, from numeric integration along the length of generation field coils. Figure 2 shows the geometric details of this calculation. Discretizing the contour of the helicoidal curve of the coil in Nc divisions and forgetting the transversal section of wire, we obtain the expression for the calculation of the primary field $\boldsymbol{B_p}$ in a space position $\boldsymbol{r}$:

$$\boldsymbol{B}_p(\boldsymbol{r}) = \frac{\mu_o I_s}{4\pi} \sum_{n=1}^{N_c} \frac{\Delta \boldsymbol{L}_n \times (\boldsymbol{r} - \boldsymbol{s}_n)}{|\boldsymbol{r} - \boldsymbol{s}_n|^3}, \tag{1}$$

where μ_o is a magnetic permeability in vacuum, I_s is the peak value of the sinusoidal current flowing in the solenoid, $\Delta \boldsymbol{L}_n$ is the elementary displacement along the solenoid, $\boldsymbol{s}_n$ is the vector that expresses the position of the element $\Delta \boldsymbol{L}_n$, and $\boldsymbol{B}_p$ is the magnetic induction produced by the action of the solenoid at the point P.

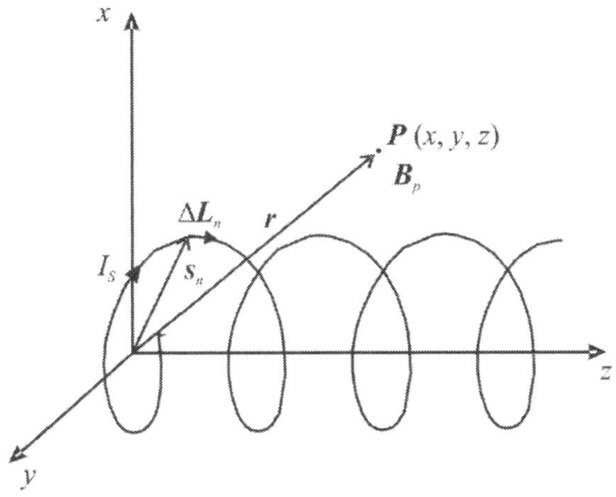

Figure 2 – a) Geometric representation to calculate the primary and secondary fields.

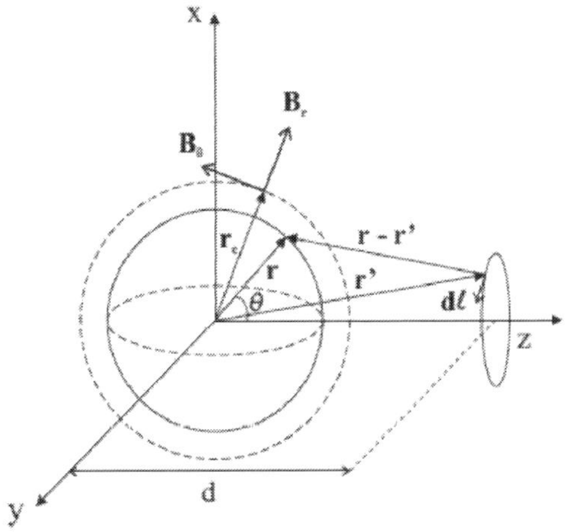

Figure 2 – b) Geometric representation to calculate the primary and secondary fields.

4. Calculation of Eddy Currents

In order to obtain the values of the secondary field, you must first calculate the induced currents. The impedance method was used by dividing the region of interest in a regular grid of $N_x \times N_y \times N_z$ cubic volume elements, considered homogeneous, connected to their neighbors by impedance dependent on the size of the element and their electrical characteristics. The resulting circuit is shown in Figure 3.

The spatial distribution of the magnetic potential vector $\boldsymbol{A}$ is calculated from the contributions of the elementary field sources $I_s\, \boldsymbol{\Delta L_n}$. In others words,

$$\boldsymbol{A}(\boldsymbol{r}) = \frac{\mu_0 I_s}{4\pi} \sum_{n=1}^{N_c} \frac{\Delta \boldsymbol{L}_n}{|\boldsymbol{r} - \boldsymbol{s}_n|}. \qquad (2)$$

Once the discrete distribution of the magnetic potential vector $\boldsymbol{A}$ in the various vertices of the three-dimensional grid is known, it is possible to calculate the induced voltage in the primary field at each grid vertice using the equations (3), (4) and (5), that are based on Faraday's Law:

$$V_X(i,j,k) = j\omega\{[A_Y(i,j,k+1) - A_Y(i,j,k)]\Delta y + [A_Z(i,j,k) - A_Z(i,j+1,k)]\Delta z\}; \quad (3)$$

$$V_Y(i,j,k) = j\omega\{[A_X(i,j,k) - A_X(i,j,k+1)]\Delta x + [A_Z(i+1,j,k) - A_Z(i,j,k)]\Delta z\}; \quad (4)$$

$$V_Z(i,j,k) = j\omega\{[A_Y(i,j,k) - A_Y(i+1,j,k)]\Delta y + [A_X(i,j+1,k) - A_X(i,j,k)]\Delta x\}, \quad (5)$$

where Δx, Δy and Δz are, respectively, the length in directions x, y and z of the volume elements. $\boldsymbol{A}_x$, $\boldsymbol{A}_y$ and $\boldsymbol{A}_z$ are the components x, y and z of the magnetic potential vector calculated by Equation (2).

Eddy currents can be calculated using traditional mesh analysis. Each branch of the equivalent circuit is shared by four meshes in two perpendicular planes, and four mesh currents must be algebraically added to each branch.

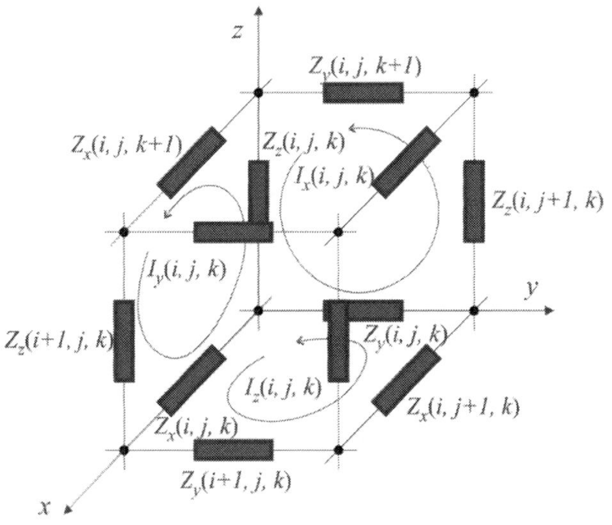

Figure 3 – Equivalent circuit for connection between neighboring elements.

Assuming that in each volume element, the value of the parameters σ and ε are constant. However, these parameters may be different in adjacent elements, which enables modelling heterogeneous media and anisotropic shape.

As each volume element is considered homogeneous, the value of the impedances can be calculated by Equations (6) to (8).

$$Z_X(i,j,k) = \Delta x\{[\sigma_x(i,j,k) + j\omega\varepsilon_o\varepsilon_x(i,j,k)]\Delta y\Delta z\}^{-1} \quad (6)$$

$$Z_Y(i,j,k) = \Delta y\{[\sigma_y(i,j,k) + j\omega\varepsilon_o\varepsilon_y(i,j,k)]\Delta z\Delta x\}^{-1} \quad (7)$$

$$Z_Z(i,j,k) = \Delta z\{[\sigma_z(i,j,k) + j\omega\varepsilon_o\varepsilon_z(i,j,k)]\Delta x\Delta y\}^{-1} \quad (8)$$

where σ is the conductivity, ε is the dielectric constant and ε_o is the vacuum permittivity.

Applying Kirchhoff's Voltage Law to the element meshes involving (i, j, k) results in Equations (9) to (11).

$$\begin{aligned} &Z_y(i,j,k)\left[I_x(i,j,k)-I_x(i,j,k-1)+I_z(i-1,j,k)-I_z(i,j,k)\right]+\\ &Z_z(i,j,k)\left[I_x(i,j,k)-I_x(i,j-1,k)+I_y(i-1,j,k)-I_y(i,j,k)\right]+\\ &Z_y(i,j,k+1)\left[I_x(i,j,k)-I_x(i,j,k+1)+I_z(i,j,k+1)-I_z(i-1,j,k+1)\right]+\\ &Z_z(i,j+1,k)\left[I_x(i,j,k)-I_x(i,j+1,k)+I_y(i,j+1,k)-I_y(i-1,j+1,k)\right]=V_x(i,j,k) \end{aligned} \quad (9)$$

$$\begin{aligned} &Z_x(i,j,k)\left[I_y(i,j,k)-I_y(i,j,k-1)+I_z(i,j-1,k)-I_z(i,j,k)\right]+\\ &Z_z(i,j,k)\left[I_y(i,j,k)-I_y(i-1,j,k)+I_x(i,j-1,k)-I_x(i,j,k)\right]+\\ &Z_x(i,j,k+1)\left[I_y(i,j,k)-I_y(i,j,k+1)+I_z(i,j,k+1)-I_z(i,j-1,k+1)\right]+\\ &Z_z(i+1,j,k)\left[I_y(i,j,k)-I_y(i+1,j,k)+I_x(i+1,j,k)-I_x(i+1,j-1,k)\right]=V_y(i,j,k) \end{aligned} \quad (10)$$

$$\begin{aligned} &Z_x(i,j,k)\left[I_z(i,j,k)-I_z(i,j-1,k)+I_y(i,j,k)-I_y(i,j,k-1)\right]+\\ &Z_y(i,j,k)\left[I_z(i,j,k)-I_z(i-1,j,k)+I_x(i,j,k-1)-I_x(i,j,k)\right]+\\ &Z_x(i,j+1,k)\left[I_z(i,j,k)-I_z(i,j+1,k)+I_y(i,j+1,k-1)-I_y(i,j+1,k)\right]+\\ &Z_y(i+1,j,k)\left[I_z(i,j,k)-I_z(i+1,j,k)+I_x(i+1,j,k)-I_x(i+1,j,k-1)\right]=V_z(i,j,k) \end{aligned} \quad (11)$$

To specify the boundary conditions, all currents out of the mesh discretization are nil. Here, Ix, I_y and I_z are the induced currents in the mesh. The set of equations $3N$, where N is the number of cubic elements, must be resolved to obtain the mesh currents in the equivalent circuit. For the secondary field produced by currents induced in the object, you must obtain the branch currents using Equations (12) to (14).

$$\begin{aligned} I_{bx}(i,j,k) &= I_z(i,j,k)-I_z(i,j-1,k)-\\ &\quad I_y(i,j,k)+I_y(i,j,k-1); \end{aligned} \quad (12)$$

$$\begin{aligned} I_{by}(i,j,k) &= I_x(i,j,k)-I_x(i,j,k-1)-\\ &\quad I_z(i,j,k)+I_z(i-1,j,k); \end{aligned} \quad (13)$$

$$\begin{aligned} I_{bz}(i,j,k) &= I_y(i,j,k)-I_y(i-1,j,k)-\\ &\quad I_x(i,j,k)+I_x(i,j-1,k). \end{aligned} \quad (14)$$

Thus, the secondary field is calculated using the Biot-Savart Law:

$$\boldsymbol{B}_s(\boldsymbol{r}) = \frac{\mu_0 h}{4\pi}\sum_{n=1}^{N}\frac{(I_{bx}\boldsymbol{u}_x + I_{by}\boldsymbol{u}_y + I_{bz}\boldsymbol{u}_z)_n \times (\boldsymbol{r}-\boldsymbol{r}_n)}{|\boldsymbol{r}-\boldsymbol{r}_n|^3}, \quad (15)$$

where the mesh parameter $h = \Delta x = \Delta y = \Delta z$, $N = N_x N_y N_z$ is equal to the number of volume elements used in the discretizing, $\boldsymbol{u}_x$, $\boldsymbol{u}_y$ and $\boldsymbol{u}_z$ are the unit vector in directions x, y and z, and $\boldsymbol{r}_n$ is the position vector of element n.

5. Inverse Problem

5.1.1. Sensitivity Computing

With the topology used, as can be seen in Figure 2b, the sensor coils are excited by the radial component of the magnetic field resulting in the radial direction, which is calculated using Equation (16).

$$B_r = B_y \sin\theta + B_z \cos\theta, \tag{16}$$

where θ is the angle measured from the positive direction of the axis z in direction of the axis y in the positive direction. When a conductive object is introduced into the tomograph, it induces a secondary field at the time quadrature with respect to the primary field, the signal induced in the sensors is shifted in phase by the following quantity:

$$\Delta\phi_{m,n} = \tan^{-1}\left(\frac{B_{s\,m,n}}{B_{p\,m,n}}\right) \approx \frac{B_{s\,m,n}}{B_{p\,m,n}}, \tag{17}$$

where m denotes the generation position and n the measuring position (Figure 1). The approach used is valid because the secondary field has much smaller amplitude than the primary field. As the phase shift is directly related to the sensitivity to the conductivity, it is possible to obtain the distribution of conductivity in the region of interest from this measurement.

The calculation of sensitivity is straightforward using the Biot-Savart Law and the impedance method.

Employing N_f sources and N_s field sensors, it is possible to obtain $N_f N_s$ independent measurements of phase, and with this to obtain the conductivity values in $N_f N_s$ positions. In this work, the interest is to obtain the image of the transversal section of the conductivity. In the idealized image, it is expected that the conductive region will appear white in color, while the region containing air will appear to be black. For this reason, the area of interest is divided into a two-dimensional regular grid of area elements and a system of equations is defined relating the measured phase of the signal in the set of sensors with the distribution of conductivity in the area of the discretized object:

$$[S]_{N_f \cdot N_s \times N_f \cdot N_s} \cdot [\sigma]_{N_f \cdot N_s \times 1} = [\Delta\phi]_{N_f \cdot N_s \times 1}, \tag{18}$$

where $[\Delta\phi]$ is the vector of phase differences defined in equation (19). The sensitivity matrix or Jacobian matrix, denoted by $[S]$ relates the vector of electrical conductivity $[\sigma]$ in the object with the vector phase difference $[\Delta\phi]$ that will be generated when excited by field sources:

$$[\Delta\phi] = \left[\Delta\phi_{0,0}\ \Delta\phi_{0,1} \cdots \Delta\phi_{0,N_s-1} \cdots \Delta\phi_{N_f-1,N_s-1}\right]^T, \tag{19}$$

where $\Delta\phi_{m,n}$ is defined in equation (17).

We can define the sensitivity matrix so that $S_{i,j}$ is the phase difference to the source-sensor combination $i = m \cdot N_s + n$ per unit of electrical conductivity in the area element $j = l \cdot N_c + c$, where l is the row index and c is the column index of the element area, N_c is the total number of columns. In other words, to obtain the sensitivity matrix $N_f N_s$ numerical simulations are performed which calculate the phase difference for each source-sensor combination. Each simulation is referring to an element of the image area and provides a column of the matrix of sensitivity. The only parameter which changes from one to another simulation is the conductivity of the mesh so that the conductivity is one for all volume elements contained in the region perpendicularly above or below the element area j. The remaining elements of volume conductivity are zero. The process of calculating the matrix of sensitivity would appear time-consuming, but as it is known *a priori* that eddy currents are zero throughout the volume where the conductivity is zero, the system to be solved for calculating the current has reduced their order. In addition, once the electrical potentials in the loop for all the field sources are calculated, they do not change when the conductivity of the fabric is altered.

5.2. Regularization of the sensitivity matrix

Once the matrix $[S]$ is calculated, it is possible to obtain the distribution of conductivity of a given object as the test vector phase difference $[\Delta\phi]$ accordingly. However, the matrix sensitivity is extremely ill-conditioned due to the symmetry in this system, a fact which makes the traditional methods of resolution of linear systems inadequate. A widely used method in these cases is the Tikhonov regularization technique that, for a linear system of the form:

$$Ax = b \tag{20}$$

is to obtain the value x such that:

$$x \to \min\left\{\|Ax-b\|_2^2 + \lambda^2 \|Lx\|_2^2\right\} \tag{21}$$

where L is the matrix regularization, usually equal to the identity matrix, and λ is the regularization parameter. There are some ways to get the regularization optimal parameter λ. One is the curve L which is to find the inflection point of the parametric curve in relation to λ:

$$L : C_\lambda \to \left(\log\|Ax-b\|_2, \log\|Lx\|_2\right). \tag{22}$$

Good results were obtained using the technique of Tikhonov regularization, however, due to the need to find the optimal parameter λ, the process becomes slow. To solve this problem we developed a new relaxation technique for solving linear systems called the *technique of successive approximations*, briefly described in the Appendix.

6. Results

The imaging algorithm was tested for various configurations of the test object. All results presented are for the following series discretization with a grid with an edge of 24 cm and 48 divisions in each coordinate axis. In the simulation, 16 sources and 16 field sensors were modeled, which totals 256 area elements of the image obtained. Figure 4 shows the calculation area. These are distributed in a regular two-dimensional 16×16 element (pixel) grid. After the image is obtained, it is interpolated for a 48×48 pixel grid which is then filtered to remove high frequency components of the signal.

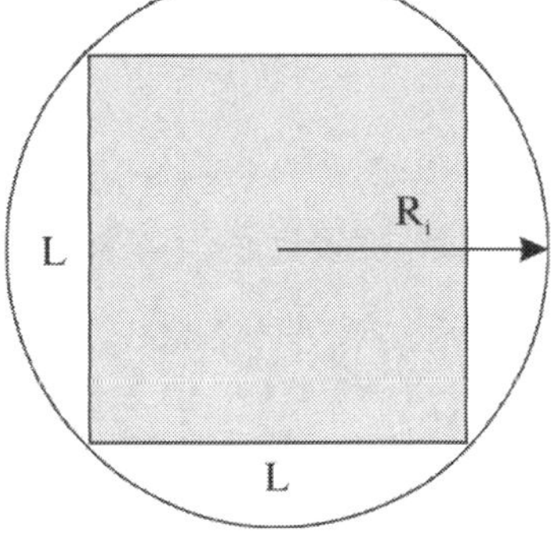

Figure 4 – Top view of the calculation area; R_i=17,81 cm is the internal radius of the tomograph; L=24 cm is the length of the discretized field for implementing the impedance method.

TABLE I – Mean squared errors.

Field	Error (%)
Primary radial B_{pr}	0,11
Primary polar $B_{p\theta}$	2,39
Secondary radial B_{sr}	3,17
Secondary polar $B_{s\theta}$	8,67

The figures illustrations relating to the simulated images are presented in the following. They describe the problems of testing and then display the results, which are images.

Figure 5 shows the real images reconstructed using the Tikhonov regularization technique (TRT), and using the technique of successive approximations (TSA) for a cylindrical object of radius 4 cm , 24 cm of height centered on the position y = 7 cm and z = -7 cm relative to the center of the tomograph. In Figure 7 the images for two cylindrical objects are presented , one identical to the previous (Fig. 6) and the other of radius 7.5 cm, 24 cm of height and center in y = 3 cm and z = -3 cm. Figure 8 shows images for a circular ring centered with the inner radius of 4.5 cm, 9 cm of outer radius. For analysis of the quality of the images obtained, indices defined by Wang and Bovik were used [9]. These indices measure the similarities between the images with respect to average luminance [0, 1], the linear correlation [-1, 1] and contrast [0, 1]. Figures in brackets indicate the limits for each index, where the lower limit indicates the maximum dissimilarity and the upper limit indicates the maximum similarity between the images. The image quality is calculated by multiplying the four previous indices. Table 2 presents these indices for each of the objects.

TABLE II – Indices of image quality.

Index	Object					
	1		2		3	
	TRT	TAS	TRT	TAS	TRT	TAS
Average luminance	0.998	0.850	0.990	0.908	0.996	0.917
Linear Correlation	0.550	0.781	0.442	0.632	0.126	0.241
contrast Image Quality	0.209	0.157	0.604	0.516	0.659	0.585
Contrast Image Quality	0.115	0.104	0.264	0.297	0.0824	0.129

Figures 8 to 10 show the comparison between the ratios for the images obtained using the Tikhonov and successive approximations techniques for cylindrical objects 24 cm in height, varying their radius Δx to ½NxΔx where the radius value is normalized to the number of area elements (pixels) of the image.

7. Conclusion

It can be observed in images reconstructed in the figures (Figs. 5-7) that the technique of Tikhonov regularization (TRT) leads to lower quality images compared to images obtained using the technique of successive approximations (TAS), This fact is confirmed by the levels of image quality shown in Table 2. These results show the prevalence of TAS in relation to linear correlation between the real image and obtained image but also indicate the prevalence of TRT with respect to contrast and similarity between the indices of average luminance for the two methods. These quality characteristics are also shown in Figures 8 to 11. However, they show that TAS prevails for smaller objects and TRT for larger objects, showing the greater robustness of TAS. It also shows the limit of detectability of the techniques relative to the size of the test object which is about 3 pixels radius for a cylindrical object.

7.1 Obtained images

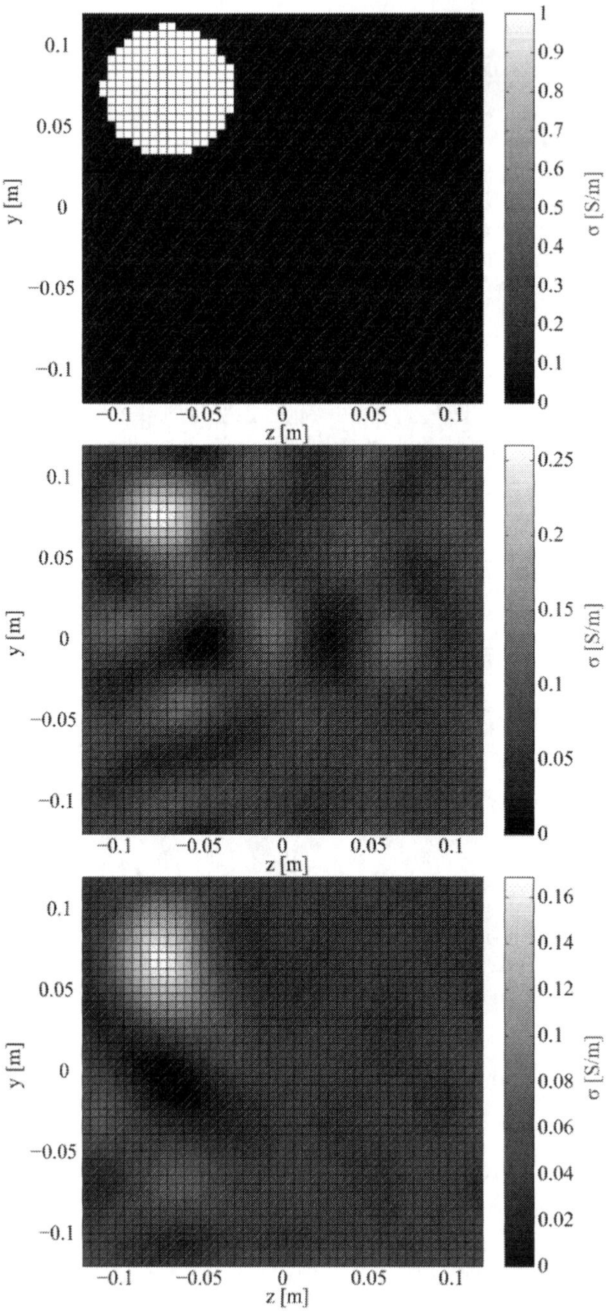

Figure 5 – Real and obtained images by Tikhonov and successive approximations techniques for a cylindrical object of radius 4 cm, 24 cm of height with center at coordinates (0.07, -0.07).

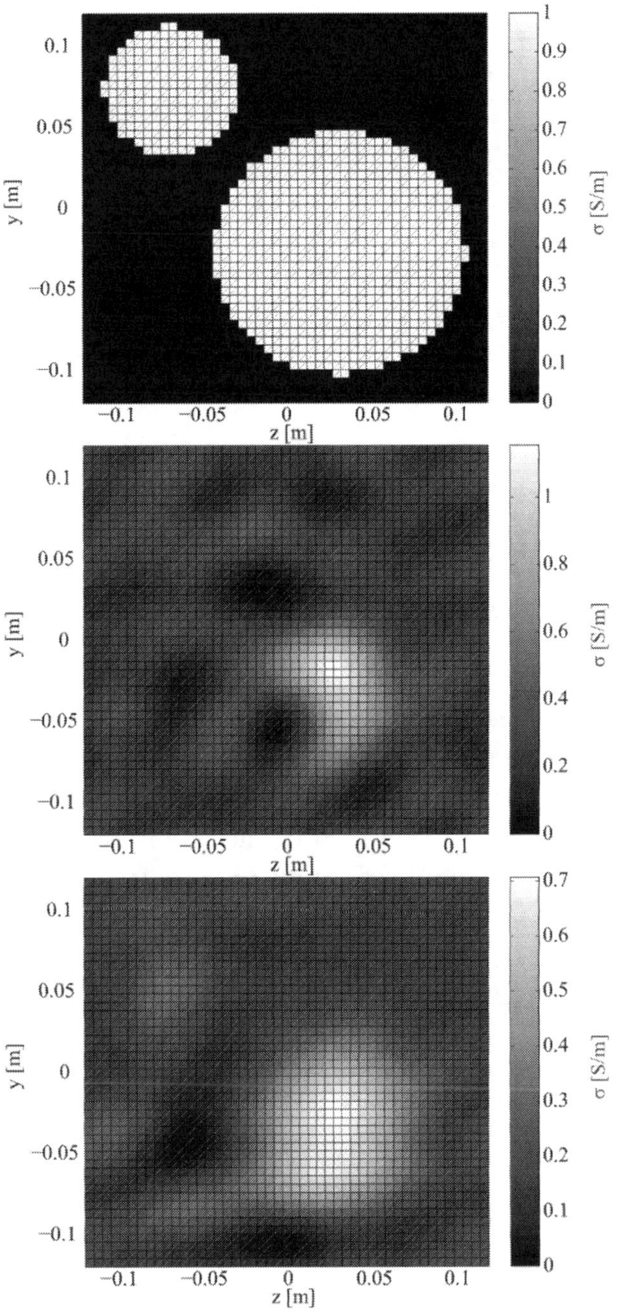

Figure 6 – Real and obtained images by Tikhonov and successive approximations techniques for two cylindrical objects, the first identical to Figure 5 and the second of 7.5 cm of radius, 24 cm of height with center at coordinates (-0.03, 0.03).

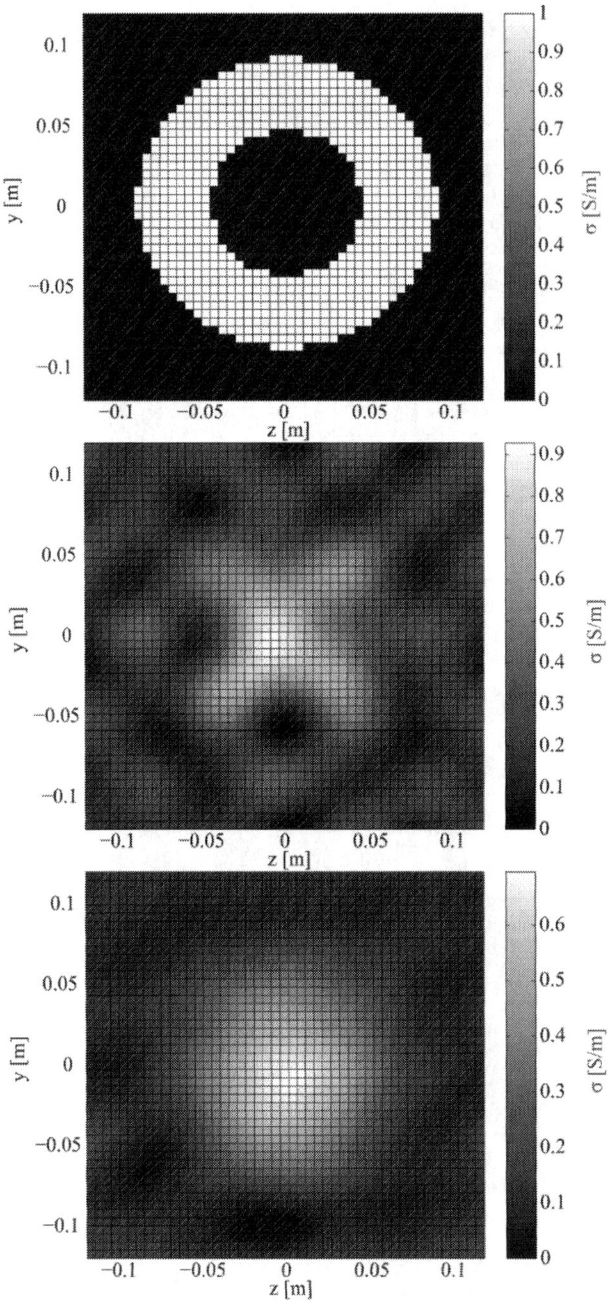

Figure 7 – Real and obtained images by the Tikhonov and successive approximations techniques for a circular ring with 4.5 cm of inside radius, 9 cm of external radius and 24 cm of height, concentric to tomograph.

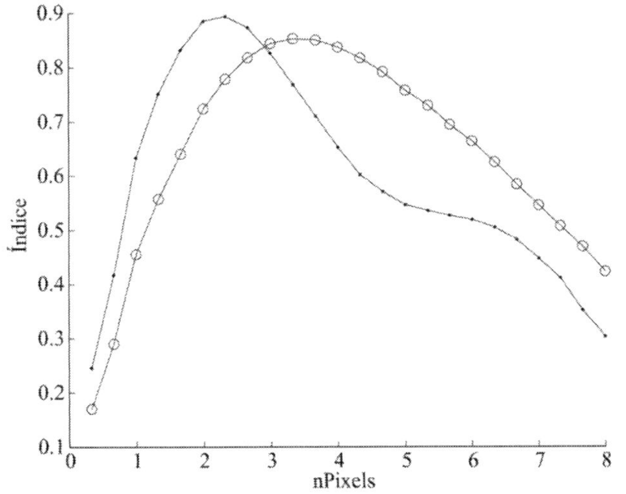

Figure 8 – Comparison between the linear correlation coefficients for Tikhonov technique (•) and successive approximations (○) according to the number of pixels of the object beam.

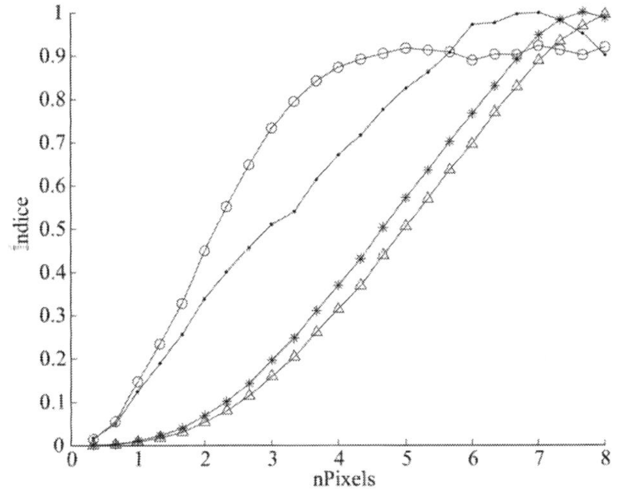

Figure 9 – Comparison between the average luminance ratios for Tikhonov technique (•) and successive approximations (○) and contrast for Tikhonov technique (⁎) and successive approximations (△) according to the number of pixels of the object beam.

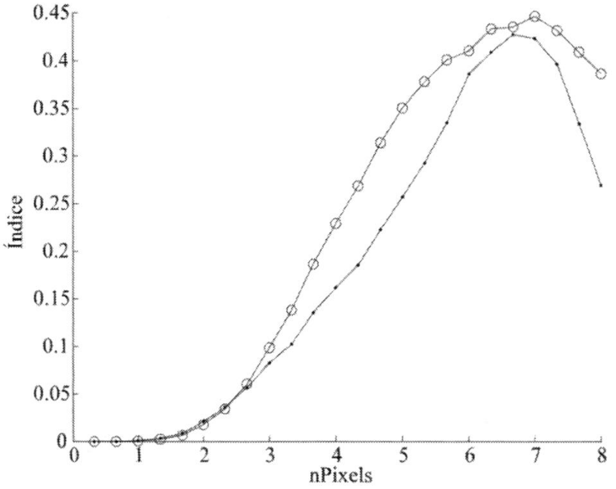

Figure 10 – Comparison between the image quality scores for Tikhonov technique (•) and successive approximations (○) according to the number of pixels of the object beam.

8. Appendix

The technique consists basically in carrying out successive increments to each position of the vector x so that the Euclidean norm of the residual vector r is minimized. The first method assumes that, given an initial approximation x_0 to x, it is possible to obtain an increased δx_i that minimizes the norm of residual vector:

$$r_1 = b - Ax_1, \qquad (23)$$

where the vector x_1 is the vector x_0 after the increment in the element i. It is easily verifiable that the residual vector r_0 for the initial approximation x_0 and the vector r_1 for the vector x_1 are related as follows:

$$r_1 = r_0 - \begin{bmatrix} a_{1,i}\delta x_i & a_{2,i}\delta x_i & \cdots & a_{n,i}\delta x_i \end{bmatrix}^T, \qquad (24)$$

where $a_{i,j}$ denotes the element in row i and column j of the matrix A. Thus, it is possible to obtain more accurate solutions by carrying out the increments δx_i successively in each of the n elements of x_n, where x_n is the vector of responses of the iteration n.

In order to obtain the optimum value of δx_i, consider the expression for the norm of the residual vector for the iteration $m+1$, depending on the residual vector of the iteration m:

$$\text{res}_{m+1}(\delta x_i) = \left\{ \sum_{k=1}^{n} \left[r_m(k) - a_{k,i}\delta x_i \right]^2 \right\}^{1/2}. \qquad (25)$$

We are interested in the minimum point of the function. At this point the first derivative of $\text{res}_{m+1}(\delta x_i)$ is zero. Thus, the critical points are obtained from the following equation:

$$\frac{\mathrm{d}\,\text{res}_{m+1}(\delta x_i)}{\mathrm{d}\delta x_i} = \frac{1}{2} \underbrace{\left\{ \sum_{k=1}^{n} \left[r_m(k) - a_{k,i}\delta x_i \right]^2 \right\}^{-1/2}}_{} \cdot \sum_{k=1}^{n} \left[a_{k,i}\left(r_m(k) - a_{k,i}\delta x_i \right) \right] = 0. \qquad (26)$$

As the highlighted word in the last equation is non-zero, it must be the case that:

$$\sum_{k=1}^{n} \left[a_{k,i}\left(r_m(k) - a_{k,i}\delta x_i \right) \right] = 0. \qquad (27)$$

Thus, we obtain the critical value of δx_i, as:

$$\delta x_{i_c} = \frac{\sum_{k=1}^{n}\left[a_{k,i} r_m(k)\right]}{\sum_{k=1}^{n} a_{k,i}^{2}}. \tag{28}$$

So taht the value obtained by equation (28) is a minimum point, it is necessary that:

$$\lim_{\delta x_i \to \delta x_{i_c}^-} \frac{\mathrm{d}\,\mathrm{res}_{m+1}(\delta x_i)}{\mathrm{d}\delta x_i} < 0; \tag{29}$$

$$\lim_{\delta x_i \to \delta x_{i_c}^+} \frac{\mathrm{d}\,\mathrm{res}_{m+1}(\delta x_i)}{\mathrm{d}\delta x_i} > 0. \tag{30}$$

As the function $\mathrm{res}_{m+1}(\delta x_i)$ has only one critical point, you can use limits $\delta x_i \to -\infty$ and $\delta x_i \to +\infty$ in the above limits. Substituting Equation (26) for (30), and considering that the term in focus is always positive and nonzero, we obtain:

$$\begin{gathered}
\lim_{\delta x_i \to -\infty} \left\{ \delta x_i \sum_{k=1}^{n} a_{k,i}^2 - \sum_{k=1}^{n} \left[a_{k,i} r_m(k) \right] \right\} < 0; \\
\lim_{\delta x_i \to +\infty} \left\{ \delta x_i \sum_{k=1}^{n} a_{k,i}^2 - \sum_{k=1}^{n} \left[a_{k,i} r_m(k) \right] \right\} > 0.
\end{gathered} \tag{31}$$

It appears that the conditions imposed in (31) are satisfied for any values of the matrix A and vector rm. Thus it is concluded that the point δx_{ic} given by equation (28) is always a minimum as desired.

References

[1] Orcutt N, Gandhi OP (1990) "Use of the impedance method to calculate 3-D power deposition patterns for hyperthermia with capacitive plate electrodes". *IEEE Transactions on. Biomedical Engineering*, 37(1):36-43.

[2] Yin W, Peyton AJ (2006) "A planar EMT system for the detection of faults on thin metallic plates". *Measurement Science and Technology*, 17(8):2130-2135.

[3] Han Y, Lu Y, Zhang J, Yang B (2005) "A 3-D impedance method to compute current inducted in human body exposure to power lines". *IEEE International Symposium on Microwave, Antenna, Propagation and EMC Technologies for Wireless Communications*, 1:470-473.

[4] Orcutt N, Gandhi OP (1988) "A 3-D impedance method to calculate power deposition in biological bodies subjected to time varying magnetic fields". *IEEE Transactions on. Biomedical Engineering*, 35(8):577-583.

[5] Ramos A, Raizer A, Marques JLB (2003) "A new computational approach for electrical analysis of biological tissues". *Bioelectrochemistry*, 59:73–84.

[6] Nadeem M, Thorleif T, Gandhi OP (2003) "Computation of electric and magnetic stimulation in human head using the 3-D impedance method". *IEEE Transactions on Biomedical Engineering*, 50(7):900-907.

[7] Peyton AJ, *et. al* (1999). "Development of Electromagnetic Tomography (EMT) for Industrial Applications. Part 1: Sensor Design and Instrumentation". *1st World Congress on Industrial Process Tomography*.

[8] Pham MH, Peyton AJ (2008) "A model for the forward problem in magnetic induction tomography using boundary integral equations". *IEEE Transactions on. Magnetics*, 44(10):2262-2267.

[9] Ramos A, Wolff JGB (2011) "Numerical modeling of magnetic induction tomography using the impedance method". *Medical & Biological Engineering & Computing*, 49(2):233-240.

Spatial filtering in electrical impedance tomography

M Fernández-Corazza[1,2,3], N von Ellenrieder[1,3] and C H Muravchik[1,4]

[1] Laboratorio de Electrónica Industrial, Control e Instrumentación (LEICI)
Facultad de Ingeniería, Universidad Nacional de La Plata (UNLP), Buenos Aires, Argentina.
[2] Departamento de Ciencias Básicas, Facultad de Ingeniería, UNLP, Buenos Aires, Argentina.
[3] Consejo Nacional de Investigaciones Científicas y Técnicas (CONICET), Argentina.
[4] Comisión de Investigaciones Científicas de la Provincia de Buenos Aires (CICpBA), Buenos Aires, Argentina.

E-mail: `marianof.corazza@ing.unlp.edu.ar`

Abstract. We propose a new method to localize electrical conductivity changes in the human head using Electrical Impedance Tomography (EIT). In EIT, a localized conductivity change produces a change in the electric potential distribution which is equivalent to a potential generated by a dipole at the same location. We propose to estimate the location of conductivity changes with the same techniques used to solve the Electroencephalography source localization problem. In particular, we show an adaptation of the Linear Constrain Minimum Variance Beamforming technique to perform this estimation. We simulate a localized 10% conductivity change in a realistic model of the human head to apply the method. Results show that depending on the noise level, the method can accurately localize the conductivity change.

1. Introduction

Conductivity changes in the human head are of interest, for example, in the study of epileptic seizures and brain function. Electrical Impedance Tomography (EIT) can image these conductivity changes when applied to the human head. In [1] it is shown that there is an equivalence between a localized conductivity change and a dipolar source. To our knowledge, the use of this equivalence to image or localize conductivity changes has not been fully explored yet. In this work, we study the possibility of applying the Beamforming technique used in the Electroencephalography (EEG) source localization problem to localize and image conductivity changes from EIT measurements.

EIT is a technique that can be used to estimate the electrical conductivity distribution of an object based on the electric potential measured on its surface when electric current is applied to it. Its clinical application is considered to have a great potential in diagnosis as EIT is a relatively low-cost and minimally invasive method [2]. While applied to the human head it can be used to estimate a relatively small number of parameters such as the electrical conductivity of the main tissues of the head [3, 4], or to image the internal conductivity distribution of the head, a problem known as EIT reconstruction [5]. The parametric approach is of interest for improving the electrical head models needed in the EEG source localization problem [6, 7] and in the EIT reconstruction problem. The conductivity changes that EIT can image are mostly produced by blood flow and volume changes known to occur during evoked activity and epileptic seizures [8, 2]. Several algorithms were designed for solving the EIT reconstruction problem, and a good review of them can be found in [1]. As the problem is ill-conditioned, most of the algorithms are based on a regularized least squares minimization with different approaches for the regularization term.

Spatial Filtering, also known as Beamforming, was originally developed in the field of sensor-array signal processing, and its application in the field of electromagnetic brain imaging was first introduced by Robinson [9]. It applies a linear operator to the measurements of a sensor array such that it points to a particular location. The measurements are typically obtained from EEG and/or Magnetoencephalograpy (MEG) records, but in this work we introduce the application of Spatial Filtering to EIT measurements. The output of the filter would be an estimation of the conductivity at the pointing location. As the pointing location is controllable, the entire head (or a particular zone of interest) can be scanned producing a 3D map of the internal conductivity of the head. One of the main issues of Beamforming in brain imaging is that the filter output is less sensitive to deeper sources so that they are harder to detect leading to false positives in deep regions. To overcome this problem, normalization is required. The Linear Constrained Minimum Variance (LCMV) Beamformer introduced by Van Been includes a possibility for normalization [10]. This Beamformer assumes the knowledge of the resulting potential on the electrodes for a dipolar current source of unit norm, oriented in the three canonical directions, and localized at each point of the region of interest. The output of this filter is an estimate of the amplitude and orientation of the sources [10]. The method can be improved if an orientation scanning is also performed which is called Synthetic Aperture Magnetometery (SAM) [11], but it increases the computational cost. An alternative is to estimate the dipole orientation at each spatial point and then to perform the volumetric scanning only for that orientation. This beamformer is known as the Minimum Contrast Beamformer (MCB) [12].

In EIT, a change of the electric potential produced by a conductivity change in a specific location is equivalent to a voltage distribution produced by a dipolar source localized at the same location and with its strength proportional to the conductivity change [1]. We propose the use of the Beamforming technique to estimate the conductivity change as used in EEG source localization problems. Two key issues particularly affect EIT beamforming: the noise (because the difference of the potential is of the order of the electric noise levels) and the combination of different measurements when several electrode pairs are used for the current injection. This work is mostly focused on these two issues.

2. Materials and Methods

2.1. Principle of equivalence

Let $\sigma(\vec{x})$ be the conductivity at each point of the space $\vec{x}$ of a volume Ω. When an electric current is applied, an electric potential distribution $\Phi(\vec{x})$ is generated. If a conductivity change $\delta\sigma(\vec{x})$ is produced, the electric potential also changes $\delta\Phi(\vec{x})$. The equation that governs the physics of the problem is:

$$\nabla \cdot (\sigma(\vec{x})\nabla\Phi(\vec{x})) = 0, \tag{1}$$

in all $\vec{x}$ except from the current injection points. Relation (1) must also be satisfied when the conductivity change is produced:

$$\nabla \cdot ((\sigma(\vec{x}) + \delta\sigma(\vec{x}))\,\nabla\,(\Phi(\vec{x}) + \delta\Phi(\vec{x}))) = 0. \tag{2}$$

Applying properties of the gradient operator, (2) becomes

$$\nabla \cdot (\sigma\nabla\Phi) + \nabla \cdot (\sigma\nabla\delta\Phi) + \nabla \cdot (\delta\sigma\nabla\Phi) + \nabla \cdot (\delta\sigma\nabla\delta\Phi) = 0, \tag{3}$$

where de dependence with $\vec{x}$ has been omitted for clarity. The first term of (3) is zero because of (1), and the last term of (3) can be neglected for small conductivity changes as it is $O(\delta\sigma^2)$ [1]. Then, (2) can be approximated by

$$\nabla \cdot (\sigma\nabla\delta\Phi) \approx -\nabla\,(\delta\sigma\nabla\Phi)\,. \tag{4}$$

This relation is equivalent to the equation

$$\nabla \cdot (\sigma \nabla \Phi) = \nabla J_p, \tag{5}$$

which governs the EEG source localization problem with J_p being the primary current density of a dipolar source. Then, the EIT problem is equivalent to the EEG source localization problem where the difference of the potential is the known or measured data while the equivalent current density at a point of the space is the conductivity change multiplied by the gradient of the potential at that point. This enables the possibility of using the EEG/MEG source localization algorithms and tools to localize conductivity changes with EIT. To our knowledge, this is an interesting possibility that has not been explored yet. In this work we adapt the LCMV spatial filter to this new application, but some other well known algorithms such as LORETA, or min-norm could be studied as well.

2.2. Spatial Filter design

General concepts about the design of filters for Brain Imaging can be found in [9], while the LCMV and MCB filters are described in [10] and [12] respectively. In this section we focus on the design of the filter for its application in EIT.

We assume that two electrodes of a set of L electrodes are dedicated to the current injection, while the remaining $L-2$ electrodes are used to measure the potential. The measurement model is

$$y = v_1 - v_2 + n_1 - n_2 = \delta v + \delta n, \tag{6}$$

where vectors v_1 and v_2 are the $L-2$ expected values of the electric potential measured at the electrodes with and without the conductivity change, and n_1 and n_2 account for the electronic noise due to the amplifiers and the electrode-skin impedance.

The Lead Field (LF) matrix $LF(\vec{x})$ is an $L \times 3$ matrix where each column has the expected potential at the measurement electrodes generated by a unit-magnitude dipole source localized at the point $\vec{x}$ and directed by each of the three canonical directions (x, y, and z). As two electrodes are used for the current injection, the LF has the two rows corresponding to those electrodes removed. From the equivalence between (4) and (5), the expected δv is just the Lead Field matrix multiplied by the equivalent current density J_p:

$$y = -LF(\vec{x})\delta\sigma(\vec{x})\nabla\Phi(\vec{x}) + \delta n = -\delta\sigma(\vec{x})l_f(\vec{x}) + \delta n, \tag{7}$$

where l_f is the Lead Field vector which is the Lead Field matrix multiplied by the gradient field.

On EIT, the injection pair can be changed to another two electrodes and a new set of $L-2$ measurements can be obtained. Then, if M is the number of pairs used, there will be M sets of $L-2$ measurements. In this work, we assume that the electrodes used for the current injection are used to measure when the injection pair is changed. Then the LFs are different for each set. Assuming that there is no conductivity change between measurements with different sets, they can be arranged in a column resulting in an $M(L-2)$ vector $y_T = -\delta\sigma l_{fT} + \delta n_T$, where l_{fT} and δn_T are the arranged vectors with the l_f and δn of each set.

The Beamforming method proposes a weight vector w such that $w^T y_T$ is an estimator of the parameter of interest, in this case, $\delta\sigma$. The weight vector for the unit-gain constraint LCMV Beamformer is

$$w^T = \left(l_{fT}^T C^{-1} l_{fT}\right)^{-1} l_{fT}^T C^{-1}, \tag{8}$$

where C is the covariance matrix of the signal [10, 9]. If the LCMP filter is used instead, the matrix C is replaced by the correlation matrix R. In this work we used the LCMP filter since it performed better. Then, the output of the filter is an estimator of the conductivity change, i.e.

$$\widehat{\delta\sigma} = w^T y_T = \left(l_{fT}^T R^{-1} l_{fT}\right)^{-1} l_{fT}^T R^{-1} y_T, \tag{9}$$

Journal of Physics: Conference Series **407** (2012) 012023 doi:10.1088/1742-6596/407/1/012023

where

$$R = (\delta\sigma l_{fT})(\delta\sigma l_{fT})^T + C_n, \tag{10}$$

and C_n is the noise covariance matrix. As $\delta\sigma$ is unknown, the estimator of the correlation matrix used in this work is

$$\widehat{R} = y_T y_T^T + C_n. \tag{11}$$

We followed the normalization denominator used in [12], which is the variance of the filter output when only noise is being measured (y_n). We define a Conductivity Change Index (CCI) as:

$$CCI = \frac{\left(\widehat{\delta\sigma}\right)^2}{\mathrm{Var}(w^T y_n)} = \frac{w^T y_t y_t^T w}{w^T C_n w}. \tag{12}$$

Note that in (8) the inverse of C (R in our work) has to be computed. For a fast computation of the CCI, the matrix inversion lemma was used to decrease the size of the matrix to be inverted.

2.3. Simulation Setup

A 10% conductivity increase in a sphere of radius 1cm localized at the right side of the brain in the union of the parietal and frontal lobes was simulated. A three layer and realistically shaped head model was employed. An electrode cap with 64 electrodes was assumed, where the electrode positions correspond to the $10-10$ EEG standard placement system. The assumption is the use of the same electrodes used in EEG where the surface of the electrodes is much lower than the total surface of the head and then, the electrodes can be modeled as a point with no surface [13]. The EIT forward problem was solved using the Finite Element Method (FEM), both for the simulation of the measurements and for the localization problem solution. However, a different tessellation (314798 tetrahedrons for the brain) was used for the localization problem where the tetrahedron distribution inside the brain is uniform. An amplitude of 100μA was adopted for the injected current strength as allowed by the IEC60601 standard [14, 15]. For the scalp and skull, anisotropic conductivity values were adopted with radial:tangential ratios of 1:1.5 (0.3 and 0.45 S/m) and 1:10 (0.0015 and 0.015 S/m) respectively [16, 17], whereas for the brain, an isotropic value of conductivity (0.4 S/m) was adopted.

The electronic noise due to the amplifiers and the electrode-skin contact impedance was modeled as white gaussian noise (WGN). Independence between channels was assumed. Another source of noise is the electrical activity of the brain. There are at least two possibilities to mitigate its effect: taking advantage of the knowledge of the waveform of the injected current [3, 18] or the use of frequencies greater than $\sim$ 200Hz knowing that brain activity is mostly concentrated at low frequencies, but lower than $\sim$ 1kHz so that the electromagnetic quasi-static approximation remains valid. We adopt $C_n = 2\sigma_n^2 I_{(L-2)M}$, where σ_n is the standard deviation (SD) of the noise, the 2 factor is a consequence of the difference between n_1 and n_2, and $I_{(L-2)M}$ is the identity matrix. Adopting 0.1μV/$\sqrt{\text{Hz}}$ for the noise SD and assuming slow conductivity changes of bandwidth 1/4Hz, the noise SD would be 50nV. For 100 independent repetitions of the experiment the SD can be reduced up to 5nV. We performed simulations for three different values of σ_n: 1×10^{-7}V, 1×10^{-8}V, and 1×10^{-9}V.

In order to study the influence of the number of sets used in the inverse problem, two injection patterns were simulated. The first one consisting of 63 pairs where the Cz electrode is fixed as one injection node and the other node varies between the other electrodes, and the second one consisting of $M = 125$ combinations where the fixed electrodes are the Pz and the Fz.

3. Results

The reconstruction problem was simulated for both patterns and for the three levels of noise. Figure 1 shows the normalized CCI ($CCI(\vec{x})/max\{CCI\}$) for both patterns ($M = 63$ and

$M = 125$), and for different noise levels (1×10^{-7}V, 1×10^{-8}V, and 1×10^{-9}V of SD). Three cross sections of the head are shown and the colorbar indicates the normalized CCI level.

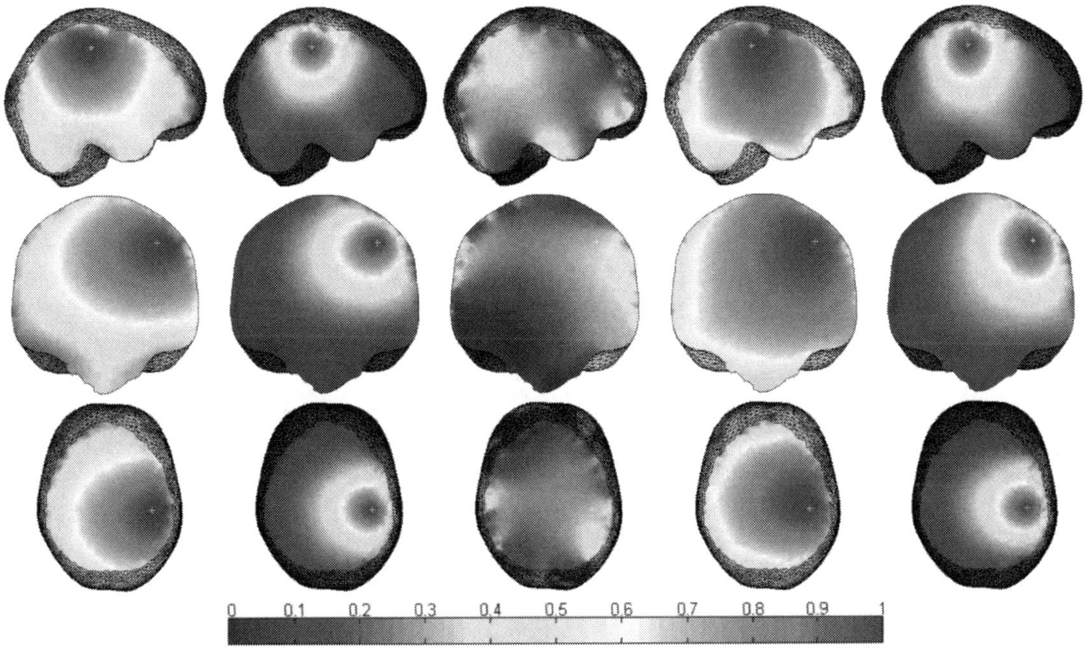

Figure 1. Cross sections of the head depicting the CCI obtained with 63 (first two columns) and 125 (last three columns) electrode pairs for the current injection and with a noise SD of 1×10^{-7}V (third column), 1×10^{-8}V (first and fourth columns), and 1×10^{-9}V (second and last columns). White crosses indicate the true centre of the sphere.

As an indicator of performance, the Localization Error (LE) was also computed. It is defined as: $LE = ||\vec{x}_{est} - \vec{c}||$, where $\vec{x}_{est}$ is the location of the maximum CCI value and $\vec{c}$ is the centre of the sphere. Twenty different noise realizations where simulated for each pattern and for each noise level. The LE is shown in Figure 2, where the simulations with 1×10^{-7} of noise SD were excluded for clarity.

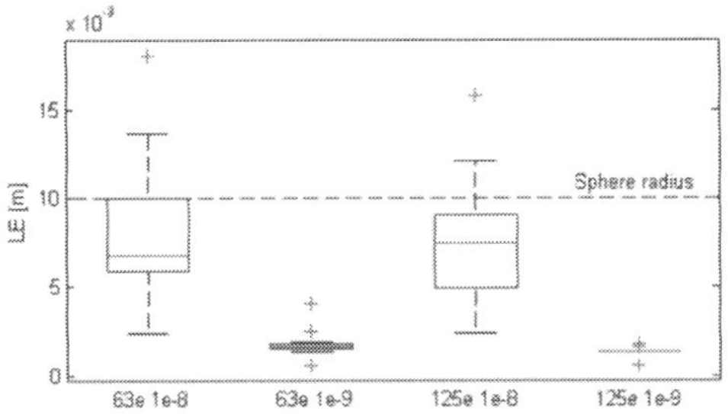

Figure 2. Localization Error for 63 and 125 electrode pairs used for the current injection with 1×10^{-8}V and 1×10^{-9}V for the noise SD of the simulated measurements. The central mark is the median, the edges of each box are the 25th and 75th percentiles, and the whiskers extend to the most extreme data points not considered outliers. The red crosses are considered outliers.

4. Discussion and Conclusions

The equivalence between conductivity changes in EIT and dipolar sources in EEG appears to be promising as it enables the use of the techniques and algorithms of EEG and MEG source localization problems. Results of our work suggest that one of those techniques, the spatial filtering, could reveal the conductivity change location within the brain, based on EIT measurements. Other techniques should be studied and a comparison with the EIT linear reconstruction algorithms described in [1] must be performed. However, the Beamforming technique does not need the inversion of full Jacobian matrices and the CCI may be simpler to compute. Note that the reconstructed image indicates the location of conductivity changes that best fit with the difference of the measured potentials.

We believe that the noise effect is a major technical issue of the method presented here. However, it is an inherent limitation in EIT as the conductivity changes produced in the brain generate small changes in the potential measured at the scalp. Averaging, repeating experiments, a smart selection of electrode pairs for the current injection, and some prior information related to the problem under study may improve the results. Note that for equal noise levels, the use of 125 pairs instead of 63 pairs for the current injection does not appear to significatively improve the performance of the localization estimation. It can be a consequence of the sphere position as the fixed electrode for the pattern with 63 pairs (Cz) is nearer to the sphere than the fixed electrodes used for the injection pattern of 125 pairs (Fz and Pz). Further simulations where the conductivity change is localized at different positions are needed.

The introduction of Spatial Filtering to EIT to localize conductivity changes looks promising as it is a method known to have a excellent performance while applied to EEG/MEG source localization problems. The method should be validated with real measurements.

Acknowledgments

This work was supported by ANPCyT PICT 2011-0909, UNLP I166, CONICET and CICpBA.

References

[1] Lionheart W, Polydordes N and Borsic A 2004 *Electrical Impedance Tomography: Methods, History and Applications* ed Holder D S (Inst. Phys.) chap The reconstruction problem, pp 3–64
[2] Bayford R 2006 *Annu. Rev. Biomed. Eng.* **8** 63–91
[3] Gonçalves S, de Munck J, Verbunt J, Bijma F, Heethaar R and da Silva F L 2003 *IEEE Trans. Biomed. Eng.* **50**(6) 754–767
[4] Fernández-Corazza M, Beltrachini L, von Ellenrieder N and Muravchik C H 2011 *XIV Reunión de Trabajo Procesamiento de la Información y Control RPIC 2011* (Oro Verde) pp 845–850
[5] Abascal J F P, Arridge S R, Atkinson D, Horesh R, Fabrizi L, Lucia M D, Horesh L, Bayford R H and Holder D S 2008 *NeuroImage* **43** 258 – 268
[6] Vanrumste B, Van Hoey G, Van de Walle R, D'Havé M, Lemahieu I and Boon P 2000 *Med Biol Eng Comput* **38**(5) 528–534
[7] Wolters C, Anwander A, Tricoche X, Weinstein D, Koch M and MacLeod R 2006 *NeuroImage* **30** 813 – 826
[8] Holder D 2008 *World Automation Congress* (Hawaii)
[9] Sekihara K and Nagarajan S S 2008 *Adaptive Spatial Filters for Electromagnetic Brain Imaging* (Berlin: Springer)
[10] van Been B D, van Drongelen W, Yuchtman M and Suzuki A 1997 *IEEE Trans. Biomed. Eng.* **44** 867–880
[11] Vrba J and Robinson S 2000 *Conference Record of the Thirty-Fourth Asilomar Conference on Signals, Systems and Computers, 2000.* vol 1 pp 313–317
[12] Chen Y S, Cheng C Y, Hsieh J C and Chen L F 2006 *IEEE Trans. Biomed. Eng.* **53** 1765–1774
[13] Paulson K S, Breckon W R and Pidcock M K 1992 *SIAM Journal of Applied Mathematics* **52** 1012–1022
[14] Gilad O, Horesh L and Holder D 2007 *Med. Biol. Eng. Comput.* **45** 621–633
[15] IEC60601 2005 ed 3.0 (Geneva: International Electrotechnical Commission)
[16] Rush S and Driscoll D 1968 *Anesth. Analg.* **47**(6) 717–723
[17] Horesh L 2006 *Some Novel Approaches in Modelling and Image Reconstruction for Multi-Frequency Electrical Impedance Tomography of the Human Brain* Ph.D. thesis University College London
[18] Fernández-Corazza M, Beltrachini L, von Ellenrieder N and Muravchik C H 2012 *ArgenCon 2012* 1–6

First Latin-American Conference on Bioimpedance (CLABIO 2012) IOP Publishing
Journal of Physics: Conference Series **407** (2012) 012024 doi:10.1088/1742-6596/407/1/012024

Electrical Impedance Tomography for pulmonary oedema extent monitoring: review and updated design

Eduardo Santos, Franco Simini

Núcleo de Ingeniería Biomédica, Facultades de Medicina e Ingeniería, Universidad de la República, Montevideo, Uruguay

E-mail: esantos@fing.edu.uy

Abstract. Monitoring pulmonary oedema would be greatly facilitated by the availability of a graphical representation of its size and density to guide therapeutic interventions. Currently the clinician has only indirect estimations because X-ray imaging or computed tomography can not be repeated often. To avoid the transfer of critically ill patients and to have continuous information Electrical Impedance Tomography (EIT) is suggested. Circuit designs for EIT are reviewed, including wave generators, current sources, differential amplifiers, synchronous voltmeters, multiplexers and control modules. Nine designs are studied, the characteristics of five of them are presented (England 2005, Uruguay 2002, Iran 2006, China 2007 and Switzerland 2012). Three different solutions are compared (components cost range from) and an optimal design is proposed which includes a direct digital synthesizer (DDS) for signal generation, a modified Howland configuration for current source, 16 bits for the analog to digital conversion, and a digital signal processor (DSP) for the synchronous demodulation as well as to process the measurements for the reconstruction algorithm. This allows us to design low cost, gross graphical representations for oedema extent monitoring, with little anatomical accuracy.

1. Introduction

Treatment of pulmonary oedema is based among other parameters upon the estimation of alveolar volume occupied by liquid. The injection of high frequency current ($> 20kHz$) at amplitudes not perceived by the human body (of the order of $2mA$) and the resulting measurements of voltages on the chest's skin, allow to estimate the electrical impedance. This estimation can be dispayed as a tomographic image. The method, known as Electrical Impedance Tomography (EIT), low-cost, non-invasive and prolonged application to show a low resolution image of the distribution of fluid in the lungs and its evolution. Is thus an attractive alternative to X-rays and other cumbersome procedures

Since the pioneering work of Barber and Brown in 1984 [1], the last two and a half decades have seen a considerable growth of EIT applications in research centers, with few commercial offers. Since 1995 the Núcleo de Ingeniería Biomédica (nib) has developed circuits [2], reconstruction softwares [3] and complete prototypes [4, 5] under the name of IMPETOM (**impe**dance **tom**ography) with test results in phantoms and healthy volunteers.

This paper presents the evolution of EIT designs adopted, as published in English literature.

2. EIT hardware architecture

The different blocks of an EIT system are the following:

- Waveform synthesis
- Current source
- Differential amplifier
- Synchronous voltmeter
- Multiplexers
- Control block

Figure 1 shows the architecture of a typical EIT system.

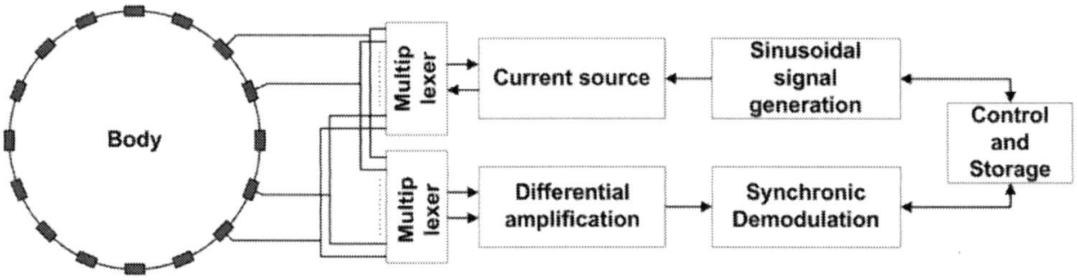

Figure 1. EIT system basic block structure with a differential current source. There are 16 electrodes affixed onto the skin of the patient's body

2.1. Waveform synthesis

The wave generator makes the reference signal for the sinusoidal current source,besides generating the synchronizing signal to be used by the demodulator. There are analog and digital solutions, the latter being the most commonly used in recent developments such as the circuit described by Saulnier [6]. These solutions feed a digital-analog converter with stored samples. The performance of this unit is measured in terms of spectral purity and its signal to noise ratio (SNR).

2.2. Current source

The current source is driven by the wave generator signal to convert it into a current to be injected into the body through the electrodes. You can use single-ended or floating sources. The former are ideal for systems with multiple sources, while the latter are used on systems with a single source [6]. The proposed solutions for the current source are varied, mostly using discrete op amps or transistors, but all eventually face the same problem: the output impedance is difficult to maintain high [6]. This impedance consists of an output resistance and stray capacitance: the output impedance thus affects the value of the injected current, but not always in the same way, as it depends on the load seen by the source, in this case the human body (and the electrode interface skin), which is constantly changing. That is why the design of the source must take into account load variability in order to obtain an output impedance that allows the current to vary at most within an acceptable range. Ideally, this range should not exceed the precision of the system in terms of A/D bits [6]. One may also measure the actual current being injected and to reflect it in the reconstruction algorithm.

2.3. Differential amplifier
Either differential or single-ended voltages may be measured in principle, but the vast majority of systems adopts differential configurations that result with a reduced dynamic range of the voltages [6]. Designers face here a second problem to solve: the common mode voltage of the amplifiers. To reduce it, the components and the operating point must be chosen very carefully. In addition to this, one may use specific techniques, such as an additional electrode located away from the area where the measurement is taken to reintroduce feedback current much in the same way as the ECG driven-right-leg circuit (Webster) [6].

2.4. Synchronous voltmeter
To reconstruct a tomographic image it is necessary to take phase-sensitive voltage measurements, either to measure the resistivity and permittivity, or just the resistivity. This is why it is necessary to have a coherent signal from the wave generator. One can perform this demodulation in either the analogue or digital domain, the latter being the most used recently [6]. One way to perform the digital demodulation is via an A/D which converts the input analogue signal to digital samples which are then multiplied by sine or cosine functions of the same frequency. Then the samples are integrated over a number of cycles of the original wave as described by Saulnier [6], to obtain a value used in the algorithm.

2.5. Multiplexers
The multiplexers are needed in systems with a single current source or with fewer voltmeters than electrodes. Multiplexers are used to select channels through which current is injected and where the voltage is measured. The multiplexers have non ideal characteristics that create problems in the system, as the on-resistance, and above all, the input and output capacitances.

2.6. Control and data storage block
Control is necessary to synchronize all other blocks, to store the measured data and to transfer it to a computer that will perform the image reconstruction. Depending on the complexity of this block, the options are varied, from basic microcontroller [7] to high performance DSPs (digital signal processors) [8, 9, 10, 11] in order to make a more complex pre-processing of the data.

3. Literature review method
We searched the literature on EIT systems, excluding publications solely related to reconstruction algorithms, therefore not including hardware information. Selected publications contain descriptions of circuits used, and were published from 2004 to 2012. The previous study by NIB was concluded in 2004, and therefore the publications prior to this date are analyzed in the documentation of these projects [2, 3, 4, 5]. In the present analysis we also included the characteristics of IMPETOM [2, 3, 4, 5] to compare it with newer equipment.

The search yielded seven publications by five groups, that along with the solutions for IMPETOM, allow us to identify trends in design and performance of EIT systems.

4. EIT projects
Ferreira, Rodriguez and Simini proposed in 2002 [2] a circuitry for electrical impedance tomography with sinusoidal signal generation using the Function Generator MAX038 (Maxim). The current source includes the operational amplifier AD844 (Analog Devices). Two multiplexers MAX336 (Maxim) select the electrodes that carry the current. In the measurement stage, 16 identical circuits avoid the use of a multiplexer. The INA114 (Burr Brown) functions as an instrumentation amplifier and the output is fed to a bandpass filter made with the MAX274

chip (Maxim). A sample and hold circuit captures the filter output with the LTC1043 (Linear Technology) IC.

Gonzalez, Liguori, and Simini in 2005 [4] designed a prototype to integrate with the previous project, developing the control stage and the analog to digital conversion through the PC-LPM-16PnP (National Instruments) board, which has 16 input channels and 12-bit resolution converters. The results measured by [2] show a SNR (signal to noise ratio) of $40dB$, CMRR (common mode rejection ratio) in the order of $50dB$ and a current source output impedance of $560k\Omega$ at $50kHz$.

Wang, Liu and Wang in 2005 used a DSP [8, 9] for control and data preprocessing, the DSP used is the TMS320F206 (Texas Instruments). A VCCS (voltage controlled current source) generates the signal that is injected by the electrodes. For measuring voltages, a multiplier is implemented with the AD734 (Analog Devices) IC, A low pass filter with the MAX275 chip (Maxim), and the ADC used is the AD1674 (Analog Devices). The results are oriented to rebuild times, reaching a rate of more than thirty frames per second.

Soleimani in 2006 [12] showed a low-cost equipment for reconstruction in 2 dimensions. The signal is digitaly generated, using an EPROM (27C258) to storage a sinusoidal of $23kHz$. A counter reads the data from the EPROM and feeds a DAC (DAC-0808 National Semiconductor). Harmonic distortion was measured in 1.3%. This signal is fed to a buffer which in turn is connected to a VCCS, performed with the operational AD644 (Analog Devices). For measuring the voltage, synchronous demodulator is used due to its ability to remove noise. It uses the AD625 instrumentation amplifier (Analog Devices) as the input stage before the acquisition card PCL-812PG I/O (Advantech).

Xu et al. created in 2007 a system [10] using 128 electrodes for three-dimensional images of the human thorax. A multiple frequency current is injected using the IC AD9852 (Analog Devices). The frequency range is from $1kHz$ to $2MHz$. The control module is implemented with the chip TMS320F2812 (Texas Instruments). The multiplexer MAX306 (Maxim) is used. For the acquisition of data the integrated AD624 (Analog Devices) is used as a pre-amplifier, then a fourth-order Butterworth filter is implemented with the integrated MAX275 (Maxim). The synchronizing signal is taken from the AD9852, and a synchronous demodulator is implemented in the DSP. As conclusion, reconstructed images are shown, but they don't provide hardware performance data in the form of SNR or any other form.

Bera and Nagaraju developed in 2009 [13] a system to study calibration. For current injection they use a VCO (voltage controlled oscillator) built with the MAX038 (Maxim), which feeds a modified Howland type current source [6]. This module is constructed with two operational amplifiers AD811 (Analog Devices) and generates current of $1mA$ and $50kHz$. A differential amplifier stage and a filter is used for the voltage measurement. Then the resulting voltage signal is measured with a multimeter and a digital oscilloscope. A central electrode is used to reduce the common mode signal. The results show that the common mode is reduced to a maximum of $67mV$.

Hamidi et al. implemented in 2010 [11] a synchronous demodulator with a DSP (Texas Instruments MS320C6713) which is mounted on the development board TLV320AIC23(Texas Instruments). This board has two 16 bits ADCs (AIC23 Texas Instruments) with a sampling rate of 96 kbps. With simulated data, the system had an error in the phase of 0.12 degrees and a signal to noise ratio (SNR) of $130dB$. Using a phantom, the results were worse, but still significantly good.

Gaggero et al. in 2012 [14] developed a system to address two important problems affecting the reliability and usability of electrical impedance tomography. These are: a) EIT is very sensitive to the contact impedance of the electrodes, which can be very large and vary over time due to movement. b) The difficulty of placing the electrodes individually. To solve these problems they implemented a method of active electrodes, which uses a voltage buffer physically close to the electrodes. This stabilizes the contact impedance, and also using a multiplexing method reduces the wires from the electrode belt to the central block. In the central block of the equipment is the signal generator, power source, the electrode handling and the communication with the PC. For this, two boards are used: an Altera Stratix II development kit (Altera) that has integrated the AD9433 ADC (TI), the DAC904 DAC (TI) and Ethernet inteface. Another custom board is responsible for the analog stage, it has a differential amplifier and high pass filter implemented with an AD8221 (Analog Devices) then converts the signal into a differential signal and performs a low pass filter with the integrated THS4502 (IT). The sinusoidal signal is generated in the FPGA through a numerically controlled osciliador (NCO), and a modified Howland circuit is used as a current source. The results shown are the output resistance of current source, and provides that for lower frequencies, this resistance will be lower.

Dixtal Biomedical developed a commercial system, and seen in operation in Montevideo in 2010, called DX 1800, which provides real-time chest images. It has two modules: the DX1800 itself, and a monitoring software that runs on a computer. The DX 1800 is the system responsible for the current generation, voltage measurement and data processing, this system has 32 electrodes to be placed in a line around the patient's chest. The software handles the user interface and display the reconstructed images. The connection between modules is performed via an Ethernet connection. Because it is a commercial system, it was not possible to know how the different blocks are implemented.

Another commercial system is the PulmoVista 500 of **Dräger Medical** [15] to display the distribution of ventilation. It has a processing block and a screen showing the ventilation in the lungs and various respiratory measures. A belt with one row of 16 electrodes is placed around the chest and injects currents between $80kHz$ and $130kHz$, generating images with a rate of 10 to 30 per second with a resolution of 1440 x 900 pixels.

Table 1 presents the summary characteristics outlined for the most representative systems, taking the published figures and extrapolating information wherever possible.

5. Design Options

Using the information from the previous projects, we consider three feasible options for the IMPETOM hardware, one option corresponding to a completely analog signal processing, the other two using a digital signal processor. These are analyzed based on cost factors, difficulty of assembly, performance and speed.

5.1. Discrete components without digital processing

The first option is to design a system using discrete components without a digital signal processing, this would include a basic microcontroller (eg ATmega168, Atmel, \$ 3.33) which is responsible for control and synchronization of different blocks. In this, the demodulation of the signals would be carry out using discrete components, also the filtering and gain stages. Where possible, we use analog-digital converters (A/D converters or ADCs) to digitize the signals and send them to the computer for image reconstruction. Another option derivative of this solution is to use a data acquisition card as used in the project IMPETOM [4], these cards would provide the A/D and communication with the computer. A precondition for improving system performance with respect to the previously developed is to use at least 16bit ADCs resolution. These data acquisition card (DAQs) have multiple input channels, making it possible to have a channel for each pair of electrodes, enabling measure in parallel, this also means build 16 input channels, such as those by IMPETOM C [2], increasing the cost and difficulty of design, in addition, that these DAQs use many input channels makes them very expensive, such as the NI USB-6211 (National Instruments) having a minimum cost of \$690. The main problem with this solution is to repeat the same mistakes in IMPETOM C, obtaining a system where the electrical noise

Table 1. Comparative characteristics of circuits for Electrical Impedance Tomography.

	Sheffield system England 2005 [16]	IMPETOM C Uruguay 2002 [2]	Open acces Iran 2006 [12]	128 electrodes system China 2007 [10]	Active Electrodes Switzerland 2012 [14]
Waveform synthesis	Look up table in ROM	Function generator	EPROM 27C258, VCCS(AD644)	DDS AD9852 (Altera)	Altera Stratix II (NCO) DAC904
Current source	floating with transformer	AD844	current generator $5mA$, $23kHz$		Modified Howland
Multiplexer	2 multiplexers 1 to 16 AD DG506	2 multiplexers 1 to 16	IC-4067 multiplexers	16 MAX306 multiplexers	
Differential amplifier	Instrumentation amplifier INA 110	Instrumentation amplifier INA 114	Instrumentation amplifier AD625	Pre-amplifier AD624 with 4 order low-pass filter MAX275	AD8221 + THS4502
Demodulation	Phase-sensitive voltmeter	LTC1043 sampler	Synchronized pulse demodulation	Synchronic demodulation	Digital Demodulation AD9433
Processing	External computer	IMPETOM I [3]	I/O card ADVANTECH PCL-812PG 16bits	High speed DSP TMS320 F2818	Altera Stratix II

and the imperfection of discret components dominate the measurements obtained, making it impossible to get a reasonable image. The advantage is the cost, table 2 shows the cost of the most notable components of the system, the total shown does not reflect the cost that actually could have this solution, as it ignores many minor components, like resistors, capacitors, crystals, connectors, etc. which increases the final cost.

5.2. Evaluation Board

To minimize the processing of signals in analog form, they must be passed as soon as possible for the A/D converters and perform as much processing in digital format, this would be after a basic first filtering and gain stage, the signals are then processed by a Digital Signal Processor (DSP), both in the filtering phase and in demodulation and for obtaining the amplitude and phase of the signals. This option gives a great flexibility to the system, allowing suit the system to different uses, such as measuring both resistivity and permittivity, also to reduce appreciably the errors introduced by the use of discrete components, and the processing speed ensures the final speed of the system to reconstruct the images won't be affected. By using a DSP, the system transmit to the computer the measurement vector ready to be used in the reconstruction

Table 2. Cost of EIT hardware designed with discrete components.

Block	Design	Component	Quantity	Price, each, U$S
Signal Generator	DDS	AD9832	1	11
	Multiplier DAC	DAC8811	2	13
VCCS	Operational Amplifiers	OPA602	2	10
Current Measurement	Digital potentiometer	AD7376	1	8
Multiplexers	Multiplexer	ADG406	4	12
Amps., filtering	Differential Amps.	INA114	4	12
	Programmable gain amps.	PGA203	2	17
Synch. Demod.	Discrete components	AD630	2	25
Control	Microcontroller	Atmega168	1	4
	Memory	M48Z129V	1	27
	ADC	AD7892	2	23
			Total	**322**

Notes: DDS: Direct Digital Synthesis, DAC: Digital to analog converter, VCCS: Voltage controlled current source, ADC: Analog to digital converter.

method.

Two paths can be taken to use DSPs, the first is an Evaluation Board or Starter Kit, examples of this are the Starter Kit TMS320C6713 (IT, \$ 395) or the Evaluation System ADSP-BF533 EZ-Kit (Analog, \$ 490). These systems are aimed at evaluation or initiation of the engineer in the use of a DSP model and tools developed, often for teaching purposes. It is not the purpose of these boards be used in commercial products. The advantage of using these boards is to have a complete system that contains the DSP, memory, demodulators including the A/D and programmable gain amplifiers (PGAs), communication with the computer, ports for general use, the circuit to program the DSP, this arranged in the most optimal way to get results soon from the start. Figure 2 shows the block diagram for a system implemented with an evaluation board. Figure 3 shows the block diagram of the card TMS320C6713 SDK.

The disadvantage is that being this boards a fixed set of components, some of them couldn't adapt to the problem being solved, in fact, most of these cards, at least those found in a price range between \$ 300 and \$ 600, are designed to work with audio, then the A/D and PGAs are within a codec (AIC23 IT in the TMS320C6713 [17]) specifically designed for audio, this doesn't exclude the use of this boards for EIT systems, there are systems implemented with very similar boards ([18], [11]) but it is necessary to adapt to certain limitations, the most notorious is that this codec operates at frequencies used in standard audio processing, the higher sampling frequency being $96kSPS$ (samples per second,) by the Nyquist theorem, this limits our system to work at frequencies below $48kHz$, although to get really good SNR values, operating frequency must be still smaller, below $20kHz$; this increases the total time required for the acquisition of the signals, in addition to some under-use the DSP that is designed to operate at very high speeds (the DSP within the TMS320C6713 has the same name and operates at $225MHz$ and can perform 1350 million floating point operations per second [19].) Another disadvantage is the cost, again Table 3 will present costs of the more noticeable component to use, and a estimated total cost where we have to take into account the same observations that were made for the table above, in this case the difference between this cost and the real final cost is a little lower than the case of using discrete components, since the board brings all the components needed for the entire system to function properly.

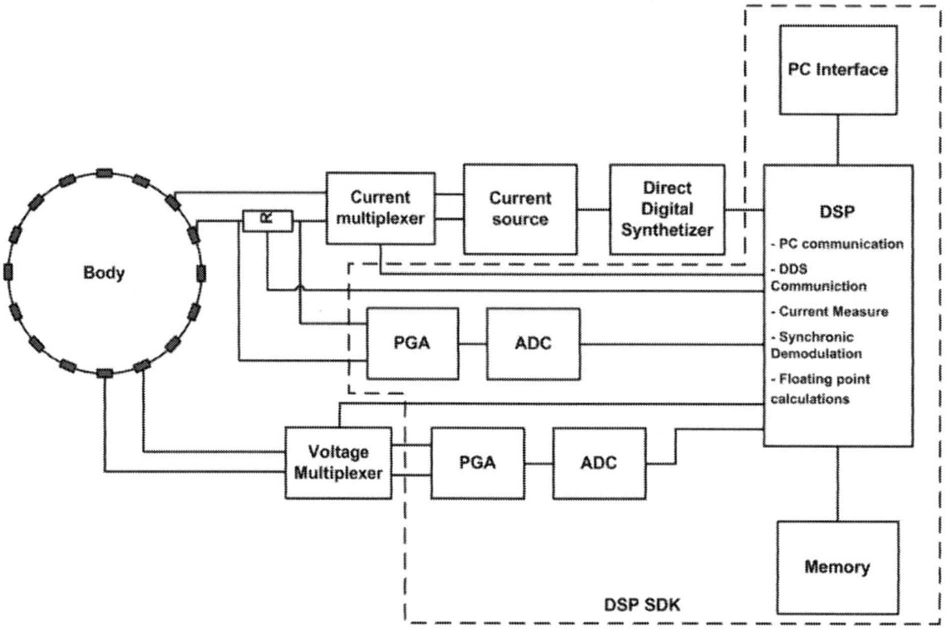

Figure 2. EIT hardware using an Evaluation Board.

Table 3. Cost of EIT hardware based on a DSP Evaluation Board.

Block	Design	Component	Quantity	Price, each, U$S
Signal Generator	DDS	AD9832	1	11
	Multiplier DAC	DAC8811	2	13
VCCS	Operational Amps	OPA602	2	10
Current Measurement	Digital potentiometer	AD7376	1	8
Multiplexers Amps, filtering	Multiplexer	ADG406	4	12
Synch. Demod. Control	DSP EB	TMS320C6713	1	395
			Total	**508**

Notes: DDS: Direct Digital Synthesis, DAC: Digital to analog converter, VCCS: Voltage controlled current source, DSP: Digital signal processor, DSP EB: DSP Evaluation board.

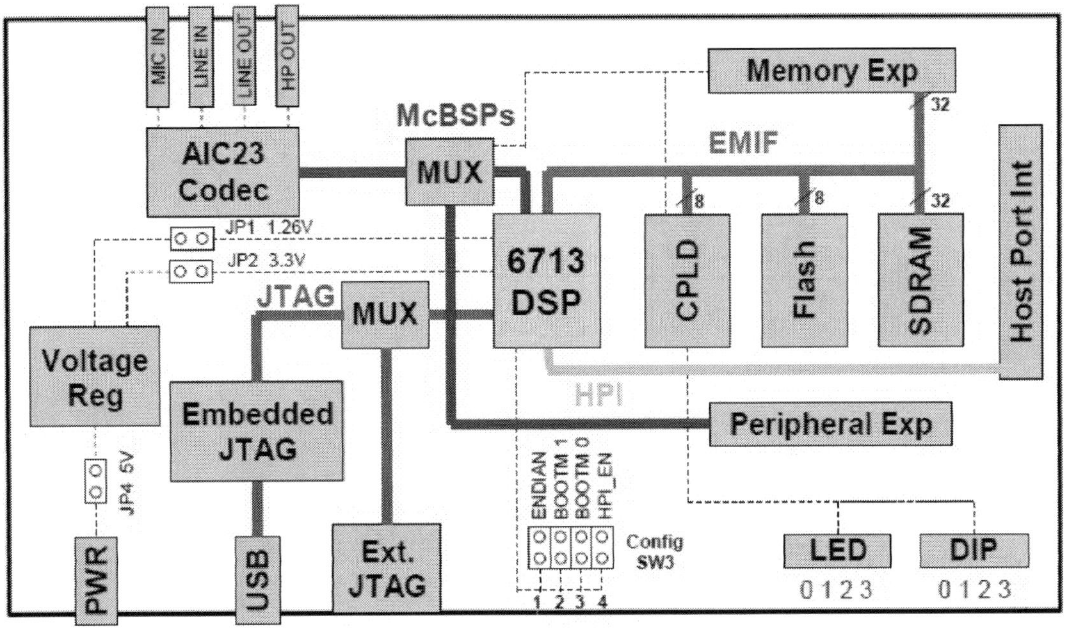

Figure 3. Block diagram for Starter Kit TMS320C671. Taken from [17].

5.3. Discrete components with digital processing

Another option to work with DSPs, overcoming the problems described of working with Starter Kits, is to set up our system with components that are most suited to our problem, that is, mainly, the use of DACs with a higher sampling frequency, such as the AD7892 (Analog, $ 23) which has 18bits of resolution and the ability to get up $1MSPS$, which theoretically would allow the system to work up to a frequency of $500kHz$. Another advantage is that you get a system more adapted to the problem at a lower cost than using a Starter Kit. One problem with this solution is the difficult to achieve a truly compatible and harmonious system, since in these types of components details such as clock speed, type of communication, supply voltages, etc. are essential for the assembled system to work properly. Another problem, not least, is the construction of the system board, since most of the components are surface mounting, and components such as DSP, memory and ADC have many contacts, it makes very difficult to assembly the components using a normal solder. Table 4 again refers to the component costs, taking into account the same considreraciones made above, the cost of this option is similar to the first.

Table 5 compares the adventages and disadvantages of the three methods discussed.

6. Proposed design

The proposed design for IMPETOM tries to improves the previous designs, overcoming the most notable problems by digitalizing the signals in early stages of the process. By doing this we can take advantage of the precision of the digital filtering and demodulation. Also, we don't have the tools needed to build a complicated board with many surface mounting components, then, our desing is based on an Evaluation Board.

The selected board is the OMAP-L137/TMS320C6747 Floating Point Starter Kit ($ 415, Spectrum Digital). The most important caracteristics for us:

- OMAP-L137 processor, with a with a C6747 VLIW DSP floating point processor and an

Journal of Physics: Conference Series **407** (2012) 012024 doi:10.1088/1742-6596/407/1/012024

Table 4. Cost of integrated and discrete components EIT Hardware.

Block	Design	Component	Quantity	Price, each, U$S
Signal Generator	DDS	AD9832	1	11
	Multiplier DAC	DAC8811	2	13
VCCS	Operational Amps.	OPA602	2	10
Current Measurement	Digital potentiometer	AD7376	1	8
Multiplexers	Multiplexer	ADG406	4	12
Amps., filtering	Operational Amps.	INA114	4	12
	Programmable gain amps.	PGA203	2	17
Synch. Demod.	DSP	TMS320C6713B	1	45
Control	DSP	TMS320C6713B	1	0
	Memory	M48Z129V	1	27
	ADC	AD7892	2	23
			Total	**313**

Notes: DDS: Direct Digital Synthesis, DAC: Digital to analog converter, VCCS: Voltage controlled current source, DSP: Digital signal processor, ADC: Analog to digital converter.

Table 5. Advantages and disadvantages of the options discussed

Option	Advantages	Disadvantages
Discrete Components without DSP	Cost	Electrical noise Components imperfections
Evaluation Board	Complete System	No flexibility for components
	Optimal Selection of components	Cost
Discrete components with DSP	System adapted to our system	Difficult that the whole system works in an harmonious way
	Cost less than a Starter Kit	Difficult build process

Notes: DSP: Digital signal processor.

ARM926EJ-S processor operating up to 300 Mhz.

- 64 Megabytes SDRAM.
- USB2 2.0 Full speed interface.
- TLV320AIC3106 Stereo Codec
- Expansion connectors for daughter card use.

The general design is simlar to the one described in 5.2. The codec has 4 input channels, considering that a channel will be used to measure the injected current, there would be three input channels to measure voltage in parallel, so at first it seems more reasonable to start with a single measurement channel and then see if expansion is needed.

This system is intended to perform two significant changes from the previous IMPETOM systems implemented [2, 4], first, measuring the current that is injected. The current source is one of the most important sources of errors in the system [6], due to imperfections in the components such as parasitic capacitances, leading to output impedance which are far from infinite. Therefore, considering that the current value does not vary with load on the reconstruction method is an important source of error, to avoid this we can measure the current that is injected and use the correct value in the reconstruction. For this measurement it is best to use a digital potentiometer, which is controlled by the control block which ensures that the signal obtained makes the most of the range of the ADC.

The other change is in principle to make a single measurement channel and to use a multiplexer to obtain all the necessary measures. This change has the disadvantages of increasing the acquisition time relative to the use of channels in parallel, in addition that the multiplexers are sources of noise and errors [6]. The reason for this change is the complexity and cost of building 16 measuring channels, plus it must have a system capable of handling these signals in parallel, the option used in the IMPETOM project [4], is that of a DAQ, which raises the cost.

7. Conclusions

The systems developed in recent years largely share some methods, such as DDS to generate the sinusoidal signal or a configuration of simple or modified Howland as current source. In the measurement stage, the instrumentation amplifier is followed by an analog to digital converter of at least 16 bit resolution, then, in the digital domain the filtering, synchronous demodulation and other data processing is performed. For control and data processing the DSP is the most popular option for its speed and calculation capacity. In addition, the fewer multiplexers are used, the better the performance of the system. This setting seems to give the best compromise between performance and cost.

The solution implemented in IMPETOM [2, 4], because the technology at the time it was originally developed, is far from what is thought to be the best implementation today. Research continues on several aspects of this development for a prototype according to the set objectives. The availability of a low-cost, noninvasive, without ionizing radiation, which has at all times the evolution of water content in lung space is an aspiration of the intensivists from around the world.

Acknowledgments

The authors are thankful to the students who developed earlier versions of IMPETOM since 2002: Adriana Ferreira, Santiago González, Raúl Hartman, Andrés Liguori, Jorge Lobo, Walter Quinteros, Alfredo Rodríguez and Mateo Ruétalo, as well as to the clinical specifications of Javier Hurtado, MD, Cristina Santos, MD and Walter Olivera, MD, and fellow EE researchers who contributed guidance and advice: Pablo Mazzara, Linder Reyes and Daniel Geido.

References

[1] Barber D C and Brown B H 1984 *J. Phys. E. Sci. Instrum.* **17** 723 – 733
[2] Ferreira A, Rodríguez A and Simini F 2002 *IMPETOM-C Tomógrafo de Impedancia Eléctrica. Academic thesis* (Universidad de la República, Montevideo)
[3] Hartman R, Lobo J, Ruétalo M and Simini F 2002 *IMPETOM-I Tomógrafo de Impedancia Eléctrica. Academic thesis* (Universidad de la República, Montevideo)
[4] González S, Liguori A and Simini F 2005 *IMPETOM. Academic thesis* (Universidad de la República, Montevideo)
[5] Quinteros W and Simini F 2007 *IMPETOM-48 Tomógrafo de Impedancia Eléctrica con tres hileras de electrodos. Academic thesis* (Universidad de la República, Montevideo)

[6] Saulnier G 2004 Eit instrumentation *Electric Impedance Tomography: Methods, History and Applications* ed Holder D (London, England: IOP Publishing) p 65
[7] Xu G, Zhang S, Wu H, Yang S, Geng D, Yan W and Wang M 2005 The acquisition hardware system with direct digital synthesis and filtered back-projection imaging in electrical impedance tomography *Engineering in Medicine and Biology Society, 2005. IEEE-EMBS 2005. 27th Annual International Conference of the* pp 7758 –7761
[8] Wang C, Liu J and Wang H 2005 Pipeline data acquisition method in the eit system *Instrumentation and Measurement Technology Conference, 2005. IMTC 2005. Proceedings of the IEEE* vol 1 pp 437 –440
[9] Wang M, Ma Y, Holliday N, Dai Y, Williams R and Lucas G 2005 *Sensors Journal, IEEE* **5** 289 – 299
[10] Xu G, Wang R, Zhang S, Yang S, Justin G, Sun M and Yan W 2007 A 128-electrode three dimensional electrical impedance tomography system *Engineering in Medicine and Biology Society, 2007. EMBS 2007. 29th Annual International Conference of the IEEE* pp 4386 –4389
[11] Hamidi S A, Jafari R, Nia A M and Soleimani M 2010 *Journal of Physics Conference Series* **224**
[12] Soleimani M 2006 *BioMedical Engineering OnLine* **5**
[13] Bera T and Nagaraju J 2009 A simple instrumentation calibration technique for electrical impedance tomography (eit) using a 16-electrode phantom *Automation Science and Engineering, 2009. CASE 2009. IEEE International Conference on* pp 347–352
[14] Gaggero P O, Adler A, Brunner J and Seitz P 2012 *Physiological Measurement* D **33** 831
[15] Dräger 2011 Dräger pulmovista 500 data sheet
[16] Smith R, Freeston I and Brown B 1995 *Biomedical Engineering, IEEE Transactions on* **42** 133 –140 ISSN 0018-9294
[17] Spectrum Digital 2003 *TMS320C6713 DSK Technical Reference*
[18] Radil T, Ramos P and Serra A C 2008 *IEEE Transactions on Instrumentation and Measurement* **57** 197–204
[19] 2011 TMS320C6713B URL `http://www.ti.com/product/tms320c6713b`

The Initial Systolic Time Interval in patients with spinal cord injury measured with impedance cardiography

Femke Hoekstra[1], Ørjan G Martinsen[2,3], Rudolf M Verdaasdonk[4], Thomas W J Janssen[1,5] and Jan H Meijer[4]

[1] VU University Amsterdam, Faculty of Human Movement Sciences, Move Research Institute Amsterdam, the Netherlands
[2] University of Oslo, Department of Physics, Oslo, Norway
[3] Oslo University Hospital, Dept. Clinical and Biomedical Engineering, Oslo, Norway
[4] VU University Medical Center, Dept. of Physics and Medical Technology, Institute for Cardiovascular Research-VU, Amsterdam, the Netherlands
[5] Rehabilitation Center Reade, Amsterdam, the Netherlands

E-mail: jh.meijer@vumc.nl

Abstract. The Initial Systolic Time Interval (ISTI), obtained from the electrocardiogram and impedance cardiogram, is considered to be a measure for the time delay between the electrical and mechanical activity of the heart. This time delay is influenced by the sympathetic nerve system. Therefore, an observational study was performed in a group of patients (SCI) with spinal cord injuries. The relationship between the ISTI and the total heart cycle (RR-interval) was established by varying the RR-interval using an exercise stimulus to increase the heart rate. The slope of this relationship was observed to be significantly higher in the SCI-group as compared with a control group, although there was no difference in ISTI in the range of common heart rates during the test between the groups. This slope and the ISTI was observed to be significantly different in an acute patient having a recent spinal cord injury at a high level. Because of the variety in injury levels and incompleteness of the injuries further, more specific research is necessary to draw decisive conclusions with respect to the contribution of autonomic nervous control on the ISTI in SCI, although the present observations are notable.

1. Introduction

When the electrical impedance of a human thorax is measured a variation synchronous with the heart activity can be observed. The time derivative of this variation is called the Impedance CardioGram (ICG). Numerous studies during the last decades have demonstrated that the origin of this ICG-signal is too complicated to interpret the *amplitude* of this signal simply as a measure of cardiac stroke volume [1, 2, 3, 4]. However, Meijer *et al* [5] have suggested that an interesting aspect of the ICG can be found in the *time relationships* of this signal with the Electro CardioGram (ECG). Regardless of the multiple sources of the signal, the ICG reflects the *mechanical aspects* of cardiac activity, while the ECG reflects the *electrical aspects*. Therefore, by recording both signals simultaneously, the time difference between electrical and mechanical events in the cardiac cycle can be studied. This concept was implemented by defining the Initial Systolic Time Interval (ISTI) as the time difference between the R-point in the ECG and by the C-point in the ICG. ISTI can be studied in various physiological conditions and clinical circumstances and may be helpful to evaluate patho-physiological processes and rehabilitation processes. The use of ISTI for diagnostic purposes is attractive: both ECG and ICG signals can be obtained fast and easily. The measurements are non-invasive, of low burden to patients and are not restricted to a clinical environment. The required technical equipment is relatively inexpensive, portable and easy to operate.

Up to date little is known about the behaviour of the ISTI in various patho-physiological and clinical circumstances. There is a need for observational research to reveal the behaviour of the ISTI under these circumstances. In theory the ISTI is influenced by at least three factors: the preload of the heart, afterload of the heart and autonomic nervous control. The influence of the preload on the ISTI was demonstrated in patients admitted to an intensive care unit after coronary artery bypass surgery [6]. The ISTI was measured before and after intravenous fluid administration. A significant relationship between the ISTI and the cardiac output was observed, reflecting the influence of preload on the ISTI. In another study [7] the influence of preload as well as sympathetic nerve activity on the ISTI was demonstrated during a breathing stimulus in a group of healthy subjects. The ISTI, and the related parameter PEP, has been reported to be controlled by the sympathetic nervous system [5,8]. These observations justify investigation of the possible use of the ISTI as a diagnostic measurement tool in rehabilitation medicine.

A spinal cord injury (SCI) leads to a partial or complete loss of function below the level of the injury. Depending on the level and (in)completeness of the lesion it leads to a wide range of different medical and health conditions including effects on cardiac function [9]. Factors such as inactivity of the 'muscle pump' in the legs, venous blood pooling in the legs can cause a reduced venous blood return to the heart resulting in a lower preload. Moreover, physical inactivity is often reported in the SCI population [10], which is a major risk factor for developing of cardiovascular diseases. Furthermore, in subjects with a lesion above level T4 problems with regulation of the heart rate and strength of contraction during exercise can occur, due to damaged sympathetic nerves [11]. Therefore, the aim of this study was to investigate whether the relationship between ISTI and RR-interval differed in subjects with a SCI compared with able bodied subjects.

2. Methods

2.1. ICG and ECG recordings

ICG recordings were made using the method described by Meijer et al. [5]. A typical example of a simultaneous recording of the ICG and ECG signal is shown in figure 1. The Initial Systolic Time Interval (ISTI) is the time interval between the R-peak in the ECG and C-wave in the ICG signal.

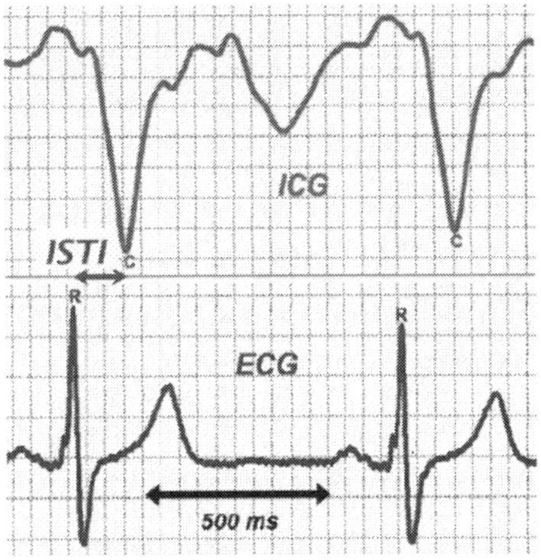

Figure 1. Example of simultaneous registration of an ICG and an ECG

Table 1. Characteristics of the subjects

	SCI group (N=7)	controls (N=9)
age (years)	46 ± 14	50 ± 11
Male/Female	2M/5F	3M/6F
Height (cm)	172 ± 7	170 ± 10
Body mass (kg)	73 ± 15	68 ± 11
BMI (kg/m2)	25 ± 5	24 ± 3

2.2. *Subjects*

Seven subjects with a motor incomplete SCI and nine age- and gender-matched able-bodied (controls) subjects participated in this study (table 1). The inclusion criteria for the SCI group were: paraplegia or tetraplegia as a result of a motor incomplete SCI (ASIA C, D) and a minimum age of 18 years. The time post-injury was at least a year (12 ± 11; 1-35 years). The level of the injury varied between the patients from L1-L2 to T1-C1. An eighth, acute patient (F, 33 years, 166 cm, 63 kg, lesion C5-T1, ASIA C) was tested for the first time 2.5 months and for the second time 7 months after injury, and is described separately. The exclusion criteria for both groups were: medical complications, such as unstable hypertension, uncontrolled cardiac dysrhythmia and other unstable cardiovascular problems, severe skeleton problems such as osteoarthritis or recent fractures of the lower limbs, severe cognitive and/or communicative disorders, other neurological and/or psychiatric disorders, or other problems which make it impossible to accomplish the tasks. The type and location of the spinal cord lesion was established by means of a clinical evaluation by a physician. After a detailed explanation of the purposes and the protocol of the experiment, all subjects signed an informed consent. The study was approved by the ethics committee of the VU University Medical Center Amsterdam and by the ethics committee of the Faculty of Human Movement Sciences of the VU University Amsterdam.

2.3. *Procedure*

The relationship between the ISTI and the RR-interval (RR) was established in each subject by varying the heart rate (HR). Each subject performed a discontinuous progressive graded exercise test on an Angio arm ergometer (Lode BV, Groningen, the Netherlands) to increase the HR. Before starting the exercise test resting values of ISTI and RR were measured during five minutes of rest in a sitting position. The exercise protocol consisted of minimal three blocks of three minutes arm pedalling with an individually adapted, increasing work load. One minute of rest after each block was included to facilitate the ICG and ECG measurements. During the first block of the exercise test, subjects started arm cranking at a low intensity (0 – 20 W) to get familiarized with the equipment. Subjects were instructed to exercise at a rate of 60 revolutions/min. Visual feedback about this cranking rate was provided. Subjects were verbally encouraged to exercise to exhaustion. The exercise test was ended when a subject was not able to continue due to exhaustion or when the subject indicated that he/she wanted to stop

2.4. *Analysis*

Linear regression analysis was used to determine the relationship between ISTI and RR according to the following equation: ISTI = slope*RR + intercept.

To investigate differences in this relationship between both groups, the regression coefficient (slope) and the intercept were used as outcome variables. To evaluate differences in ISTI in the range of the operating heart rates during the test, the ISTI was calculated from the regression relationships for each subject at an RR of 500 ms and 700 ms: ISTI(500) and ISTI(700). Differences between groups were tested using an independent *t* test and eccentricity was tested using a *z*-test.

3. Results

3.1. *Comparison of patient and control groups*

All subjects showed a significant decrease in the ISTI with a decreasing RR as a consequence of the exercise stimulus ($p < 0.005$). The mean regression coefficient (slope) for the SCI group was significantly higher as compared with the mean for the control group ($p < 0.005$), while the intercept was significantly lower in the SCI group as compared with the control group ($p < 0.025$) (table 2 and figure 2). No differences were found in ISTI(500) and ISTI(700) ($p > 0.05$).

Table 2. Relationships between ISTI and RR-interval in the two groups and in the acute patient AP. Also the ISTI-values at RR-intervals of 500 and 700 ms are given.

	SCI group (N=7)	controls (N=9)	AP first	AP second
slope (ms/ms)	0.11 ± 0.02	0.08 ± 0.01	0.29	0.26
intercept (ms)	48 ± 22	73 ± 16	-25	-33
ISTI(500) (ms)	104 ± 15	114 ± 15	119	96
ISTI(700) (ms)	127 ± 13	130 ± 15	176	148

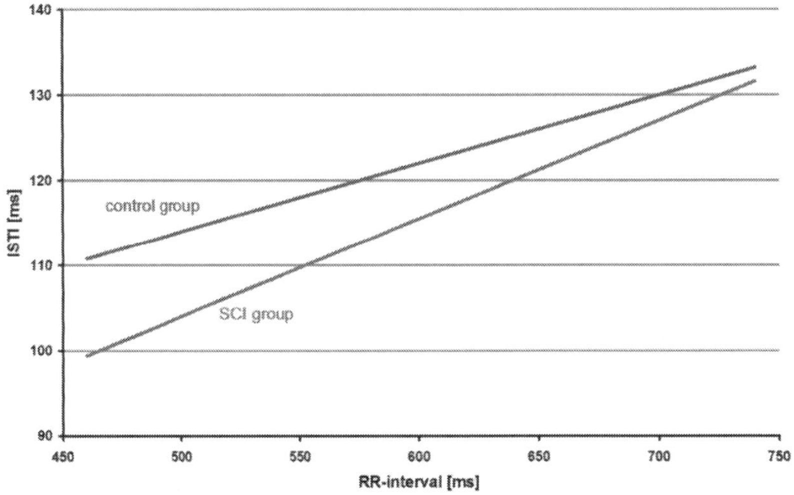

Figure 2. Mean relationship between ISTI and RR in the SCI group and in the control group.

3.2. Case report

A 33 years old acute patient having a high spinal cord injury at the level of C5-T1, ASIA C, was tested for the first time 2.5 months and for the second time 7 months after injury (see table 2). Both slope and intercept (see figure 3) of the relationship between ISTI and RR differed significantly from those in the SCI group as well the control group in the first and the second test (z-test, $p < 0.005$). ISTI(500) was not found to be significantly different on both occasions and compared with both groups ($p > 0.05$). The extrapolation to low heart rates: ISTI(700), was found to be significantly different at the first test compared with both groups ($p < 0.005$). At the second test, however, there was only a slight significance with the SCI-group ($p < 0.05$) but not with the control group ($p > 0.05$).

4. Discussion and conclusions

The present study is the first in which the relationship between the ISTI and RR-interval in patients with a spinal cord injury has been investigated. The results show that there is a significant difference in the slope of the relationship between ISTI and RR between subjects with a SCI compared with able-bodied persons, suggesting larger differences at higher heart rates. An acute patient showed a substantial difference with both groups, which was most prominent at lower heart rates. This difference in the acute phase indicate the presence of a process that develops in time. Also the development of ISTI at low heart rates during five months in this patient suggests a recovery process. The cause of the observed differences, however, may be multi-factorial. In people with spinal cord injury many control mechanisms are influenced by the lesion. An inactive lifestyle and cardiovascular adaptations can lead to a lower exercise performance [11]. Subjects with a SCI have an impaired muscle pump in the legs, which results in a lower preload to the heart. Differences in ISTI between high and low trained healthy subjects have been reported earlier [12]. In patients with a high lesion, cardiac control by the sympathetic nervous system can be impaired. The impairment depends on the level and (in)completeness of the lesion. Claydon et al. [13] reported that blood pressure dysregulation after an arm exercise stimulus is common in subjects with a cervical lesion, but not in subjects with a thoracic SCI. The sympathetic fibers controlling the heart arise at the level from the first thoracic segment (T1) to the fourth thoracic segment (T4). In subjects with a lesion at or above T4 sympathetic pathways may be interrupted which can lead to cardiovascular impairments. With a complete lesion maximal heart rate is mostly limited to approximately 120 – 130 beats per minute. In the present study only three of the total of eight subjects had a lesion above T4. However, these subjects had an incomplete lesion. During the exercise test two of these patients had a peak heart rate of respectively 143 and 155, which indicates that the sympathetic control of the heart was not complete impaired. Another limitation of the present study is that the SCI group consisted of patients with incomplete lesions at various levels. Therefore, it is not possible to draw decisive conclusions with respect to the contribution of autonomic nervous control on the ISTI in SCI. More specific research is needed to investigate the determining factors in this group of patients.

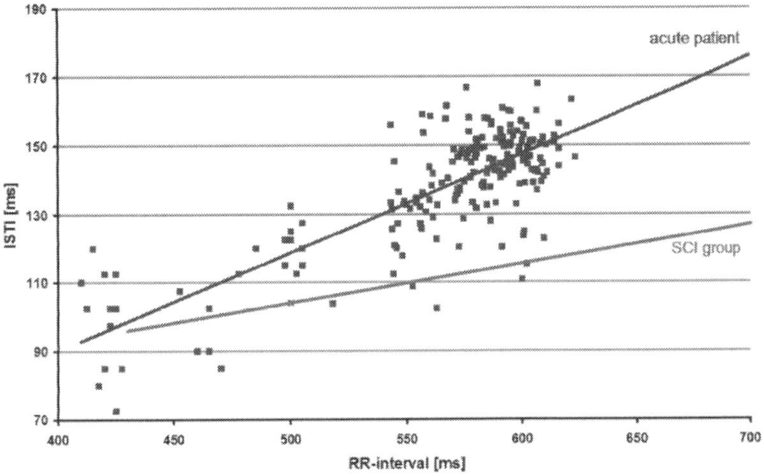

Figure 3. Relationship between ISTI and RR in the acute patient and of the mean of the SCI group.

References

[1] Visser KR, Lamberts R, Korsten HHM and Zijlstra WG 1976 *Pflügers Arch.* **366** 289-91

[2] Sakamoto K, Muto K, Kanai H and Lizuka M 1979 *Med. & Biol. Eng. & Comput.* 1979 **17** 697-706

[3] Patterson, RP 2010 *J. Phys: Conf. Ser.* **224** 012118

[4] Bernstein DP 2010 *J. Electr. Bioimp.* **1** 2-27

[5] Meijer JH, Boesveldt S, Elbertse E and Berendse HW 2008 *Physiol. Meas.* **29** S383

[6] Meijer JH, Smorenberg A, Lust EJ, Verdaasdonk RM and Groeneveld ABJ 2010 *J. Electr. Bioimp.* **1** 80-83

[7] Meijer JH, Hoekstra F, Habers E, Verdaasdonk RM and Janssen TWJ 2010 *J. Phys.: Conf. Ser.* **224** 012119.

[8] Willemsen GH, de Geus EJ, Klaver CH van Doornen LJ and Carroll D 1996 *Psychophysiology.* **33** 184-93

[9] Figoni SF 1993 *Med. Sci. Sports Exerc.* **25** 433-41

[10] Buchholz AC, Pencharz PB 2004 *Curr. Opin. Clin. Nutr. Metab. Care* **7** 635-9

[11] Phillips WT, Kiratli BJ, Sarkarati M, Weraarchakul G, Myers J, Franklin BA, et al 1998 *Curr. Probl. Cardiol.* **23** 641-716

[12] Hoekstra F, Habers E, Janssen TWJ, Verdaasdonk RM and Meijer JH 2010 *J. Phys.: Conf. Ser.* **224** 012117

[13] Claydon VE, Hol AT, Eng JJ and Krassioukov AV 2006 *Arch. Phys. Med. Rehabil.* **87** 1106-14

Journal of Physics: Conference Series 407 (2012) 012026 doi:10.1088/1742-6596/407/1/012026

A portable device to assess underwater changes of cardio dynamic variables by impedance cardiography

F Tocco[1], A Crisafulli[1], E Marongiu[1], R Milia[1], A Kalb[2] and A Concu[1-2]

[1] Department of Medical Sciences "Mario Aresu", University of Cagliari, Cagliari, Italy
[2] 2C Technologies Inc, Spin-off of the University of Cagliari, Cagliari, Italy
E-mail: concu@unica.it

Abstract. Data concerning heart rate (HR), stroke volume (SV), and cardiac output (CO) during dynamic apnoea (DA) were collected from 10 healthy male, elite divers by means of an impedance cardiograph adapted to the underwater environment (C.O.Re., from 2C Technologies Inc, Italy). Three trials were performed by the divers in a 3-m-deep pool with a water temperature of 25°C: 3-minute head-out immersion during normal breathing (A), till exhaustion immersed at the surface (B) and at 3m depth (C). Both B and C conditions did not led to changes in HR, SV and CO compared to A. Data indicate that typical diving response consisting in a reduction of HR, SV and CO was not present during DA, probably due to sympathetic activation induced by exercise during DA, which partially obscured the effects of the diving response. Moreover, this study highlights the innovative role of our portable, impedance cardiography device, i.e. the C.O.Re., in easily assessing cardiodynamic changes in subjects engaged in exercise schedules including phases of underwater, dynamic apnoea.

Introduction

The assessment of cardiovascular adjustments during underwater apnoea is rendered difficult by the technical limitations related to hemodynamic measurement in the underwater environment. Most of the studies in this field attempted to reproduce conditions similar to real diving in head-out immersed subjects [1], in simulated diving in a pressure chamber [2], and apnoeas with or without face immersion in exercising humans [3]. Furthermore, recently underwater ultrasound cardiac measurements provided exciting new data on heart rate (HR), stroke volume (SV), and cardiac output (CO) responses during diving in humans. In particular, Marabotti and co-workers [4] reported a reduction in HR and SV, and consequently in CO, during underwater, static apnoea. Therefore, this response was interpreted to induce an oxygen-conserving effect as is described in marine animals [5]. However, in the above mentioned studies apnoea was always static because of the impossibility to conduct ultrasound measurements in moving subjects since the ultrasound device was cumbersome. Thus, the only hemodynamic measurements performed during dynamic breath-hold to date were carried out during simulations in a laboratory setting [6]. This research revealed that heart rate (HR) and cardiac output (CO) decreased. The aim of the present investigation was to adapt a impedance cardiography device in such a way of measure the hemodynamic changes due to dynamic apnoea in freely moving subjects in the underwater environment. To this end, a miniaturised, portable impedance cardiograph, which was sealed in a small waterproof container, was developed to assess hemodynamics in the underwater environment. We believed such a measurement would describe better the cardiovascular changes in a situation closer to real apnoea than previous investigations had.

Methods

Subjects. Ten healthy male, elite divers (33.2±3.6 years, 172.4±7.2 cm, and 64.1±6.6 kg) were recruited. All participants received information about the aims and the procedures of the study which was performed according to the Declaration of Helsinki and conformed to Good Publishing Practice in Physiology.

Measurements. All hemodynamic measurements were performed by means of impedance cardiography, a method commonly employed to assess hemodynamic changes in resting and exercising subjects [7, 8, 9, 10]. The impedance method provides reliable, non-invasive data of SV, HR and CO. This technique assumes that, when an electrical current circulates through the thorax, the pulsatile aortic blood flow causes a proportional fluctuation in the electrical conductivity. Given this, changes in thoracic electrical impedance during systole are representative of SV [11]. By measuring several reference points on the impedance waveforms, SV can be estimated by applying the Sramek-Bernstein equation [12]. We had previously used this impedance traces processing method, which is described in detail in our previous works [8, 9, 13, 14]. Briefly, we stored derived analog traces of thorax impedance (Z_0), and the Z_0 first derivative (dZ/dt). Stored traces were afterwards analysed and signals affected by movement and respiratory artefacts were excluded from calculations. Stroke volume was calculated using the Sramek-Bernstein equation:

$$SV = (VEPT \cdot Z_0^{-1}) \cdot dZ/dt_{max} \cdot VET$$

In detail, VEPT was the volume of electrical participating tissue and was derived using a nomogram from subjects' sex, height, and weight; Z_0 was the thorax impedance measured at the end of cardiac diastole [8]; dZ/dt_{max} is the maximal Z_0 first derivative value during cardiac systole; VET was the left ventricular ejection time, calculated as the

Journal of Physics: Conference Series **407** (2012) 012026 doi:10.1088/1742-6596/407/1/012026

interval between the beginning and the minimum of the deflection in dZ/dt trace during systole. In summarising, through impedance traces it was possible to calculate the following parameters: SV, HR, and CO, the later obtained by multiplying SV•HR.
Instrumentation. To apply the impedance method during underwater apnoea, a miniaturised Cardiac Output Recorder (C.O.Re., 2C Technologies Inc, Cagliari, Italy) was utilised (Fig.1, panel A).

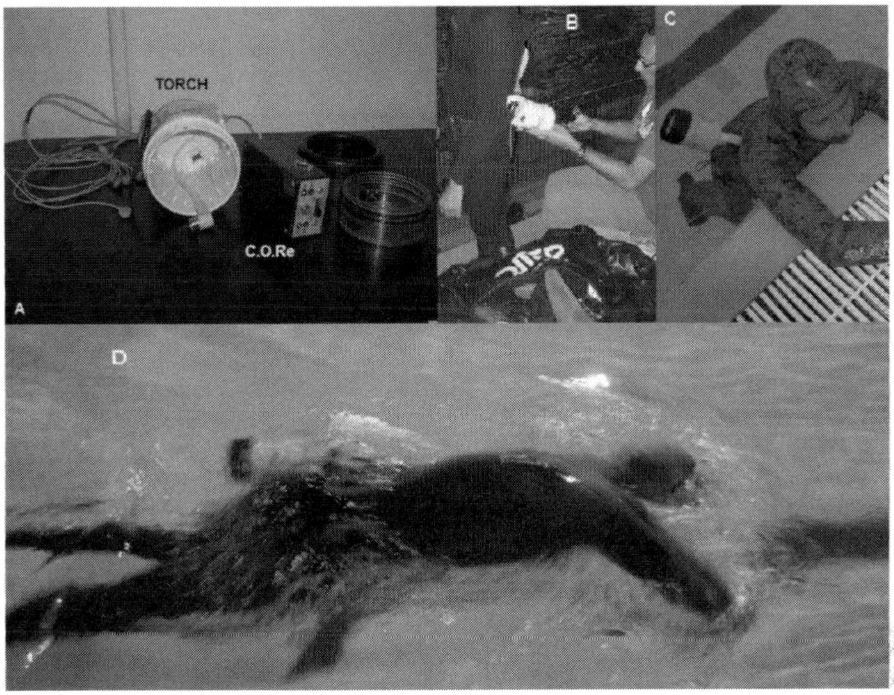

Fig. 1. Panel A: the miniaturised impedance cardiograph ("C.O.Re") contained in a waterproof torch for underwater use. Panel B: the C.O.Re is placed inside the torch before diving. Panel C: one of the divers while performing a 3-minute head-out immersion during normal breathing. Panel D: one of divers while executing dynamic apnoea fully immersed at surface.

This device consists of a analogic front-end with a A/D converter. The components of the analogic section are: a sinusoidal current generator at a constant frequency of 65 kHz, an amplifier to assess transthoracic impedance, with a demodulator, which output is connected to the digital section. The digital section consists of a 8 bit microcontroller with a A/D converter (resolution of 10 bit) which converts input signals in serial data. The C.O.Re. was placed in an underwater torch, which was waterproof to a depth of 90m (Fig.1, panel A-B) and was turned on at the beginning of each session(Fig.1, panel B). The C.O.Re. recorded impedance traces throughout the whole experimental session on a secure digital memory card. Impedance recorded traces were then analysed by employing a digital chart recorder (ADInstruments, PowerLab 8sp, Castle Hill, Australia), and hemodynamic parameters were calculated. To establish the validity of the C.O.Re., comparison with an impedance cardiograph (NCCOM3-R7, BoMed, CA Inc) previously validated in scientific literature was carried out [15, 16]. Stroke volume values were obtained by both instruments on 18 male subjects (mean ± SD of age, height, and mass were 28.8±5.1 years, 175.8±3.9 cm, and 66.8±6.1 kg respectively) during rest and exercise up to 70% of maximum HR calculated as 220-age.
Study protocol. Hemodynamic changes were assessed during dynamic apnoea in a 3-m-deep pool with a water temperature of 25°C. In detail, during one session three trials were carried out as follows by each subject.
Test A: consisted of a 3-minute head-out immersion during normal breathing. This situation was considered the baseline level (Fig.1, panel C).
Test B: involved dynamic apnoea till exhaustion with the body fully immersed at the surface (Fig.1, panel D).
Test C: during this test divers underwent dynamic apnoea till exhaustion at 3m depth.

During periods of dynamic apnoea, divers were instructed to follow a light signal placed at the bottom of the pool in order to maintain a constant speed of 0.5 m•sec^{-1}, according to the usual methods of dynamic apnoea training. The goodness of the speed control was tested by a standard chronometer. All divers had a 2 to 4 minute period of respiratory preparation before performing apnoea sessions. They were asked to begin apnoea phases after a maximal inspiration without any manoeuvres of hyperventilation. Ballast applied to the waist was used during tests B and C and the weight of ballast was chosen according to subjects' personal experience and body mass to ensure neutral buoyancy. The portable impedance cardiograph was applied to the subjects' chest by 8 electrodes [12], which were waterproofed using surgical 15x10 cm patches (Plastod, Bologna, Italy). Then, subjects simply wore a diving suit and a diving mask. At the end of each trial electrodes status was inspected to attest that the waterproof condition was maintained.

Statistical analysis. The Bland and Altman statistics [17] was used to assess agreement between the two methods of measurement to compare SV obtained from the C.O.Re. and the NCCOM3. In order to compare subjects with different body masses, data during apnoea phases were transformed into percentage changes from test A values. One-way ANOVA for repeated measurements was applied to determine significance between the various periods of the protocol and Newman-Keuls post-hoc was performed when appropriate. Significance was set at a p value <0.05.

Results

The Bland and Altman statistics show that there was positive agreement between the C.O.Re. and the NCCOM3 both at rest and during exercise as limits of agreement between the two methods were within +7.01 and -8.12 ml at rest and + 6.77 and -9.86 ml during exercise (Fig.2).

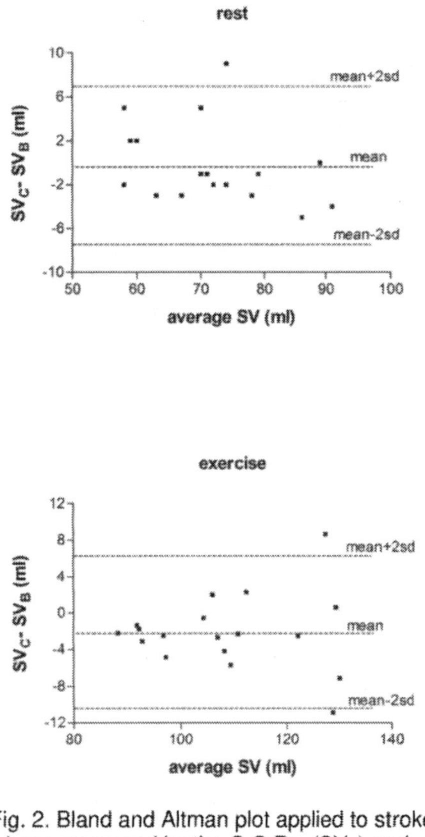

Fig. 2. Bland and Altman plot applied to stroke volume measured by the C.O.Re (SV_C) and the BoMed (SV_B) at rest (top panel) and during exercise (bottom panel).

The group mean duration of dynamic apnoea periods was 91.6±18.2 and 95±32.1 s for Test B and C respectively. Absolute values of hemodynamic data during test A preceding apnoeas were: HR = 68.8±3.9 beats•min^{-1}; SV =

Journal of Physics: Conference Series **407** (2012) 012026 doi:10.1088/1742-6596/407/1/012026

61.2±7.9 ml; CO = 4.2±0.7 l•min^{-1}. During test B the same variables showed the following values: HR = 70.2±4.8 beats•min^{-1}; SV = 62.1±6.2 ml; CO = 4.3±0.6 l•min^{-1}. During test C hemodynamic data were: HR = 69.5±4.9 beats•min^{-1}; SV = 62.7±8.2 ml; CO = 4.3±0.9 l•min^{-1}. No statistical significance was shown among hemodynamic variables assessed during A, B and C tests (Fig.3).

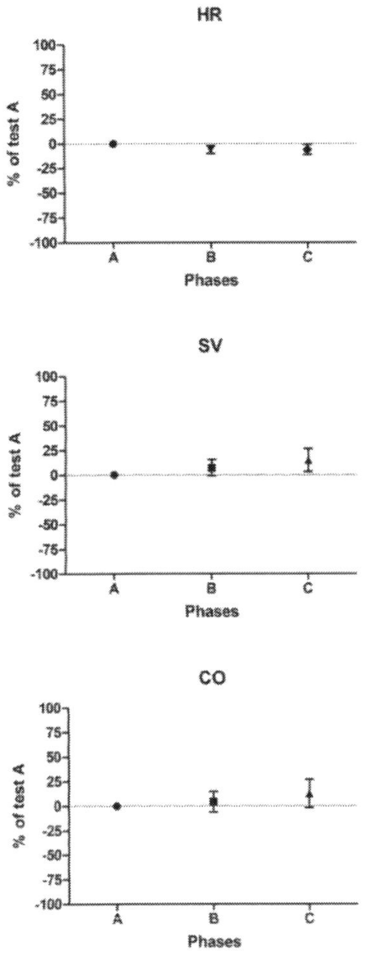

Fig 3. Heart rate (HR, top panel), stroke volume (SV, middle panel), and cardiac output (CO, bottom panel) time courses during the experimental setting. (A) Head-out immersion. (B) Dynamic apnoea at the surface. (C) Dynamic apnoea at 3m depth. Values are means ± SD percent changes form of test A.

Discussion

An important finding arising from the present investigation is that during both dynamic apnoea sessions, HR and SV were at the same level as at baseline, so that CO was stable. It is important to notice that an oxygen-sparing effect is often attributed to the classical diving response, which during exercise appears to be proportional to the degree of the bradycardia [18]. The new finding in this investigation is that hemodynamic response during static apnoea was virtually abolished when dynamic apnoea was performed. Indeed, in this setting HR and SV were substantially unmodified with respect to baseline, so that CO did not decrease. Our data consent us to speculate that the supposed

O_2-sparing effect was less evident during dynamic apnoea compared to static apnoea. Therefore, it is likely that during tests B and C the energetic demand of the exercising muscles, together with the activation of cardiovascular control areas in the brainstem, played a pivotal role in counteracting the hemodynamic effects of the diving reflex. A similar conflict between exercise and diving cardiovascular responses was hypothesized for diving ducks which usually show an increase in HR above resting values during natural dive [5].

It is well known that during exercise an increase in sympathetic tone takes place in order to guarantee adequate mean blood pressure (MBP), particularly during hypoxic exercise [19]. This increased sympathetic drive derives from the integration of two main neural mechanisms: in one, commonly termed "central command", the activation of regions of the brain responsible for motor unit recruitment also activates the cardiovascular control areas located in the medulla [20]; in the other, known as the "exercise pressor reflex", receptors within the muscles reflexively activate afferent nerves and induce sympathetic-mediated cardiovascular adjustments on the basis of the mechanical and the metabolic condition of the contracting muscle [20]. Both the central command and the exercise pressor reflex are responsible for activation of sympathetic activity, which in turn increases HR and cardiac performance, thereby raising CO [21]. These cardiovascular adjustments aim at counteracting the increase in vascular conductance due to vasodilatation in the contracting muscle, thus avoiding falls in MBP [22]. Our findings suggest that, during periods of dynamic apnoea, activation of both the central command and the exercise pressor reflex could have, at least in part, masked the typical hemodynamic response to apnoea, i.e. bradycardia and SV reduction.

We previously reported a similar cardiovascular response during exercise with face-immersed apnoea, when an increase in myocardial performance and SV occurred and obscured the cardiovascular effects of the diving response [23]. In the quoted study, we hypothesised that during simulated dynamic apnoea the exercise pressor reflex acted in opposition to the diving reflex to maintain cardiovascular homeostasis during exercise. The results of this investigation, in agreement with that hypothesis, extend the concept that exercise may also offset the effects of diving reflex for real dynamic apnoea, when hypoxia probably took place and strengthened the metaboreflex-induced sympathetic response [19].

This study also highlights the innovative role of our portable, impedance cardiography device, i.e. the C.O.Re., in easily assessing cardiodynamic changes in subjects engaged in exercise schedules including underwater, dynamic apnoea. In fact, in the light of the above findings, assessing cardiodynamic variable during underwater apnoea ought give relevant helping in optimizing training schedules of apnoea diving athletes. Moreover, underwater dynamic apnoea, with cardiodynamic assessment by the C.O.Re. device, ought be included in breath-hold exercises that often are part of respiratory rehabilitation in respiratory disease patient, since easily acquisition of cardiodynamic profile may help to prevent some cardiovascular accidents during these manoeuvres.

Keywords

Underwater Apnoea, Cardio Dynamics, Impedance Cardiograph, Portable Medical Devices.

References

1.Shiraki K *et al* 1986 *Undersea Biomed. Res.:* **13** 247
2.Ferrigno M *et al* 1997 *J. Appl. Physiol.:* **83** 1282
3.Andersson J *et al* 2004 *J. Appl. Physiol.:* **96** 1005
4.Marabotti C *et al* 2008 *Undersea Hyperb. Med.:* **35** 83
5.Butler PJ and Jones DR 1997 *Physiol. Rev.:* **77** 837
6.Lindholm P *et al* 1999 *J. Appl. Physiol.:* **87** 2122
7.Belardinelli R *et al* 1996 *Am. J. Cardiol.:* **77** 1293
8.Crisafulli A *et al* 2003 *Eur. J. Appl Physiol.:* **89** 209
9.Crisafulli A *et al* 2003 *Med. Sci. Sports Exerc.:* **35** 221
10.Crisafulli A *et al* 2011 *J. Appl. Physiol. :* **111** 530
11.Warburton DER 1999 *Sports Med.:* **7** 241
12.Bernstein DP 1986 *Crit. Care Med.:* **14** 904
13.Crisafulli A *et al* 2007 *Am. J. Physiol. :* **292** 2988
14.Crisafulli A *et al* 2008 *Exp. Physiol.:* **93** 447
15.Moore R *et al* 1992 *Chest :* **102** 448
16.Concu A and Marcello C 1993. *Eur. J. Appl Physiol.:* **66** 11
17.Bland JM and Altman DG 1986 *Lancet:* **1** 307
18.Lindholm P and Lundgren EG 2009 *J. Appl. Physiol.:* **106** 284
19.Houssiere A *et al* 2005 *Am. J. Physiol.:* **288** 1724
20.Rowell LB and O'Leary DS.1990 *J. Appl. Physiol.:* **69** 407
21.Boushel R 2010 *Acta Physiol. Scand.:* **199** 367
22.Toska K 2010 *Acta Physiol. Scand.:* **200** 211
23.Tocco F *et al* 2012 *Eur. J. Appl. Physiol. :* **112** 543

Electrodes – the challenge in electrical characterization of biological material

Andreas Barthel, Thomas Nacke, Dieter Frense, Uwe Pliquett
Institut für Bioprozess- und Analysenmeßtechnik, Heilbad Heiligenstadt, Germany
Email: Andreas.Barthel@iba-heiligenstadt.de

Abstract Electrodes are an important part of the impedance measurement chain but their influence is often underestimated. Electrochemical reactions or polarization effects, especially for galvanically coupled systems are sometimes completely neglected. The most important features of electrodes to be considered are the geometry with respect to the structure of the material to be tested and the electrochemistry at the electrode surface. Especially drift and corrosion effects may yield misleading results. If applicable, sophisticated electrode systems should be used in order to prevent the distorting influence of electrode polarization, for extending the useful frequency range of the electrodes or for enhancing the signal-to-noise ratio.

1. Introduction

Bio-impedance measurements for electrical characterization of cell based material use a wide range of frequencies. The conductivity of suspension medium in the presence of low cell concentration is usually measured at low frequency below 1 kHz while the behaviors of cell membranes overwhelm the electrical properties between 10 kHz and 10 MHz [1]. At much higher frequencies (1 – 20 GHz), water molecules can be detected which allows a simple access to humidity or total water content of a material. Despite at frequencies above 50 kHz inductive or capacitive coupling between electronics and material works well and at microwave frequencies antenna structures are favored, galvanically coupled electrodes dominate in most bio-impedance measurement setups.

The design of such electrodes ranges from simple plates or needles up to sophisticated systems with separated current injection and voltage monitor electrodes. Moreover, additional guard electrodes and special shape of the measurement chamber optimizes the sensitivity of electrode system, usually for unique measurement tasks.

Electrodes are often designed depending on the space available for the measurement system or for highest versatility of the setup. Finite element analysis is favored for prediction of the electric field distribution. However, it does not matter whether the material under test is simulated as well or not, near reality condition are achieved only to some extend owing to the complicate electrochemistry especially of metal electrodes.

Here we introduce some of the very important issues to be taken into account for designing electrodes and for interpretation of experimental results. Although the paper is not an original contribution, it should provide some thoughts especially for scientists starting in the field of bio-impedance measurement. The basic features of impedance measurement are assumed to be known.

2. Electrode geometry

The first thing to clarify is the electrode geometry. Does the entire object fit between the electrodes or does the electrode contact a small portion of the object? Is the object microscopically small (µm-size, single cell level) or is it a macroscopic object (cm-size)? In any case, the electrode geometry should be adjusted to the geometry and the structure of the material under test. In general, parallel plate electrodes are considered to be optimal, because of the homogeneous electric field between them. However, this is not true if an irregularly shaped object is placed between the electrodes. For instance, skeletal muscle or whole organs cause distortions of the electric field between the electrodes making the interpretation of the impedance measured between the clamps tricky.

Model based data processing is advantageous for retrieving material specific properties from the electrical measurement. A simple model is a cell suspension with round shaped cells. Pauly and Schwan derived equations for calculating the intra- and extracellular conductivity as well as the cell membrane capacity from the magnitude of the impedance below and above the ß-dispersion and the characteristic frequency. Although this model is used by many scientists, most of them forget about the most prominent limitations: (1) the cells are round shaped and (2) there is no interaction between them. Sometimes this model is used for cells like yeast or blood cells, departing considerably from spherical shape. In other cases, suspensions with high cell concentration above p=0.2 (cell volume fraction) are treated by using this model. Even though this is basically wrong, the departure from reality may be acceptable for practical issues.

First, one should find out what is the most important property of the material to be accessed by impedance spectroscopy [2]. As an example, detecting low cell concentration in a suspension ($<10^4$ cells/ml) works best in the low frequency region (< 10 kHz) because metabolites change the conductivity of the suspension medium. For testing the cell viability, the polarization of the whole cell becomes important. Especially in microfluidics interdigital electrodes are favored. As seen in Fig.1A, the distance between electrodes should be big enough for a field penetration sufficient for reaching a significant number of cells.

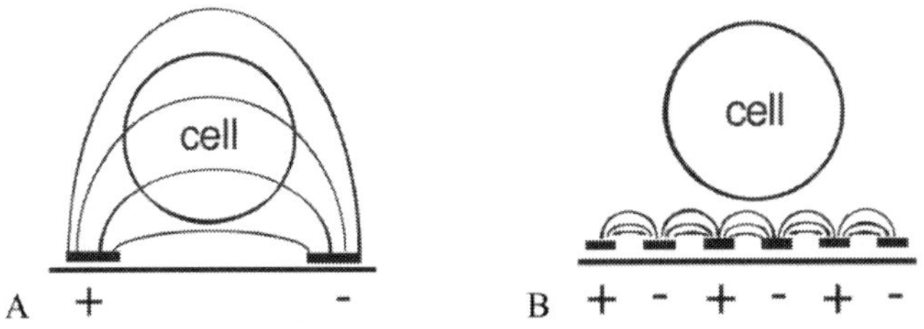

Fig1. The geometry of the electrode depends on the size of the material and the property to be measured. (A) interdigital electrodes (+ and -) with a distance (> 30 µm) for deep penetration of the electric field into the material for assessment of the polarization of cells and (B) setup to foster the measurement of the medium conductivity. The tight spacing of the electrodes (< 3 µm) greatly reduces the influence of cells on the impedance spectrum.

In opposite, when the conductivity of the suspension medium is the property of interest, an electrode geometry like in Fig.1B increases the significance of the measurement.

The electrode geometry can greatly influence the measured results for structured materials. For instance, anisotropic materials like muscle differ in their impedance spectrum between the orientation along the muscle fibers and perpendicularly to it. Inhomogeneous materials exhibit different spectrum with respect to the measurement site. Large electrodes will be used for averaged property throughout the material while for assessment of the substructure, small electrodes either scanning electrodes or electrode arrays are required.

Monopolar electrodes, i.e. usually small electrodes (µm-mm) with a large counter electrode (several cm^2) in a distant location are used for the assessment of the impedance in the immediate vicinity of the electrode itself. They are not suitable for the measurement of the impedance of large volumes. The reason is a drastic increase of the field strength at the electrode with reduced electrode dimension. Since the impedance is the ratio of voltage and current, a high electric field strength means a high voltage drop over a fixed distance which yields a high impedance for a given current. This means that most information about electrical material properties is got from regions with the highest field strength, here the immediate vicinity of small electrodes.

Despite of some applications where simple calibration to a biological quantity is sufficient, impedance measurements should be comparable and quantitatively interpretable. A simple measurement assesses the impedance between the electrodes. Besides the resistivity, ρ, also the geometry of the electrode determines the impedance: $Z = \rho\frac{l}{A} = \rho k$ where l is the distance between a pair of planar, parallel electrodes and A the surface area. The quotient l/A is usually lumped to a geometry factor k. Since for most electrode configurations other than parallel plate electrodes it is complicate or impossible to calculate distance and surface area, an experimental calibration procedure is required. Briefly, the electrode is immersed in a material with known resistivity (or conductivity, $\sigma = 1/\rho$) and the geometry factor is then calculated as $k = Z\sigma$ where Z is the measured impedance.

3. Electrode polarization

A metal in contact with electrolyte causes electrochemical reactions. Positively charged metal ions enter the solution leaving the negatively charged electrons behind. This creates a force backwards, keeping the metal ions at the metal surface or even discharging them. Thus, a current exists in both directions, for charging ions (anodic oxidation) and for discharging them (cathodic reduction). Under equilibrium condition both currents are equal resulting in zero net current. This equilibrium will be distorted by an external electric field, yielding a measurable current depending on the electrolyte and the electrode material. It is described in a lumped quantity, the exchange current density. It can range over more than 10 orders of magnitude, making an electrode polarizable or not. For instance gold in physiological electrolyte at room temperature exhibits an exchange current density on the order of 0.1-2 mA / cm^2. A higher current density charges the electrode interface, yielding a layer with immobilized ions which acts as capacitor. The real current through the electrode is the Faraday current. A diffuse layer, called Gouy-Chapman-layer, extends into the bulk of the electrolyte. Due to the rate limiting step of ionic transport by diffusion, its resistance increases with decreasing frequency. A typical locus diagram of a metal electrode is shown in Fig. 3. The line at the right side arises from the diffusive contribution of the Gouy-Chapman-layer (Warburg impedance) while the half circle at the left side is typical for a resistor (Faraday contribution + Warburg impedance) with a parallel capacity (capacity of the electric double layer).

Fig.2 (A) Locus diagram (imaginary part Z" vs. real part Z') of a typical metal electrode and (B) equivalent circuit (R_c – resistor for Faraday contribution of electrode impedance, C_{dl} – capacity of the electric double layer, Z_W – Warburg impedance, R_{bulk} – bulk resistance of the electrolyte)

Since the variability of the electrode impedance is considerably high, it becomes a serious source for measurement errors and misinterpretation [3]. Moreover, as seen in Fig.2, especially at low frequencies, its magnitude becomes very high (up to MΩ), often exceeding this of the material under test. Using the two electrode interface as in Fig.3A, means, that the measured impedance is Z_{MUT} (impedance of the material under test) plus twice the electrode impedance Z_{el}. In such case, the measured impedance mostly depends on the electrode behavior.

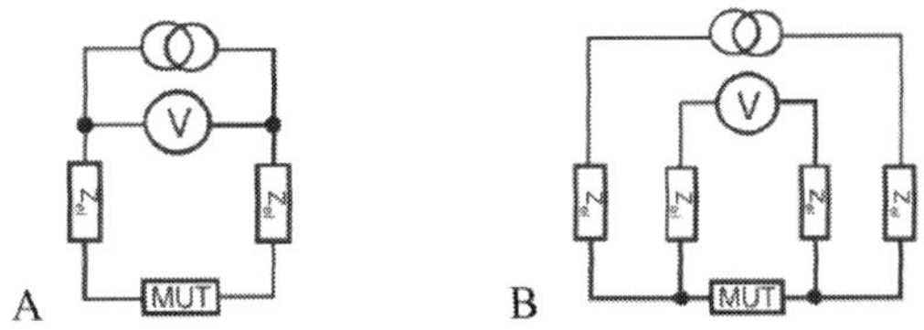

Fig.3 The principle of two (A) and four (B) electrode interface

The application of the current and measurement of voltage at separate electrodes will circumvent this problem (Fig.3B). The voltage drop at the polarized current electrodes (here outer electrodes) does not influence the measured voltage. Since the voltage monitor exhibits high impedance, no voltage drop owing to current flow through the monitor electrodes occurs. The use of monitor electrodes with equal electrochemical behavior keeps the potential difference between electrode and electrolyte equal. Therefore, it vanishes in the measured result due to the difference input of the voltage monitor [4].

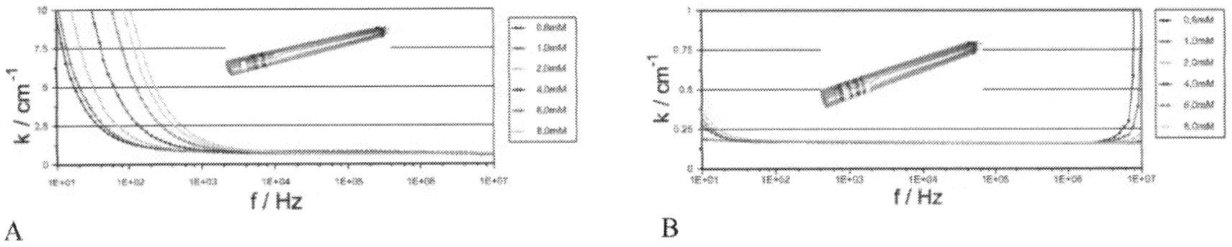

Fig.4 useful frequency range of a comparable (A) two- and (B) four electrode interface. The diameter of the ring electrodes is 6 mm while the distance between two rings is 1 mm. Each ring is machined from stainless steel with a thickness of 1 mm (inner rings) or 2 mm (outer rings).

Often electrodes are designed without testing the frequency range of reliable measurement. A simple test for electrode configuration is the measurement of the geometry factor as function of frequency at different concentrations of the test electrolyte. The most prominent test electrolyte is KCl, a 1:1 electrolyte with almost

equal size and mobility of the ionic species K^+ and Cl^-. Fig.4 gives an example for very similar electrode systems but one with two electrodes (A), the other one using a four electrode interface (B).
As seen in Fig.4B, a four electrode interface extends the useful frequency range to low frequencies by several frequency decades. Therefore, above 5 MHz a increase in k was found for the four electrode interface due to parasitic and stray capacitances. This suggests a preferred use of four electrode systems for precise measurements in the low frequency region (<10 kHz) while two electrode systems are advantageous measurements at higher frequency.

4. Non polarizable electrodes

Electrode polarization happens due to the limited exchange current density. Reversible electrodes are non-polarizable due to the electron transfer by electrochemical reaction. Typically, a metal is in equilibrium with is hardly soluble salt like silver and silver chloride. Another well-known system is Hg/HgCl (calomel). Due to the constant or determined activity of Cl^- - ions in biological systems, Ag/AgCl – electrodes predominate non polarizable electrodes in life science. The electrochemical reaction for Ag/AgCl-electrodes is: $AgCl \rightleftarrows Ag^+ + Cl^-$. At the anode, silver reacts with the chloride to AgCl depositing at the electrode surface while at the cathodic side AgCl is dissociated and metallic silver is deposit at the electrode. The potential difference of the Ag/AgCh-electrode and electrolyte depends only on temperature and the chloride activity and is for physiological condition on the order of -222 mV. In order to keep this potential constant, which is especially important for reference electrodes, chloridized silver wires are bathed by 3 M KCl in a separate compartment which has an ionic connection via glass frit to the material under test.
The key feature of non-polarizable electrodes is the high exchange current density. Besides redox electrodes, iridium sputtered titanium electrodes are used as non-polarizable or at least hardly polarizable electrodes.

5. Electrode noise

A critical property of electrodes is the generation of thermal noise. It is negligible for large electrodes but increases dramatically when electrodes became very small. The noise voltage within a frequency band of Δf is given by $u_n = \sqrt{4k_B T \Delta f R_{electrode}}$. T is the temperature and k_B the Boltzmann-constant. Since the real part of the electrode impedance (Fig.2), $R_{electrode} \propto 1/A_{electrode}$, small electrodes and especially point electrodes as often used in cell physiology exhibit a high noise level. It is even more important for electrical measurements using miniaturized electrodes since the signal level decreases with decreasing electrode surface, yielding a bad signal-to-noise ratio.

7. Conclusion

When constructing electrodes, geometry and material should be chosen with respect to the geometry of the material under test and the expected electrochemical reactions. This involves consideration with respect to aging of electrode materials as well. The simulation of electrode behavior can be only guidance. Electrodes should be tested for reliability and useful frequency range with electrolytic solutions of known conductivity.
Critical problems for impedance measurements are the distinction between electrode behavior and properties of the material under test. Four electrode interfaces can avoid distortion or the measurement in the low frequency range. Electrode aging and biofouling are important issues for long lasting measurements. This raises the question for online cleaning and conditioning of electrodes.
Since electrodes are always sources of distortions while measuring impedance at biological material, both, construction and material should keep these distortions at a minimum.

8. References

1. Foster KR, Schwan HP., Dielectric properties of tissues and biological materials : A critical review., CRC Crit Rev Biomed Eng 1989;17:25-104.
2. Pethig R, Kell DB., The passive electrical properties of biological systems: their significance in physiology, biophysics and biotechnology., Phys Med Biol 1987;32(8):933-70.
3. Zhou H, Tilton RD, White LR., The role of electrode impedance and electrode geometry in the design of microelectrode systems., J Coll Interf Sci 2006;297:819-31.
4. Bragos R, Sarro E, Fontova A, Soley A, Cairo J, Bayes-Genis A, et al., Four versus two-electrode measurement strategies for cell growing and differentiation monitoring using electrical impedance spectroscopy., Conf Proc IEEE Eng Med Biol Soc 2006;1:2106-9.

Software Tool for Analysis of Breathing-Related Errors in Transthoracic Electrical Bioimpedance Spectroscopy Measurements

F Abtahi [1], I C Gyllensten[1, 2], K Lindecrantz[1, 4] and F Seoane[1, 3]

[1]School of Technology and Health, Royal Institute of Technology, SE-141 52 Huddinge, Sweden
[2]Philips Research Europe, High Tech Campus 34, 5656AE, Eindhoven, the Netherlands
[3]School of Engineering, University of Borås, SE-501 90, Borås, Sweden
[4]Department of Clinical Science, Intervention and Technology, Karolinska Institute, SE-141 86 Stockholm, Sweden

E-mail: farhad.abtahi@sth.kth.se

Abstract. During the last decades, Electrical Bioimpedance Spectroscopy (EBIS) has been applied in a range of different applications and mainly using the frequency sweep-technique. Traditionally the tissue under study is considered to be time-invariant and dynamic changes of tissue activity are ignored and instead treated as a noise source. This assumption has not been adequately tested and could have a negative impact and limit the accuracy for impedance monitoring systems. In order to successfully use frequency-sweeping EBIS for monitoring time-variant systems, it is paramount to study the effect of frequency-sweep delay on Cole Model-based analysis. In this work, we present a software tool that can be used to simulate the influence of respiration activity in frequency-sweep EBIS measurements of the human thorax and analyse the effects of the different error sources. Preliminary results indicate that the deviation on the EBIS measurement might be significant at any frequency, and especially in the impedance plane. Therefore the impact on Cole-model analysis might be different depending on method applied for Cole parameter estimation.

Keywords: electrical bioimpedance spectroscopy, EBIS artefacts, frequency-sweep EBIS, Transthoracic Measurements, frequency-sweep spectroscopy of time-variant systems

1. Introduction

Electrical Bioimpedance Spectroscopy (EBIS) is widely used for assessing status or composition of tissue. During the last decades, EBIS estimation by using the frequency-sweep technique, i.e. sweeping through a range of different measuring frequencies, has been applied in a range of different areas e.g. Body Composition Analysis (BCA) [1], tissue characterization for skin cancer detection [2], and impedance-based systems for cardiac monitoring [3, 4]. However, the time during which the signal, from for example cardiac activity, can be considered stationary is often shorter than the measurement time for a frequency-sweep. This leads to both under sampling of important information as indicated by Sanchez et al. [5, 6] for real-time monitoring of myocardium tissue, and to measurement artefacts that might ruin the EBIS measurement.

Traditionally the tissue under study is considered to be time-invariant and dynamic changes of tissue activity are ignored and instead treated as a noise source. The error due to frequency sweep measurements under this false assumption of a time-invariant signal is illustrated in figure 1.

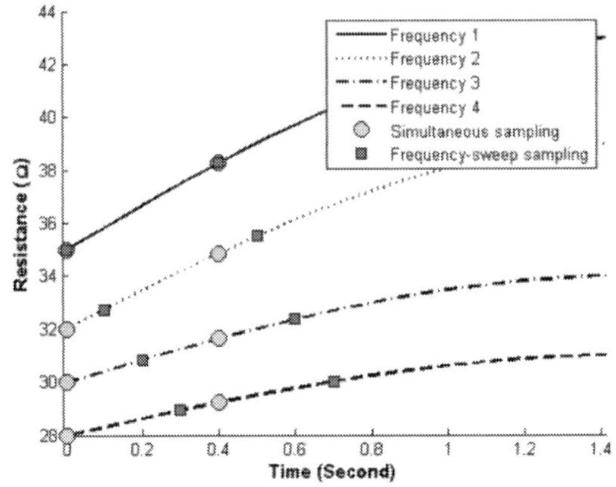

Figure 1. Resistance of thorax at four different frequencies (solid, dotted, dashdot and dashed lines); for simplification respiration is considered the only source of variance in thoracic resistance. True (simultaneously sampled) and frequency-sweep measurements are shown by green circle and red square, respectively. A respiration rate of 10 Breaths/Min and a delay of 100ms between measurements were used for this simulation.

This problem, that could have a negative impact, limiting the accuracy impedance monitoring systems and consequently impeding the proliferation of novel and emerging applications of EBIS has to date not been adequately addressed.

In this work, we present a software tool that can be used to simulate the influence of respiratory activity in frequency-sweep transthoracic EBIS measurements and to analyse the effects of the different error sources. The usefulness of the tool is demonstrated by an example of deviations and errors obtained in estimation of Cole parameters.

2. Methods

In this section, the models for creating synthetic EBIS measurements and methods for estimating the Cole parameters and calculating errors are described. EBI data was generated using the Cole equation with biologically plausible values. To model the modulation caused by the respiratory activity, the Cole-model was augmented by adding respiration related time-varying impedance.

2.1. Cole Equation

The Cole Equation (1) is a model for impedance dispersion, which was introduced by Cole in 1940 [7] to fit experimentally obtained Electrical Bioimpedance (EBI) measurements of biological tissue. The model uses four parameters to describe the impedance dispersion, R_0, R_∞ α and τ. Where R_0 and R_∞ has a direct interpretation as the impedance response at zero and infinite frequency, the α describes the models departure from a simple RC circuit model and τ is the inverse of the natural characteristic frequency ω_c [7].

$$Z_{COLE}(\omega) = R_\infty + \frac{R_0 - R_\infty}{1+(j\omega\tau)^\alpha} \quad (1)$$

The Cole equation is non-linear in the frequency domain and the impedance values generated by $Z_{COLE}(\omega)$ in (1) represent a supressed semi-circle when plotted in the impedance plane i.e. resistance vs. reactance [5].

2.2. Modelling and Synthesis of EBIS data

As mentioned earlier, generation of EBI data is simplified by considering time-invariant model for thorax and additive time-variant respiration impedance. Additive white Gaussian noise with mean μ and standard deviation σ specified by user is also included to mimic measurement noise.

$$Z_{Total} = Z_{Thorax} + Z_{Respiration} + Noise \quad (2)$$

Thoracic impedance is generated by the tool using the Cole function and user specified values for R_0, R_∞, α and f_c ($1/2\pi\tau$). Assume T as duration time of simulation, T_d as the delay between the measurements at different frequencies in the sweep (step delay) and N as the number of frequencies in the sweep. The EBIS data for the thorax during simulation time is defined as Z_{Thorax}. While each row and column shows the value at the corresponding frequency and time, respectively.

$$t = (1\ T_d\ 2*T_d\ 3*T_d\ \dots\ T)_M \quad , \quad M = \frac{T}{T_d} \quad , f_{Resp} = \frac{Respiration\ Rate}{60}$$

$$Z_{Thorax} = \begin{pmatrix} Z_{Cole\ Thorax}(1) & \cdots & Z_{Cole\ Thorax}(1) \\ \vdots & \ddots & \vdots \\ Z_{Cole\ Thorax}(N) & \cdots & Z_{Cole\ Thorax}(N) \end{pmatrix}_{N\times M}$$

According to Brown [8] the impedance change associated with respiration has an inverse relation to frequency (it decreases as the measurement frequency increases). Respiration impedance $Z_{Respiration}$ is simply modelled as a sinusoid with amplitude generated by a user specified Cole function and phase $Resp_d$ specified by user. $Resp_d$ is used to simulate spectrum measurements which are not synchronized with respiration cycle.

$$Z_{Respiration} = \begin{pmatrix} Z_{Cole\ Resp}(2\pi f_0)Sin(2\pi f_{Resp}t(1) + Resp_d) & \cdots & Z_{Cole\ Resp}(2\pi f_0)Sin(2\pi f_{Resp}t(M) + Resp_d) \\ \vdots & \ddots & \vdots \\ Z_{Cole\ Resp}(2\pi f_\infty)Sin(2\pi f_{Resp}t(1) + Resp_d) & \cdots & Z_{Cole\ Resp}(2\pi f_\infty)Sin(2\pi f_{Resp}t(M) + Resp_d) \end{pmatrix}_{N\times M}$$

Simulation of EBIS data has been done by using $Spectra_{Measure}$ and $Spectra_{True}$. True spectra are simply assigned to value of Z_{Total} for each frequency at the same time. On the other hand, $Spectra_{Measure}$ elements are assigned to value of Z_{Total} for each frequency by considering measurement delay of T_d.

$$Spectra_{True} = \begin{pmatrix} Z_{Total(1,1)} & \cdots Z_{Total(1,4)} \cdots & Z_{Total(1,M)} \\ \vdots & \ddots & \vdots \\ Z_{Total(N,1)} & \cdots Z_{Total(N,4)} \cdots & Z_{Total(N,M)} \end{pmatrix}_{N\times M}$$

$$Spectra_{Measure} = \begin{pmatrix} Z_{Total(1,1)} & \cdots Z_{Total(1,2N+1)} \cdots & Z_{Total(1,M-N)} \\ \vdots & \ddots & \vdots \\ Z_{Total(N,N)} & \cdots Z_{Total(N,3N)} \cdots & Z_{Total(N,M)} \end{pmatrix}_{N\times(M/N)}$$

2.3. *Error Analysis*

The analysis of measurement error due to the none-simultaneous measurements has been done by studying of Root Mean Square Error (RMSE) and comparing R-fitted Cole parameters for each true and measured spectrum. The RMSE of each measured spectrum compared to true spectrum has been calculated by using Equation 3. Where N is the number of frequency steps and $R(\omega)$ and $jX(\omega)$ are resistance and reactance, respectively.

$$RMSE = \sqrt{\frac{\sum(R-\bar{R})^2}{N}} + \sqrt{\frac{\sum(X-\bar{X})^2}{N}} \quad (3)$$

2.4. *Software Description*

The software is developed by using *MATLAB 7.12 (Release 2011a)* and its user interface allows us to change simulation parameters and settings, simulate the spectroscopy measurements and plot true and measured spectrum in different planes. The error analysis produces a report containing the RMSE and

R-fitted estimation of Cole parameters compared to the true spectrum. A screenshot of the software is depicted in Figure 2.

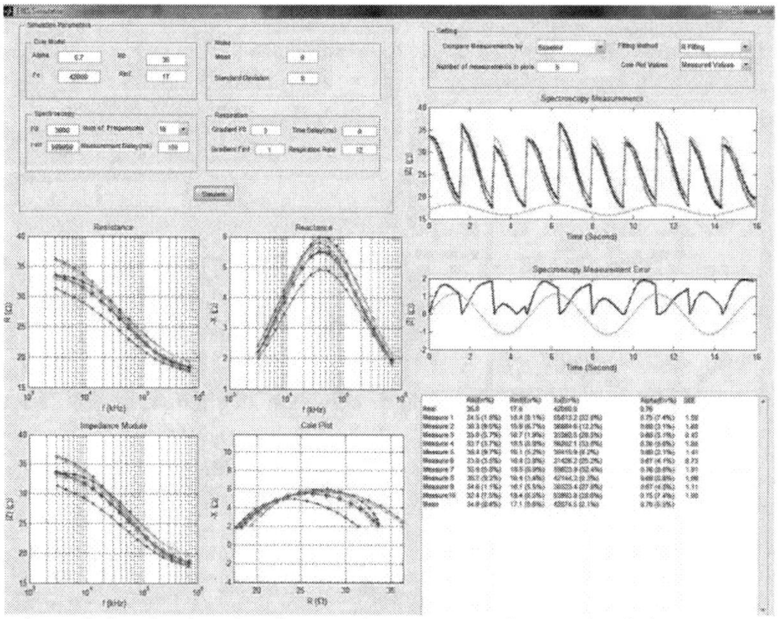

Figure 2. Screenshot of the developed software; the user can change simulation parameters and settings, simulate the spectroscopy measurements and plot true and measured spectra in different planes.

3. Results

The results for a simple simulation case are summarized in this section. The thorax and respiration impedance are generated by using Cole parameters $R_0 = 35, R_\infty = 17, f_c = 42000\ and\ \alpha = 0.7$ and $R_0 = 3\ R_\infty = 1, f_c = 42000\ and\ \alpha = 0.7$, respectively.

The simulation has been done in the frequency range of 5-1000 kHz. The spectroscopy frequency-sweep has been done at 16 frequencies with a step delay of $T_d = 100\ ms$. The respiration rate is considered at 12 per minute while mean and standard deviation of noise are set to zero.

The spectroscopy measurements and corresponding errors are shown in Figure 3. It can be clearly seen that the error is not negligible for any of the measurements and that the more prominent errors occur at maximum inhalation and exhalation. Figure 4 further illustrates the measured spectrum in different planes (impedance-plane and frequency-plane for resistance, reactance and absolute impedance). The plots in the impedance plane exhibit the variations between the measurements. It is clearly noticeable not only the Cole parameters R_0 and R_∞ but also for the characteristic frequency f_c and α.

Table 1. Summary of the Error Analysis for Spectroscopy Measurements

	R0 (Error %)	Rinf (Error %)	Fc (Error %)	Alpha (Error %)	RMSE
True Spectrum 1	35.3	17.1	42000.0	0.70	-------
Measured Spectrum 1	34.8 (1.4%)	18.3 (7.0%)	55420.8 (32.0%)	0.75 (7.0%)	1.48
True Spectrum 2	37.6	17.9	42000.0	0.70	-------
Measured Spectrum 2	38.3 (1.8%)	15.8 (11.6%)	35952.1 (14.4%)	0.68 (3.5%)	1.76
True Spectrum 3	32.5	16.2	42000.0	0.70	-------
Measured Spectrum 3	32.7 (0.7%)	16.8 (4.0%)	34421.9 (18.0%)	0.68 (2.6%)	0.41
True Spectrum 4	34.5	16.8	42000.0	0.70	-------
Measured Spectrum 4	34.0 (1.6%)	18.5 (9.5%)	56156.0 (33.7%)	0.75 (7.8%)	1.73
True Spectrum 5	37.9	18.0	42000.0	0.70	-------
Measured Spectrum 5	38.4 (1.4%)	16.0 (10.9%)	38380.0 (8.6%)	0.68 (2.5%)	1.52

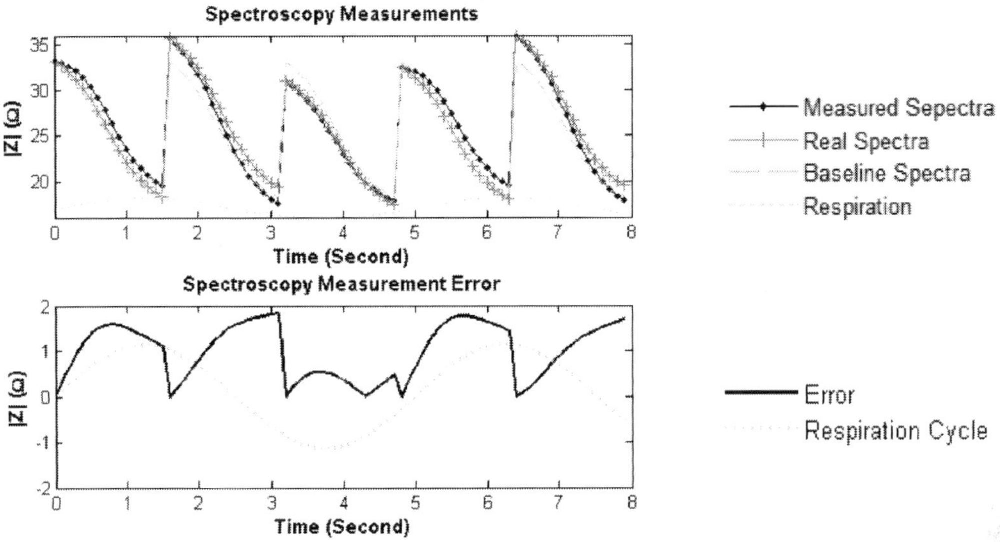

Figure 3. Simulation of spectroscopy measurements; True Spectra, Measured Spectra, Baseline (Thorax) Spectra and Respiration cycle are shown in the upper graph. Measurement error compared to the True spectra and the respiration cycle as functions of time are shown in the bottom graph.

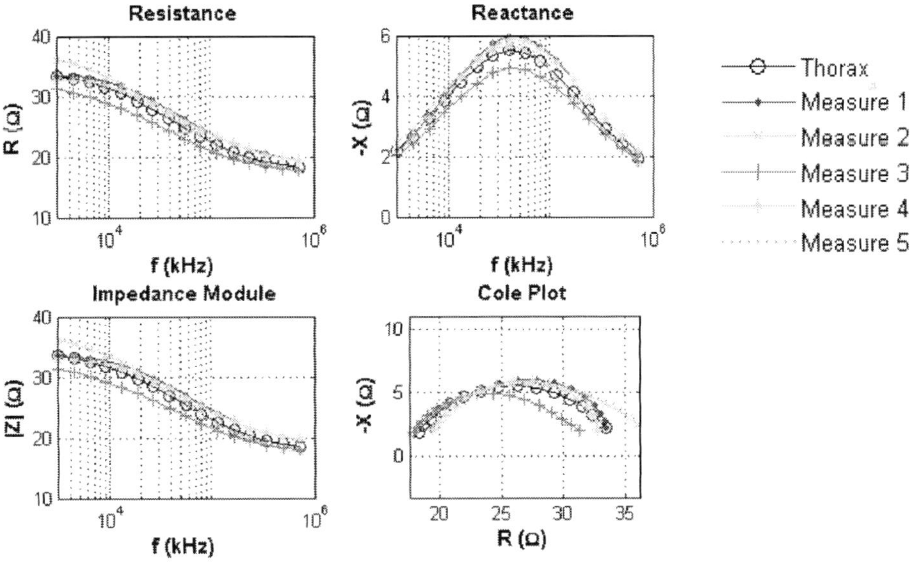

Figure 4. Graphs show measured and time-invariant Thorax spectrum in different planes **resistance** $\boldsymbol{R(\omega)}$, $\boldsymbol{reactance\, jX(\omega)}$, $\boldsymbol{impedance|Z(\omega)|}$ and the Cole Plot.

The RMSE obtained for the measured and the true spectrum and the relative error the estimation of the Cole parameters are summarized in Table 1. It is clear that even for the third measurement which has minimum RMSE; the error in estimated parameters is significant. Note that the error in the characteristic frequency is large in most of the measurements.

4. Discussion

The simulation and error analysis of the influence of the time between frequency samples on EBIS measurements indicate that Cole-based analysis done without considering measurement delay can lead to erroneous estimation of the values for the Cole parameters. All parameters suffer from erroneous estimation but the characteristic frequency f_C appears to be the most sensitive parameter.

The deviation of the EBIS measurement is significant in the impedance spectrum for all frequencies and it is more manifested in the impedance plane. Hence it is expected that the performance of Cole parameter estimation methods in different planes might be different. For instance the observed deviations of the impedance data suggest that estimating the Cole parameters from the immittance spectrum instead than in the impedance plane would produce a better result.

The simulation of the respiration-related impedance can be done by using sinusoid synthetized impedance as described in methods or by using empirically respiration impedance measurements. Currently experimental data about transthoracic EBIS measurements during breathing and different activities is collected, thus increasing the capabilities of the software tool to perform more realistic, deeper, and more complex simulations and analysis.

In addition to the search for methods and technics to enhance the analysis of EBIS measurements of the thorax, the software tool can be useful in the design of measurement protocols to be used in studies requiring EBIs transthoracic measurements.

5. Conclusion

The use of the implemented software tool has helped us to confirm our hypothesis that the measurement time delay caused by sweeping between frequencies on the acquisition of EBIS measurements influences on the measured spectrum creating a measurement artefact and consequently taint any subsequent data analysis using Cole parameters characterization. These results encourage to perform more complex studies with more realistic and advanced settings and to include experimental validation. It is expected that as a result of these studies effective tools and methods for removing or avoiding these measurement artefacts and their consequent wrong-analysis will be developed to increase the accuracy of EBIS instrumentation and fostering the proliferation of EBIS-based systems for personalized health monitoring.

References

1. van Marken Lichtenbelt, W.D., et al., *Validation of bioelectrical-impedance measurements as a method to estimate body-water compartments.* The American journal of clinical nutrition, 1994. **60**(2): p. 159-166.
2. Aberg, P., et al., *Skin cancer identification using multifrequency electrical impedance-a potential screening tool.* Biomedical Engineering, IEEE Transactions on, 2004. **51**(12): p. 2097-2102.
3. Freimark, D., et al., *Monitoring lung fluid content in CHF patients under intravenous diuretics treatment using bio-impedance measurements.* Physiological measurement, 2007. **28**: p. S269.
4. Reiter, H., J. Muehlsteff, and A. Sipila. *Medical application and clinical validation for reliable and trustworthy physiological monitoring using functional textiles: Experience from the HeartCycle and MyHeart project.* 2011. IEEE.
5. Sanchez, B., et al., *Novel Estimation of the Electrical Bioimpedance using the Local Polynomial Method. Application to in-vivo real-time Myocardium Tissue Impedance Characterization during the Cardiac Cycle.* Biomedical Engineering, IEEE Transactions on, 2011(99): p. 1-1.
6. Sanchez, B., et al. *In-cycle myocardium tissue electrical impedance monitoring using broadband impedance spectroscopy.* 2011. IEEE.
7. Cole, K.S. *Permeability and impermeability of cell membranes for ions.* 1940. Cold Spring Harbor Laboratory Press.
8. Brown, B.H., et al., *Cardiac and respiratory related electrical impedance changes in the human thorax.* Biomedical Engineering, IEEE Transactions on, 1994. **41**(8): p. 729-734.

Quantitative Label-Free Cell Proliferation Tracking with a Versatile Electrochemical Impedance Detection Platform

C Caviglia[1], M Carminati[2], A Heiskanen[1], M Vergani[2], G Ferrari[2], M Sampietro[2], T L Andresen[1], J Emnéus[1]

[1] Department of Micro- and Nanotechnology, Technical University of Denmark, Lyngby, Denmark
[2] Dipartimento di Elettronica e Informazione, Politecnico di Milano, Milano, Italy

Email: claudia.caviglia@nanotech.dtu.dk

Abstract. Since the use of impedance measurements for label-free monitoring of cells has become widespread but still the choice of sensing configuration is not unique though crucial for a quantitative interpretation of data, we demonstrate the application of a novel custom multipotentiostat platform to study optimal detection strategies. Electrochemical Impedance Spectroscopy (EIS) has been used to monitor and compare adhesion of different cell lines. HeLa cells and 3T3 fibroblasts have been cultured for 12 hours on interdigitated electrode arrays integrated into a tailor-made cell culture platform. Both vertical and coplanar interdigitated sensing configuration approaches have been used and compared on the same cell populations.

1. Introduction

Tracking bio-impedance at the microscale is becoming a ubiquitous laboratory tool to monitor in real time and with no need for labels the health and dynamic response of adherent cells to chemical or mechanical triggers, both at single cell level and for large colonies. Automatic and heavily parallel assays can be performed with beneficial impact for instance in diagnostics and drug discovery.

As first introduced by Giaever and Keese [1], cells are typically grown on planar microelectrodes and their presence and proliferation alters the interfacial impedance that is recoded over time, commonly at a single frequency. Given the amount and importance of the physiological parameters [2] that can be monitored in comparison with the intrinsically non-specific nature of impedance detection (sensitive to any environmental changes), we study the optimal detection conditions for different model cell lines with an electrochemical platform originally designed for multiparametric investigations of neural stem cells [3]. In particular, two alternative sensing configurations are experimentally compared by recording impedance spectra for both HeLa cells and mouse fibroblasts using different cell seeding densities. To our best knowledge, this is the first time two sensing configurations are applied simultaneously for the same cell populations.

2. Theory

Avoiding the addition of possibly toxic or otherwise interfering redox molecules to the cell culture medium, no Faradaic phenomena are present and the small-signal impedance equivalent of the electrochemical interface is simply constituted by the double layer capacitance (for standard *Phosphate Buffered Saline*, PBS, it can be estimated with a specific interfacial capacitance of 0.1

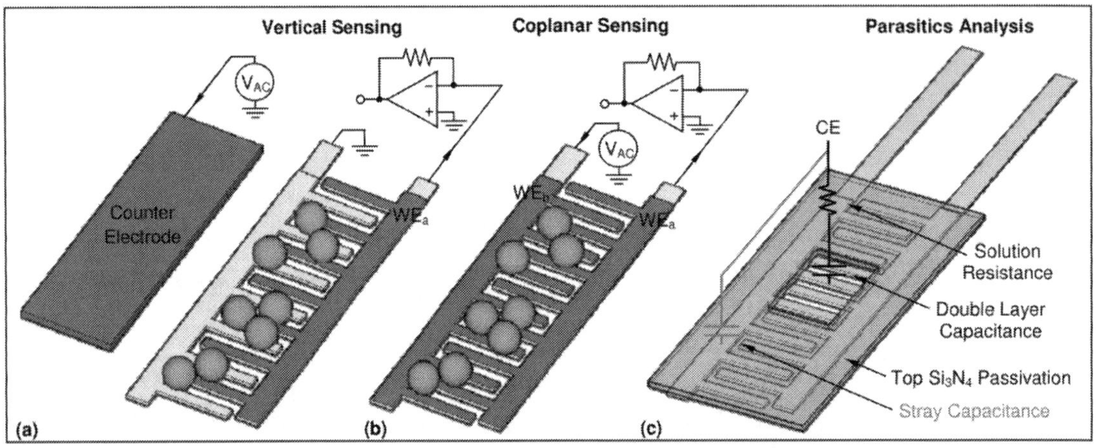

Figure 1. Simultaneous comparison for cell impedance sensing: (a) ECIS vertical configuration vs. (b) interdigitated coplanar configuration and (c) modelling of the parasitic capacitance of the electrodes.

pF/μm^2) and the resistance of the ionic solution (PBS conductivity 1.5 S/m). The presence of cells, whose membrane can be considered insulating for frequencies below 1-10 MHz since the membrane capacitance is still not bypassed, obstructs the ionic conductive paths, thus resulting in a monotonic increase in impedance magnitude with the number of cells. Several lumped-parameter equivalent models have been proposed to describe the behaviour of a cell monolayer: typically by an additive intermediate R||C couple (or R||CPE). As more complex models are valid only for a given type of cells (the amount of adhesion and the morphology), we chose to maintain the simple R-C series model also in the presence of cells and to track its "small" variations due to such a growing barrier.

As illustrated in Fig. 1a, in the classic vertical configuration, the sinusoidal voltage is applied to the large and distant counter electrode and the current is sensed at the working electrode. Instead, in the coplanar approach (Fig. 1b), the symmetrical impedance between closely-spaced interdigitated combs is measured. With the proposed platform, it is possible to switch from one configuration to the other in order to perform both measurements on the same comb (green WEa in Fig. 1) covered with the same cell population, approximately at the same time (the time shift between consecutive measurements is a few seconds, i.e. negligible with respect to the time scale of the experiments) thus quantitatively comparing their sensitivity in tracking cell proliferation. It is expected that, at a high enough frequency to enable shorting of the double layer capacitance, the coplanar scheme provides a higher sensitivity due to the higher confinement of the field lines. Unfortunately, the latter requires a higher operating frequency as the solution resistance is smaller (and capacitance is halved). Finally, the impact of parasitics has been carefully analysed and taken into account in the design of the microelectrodes. In fact, as illustrated in Fig. 1c and extensively discussed in [4], besides the connecting wires, the presence of a conductive silicon substrate and long and large metal leads covered by the Si_3N_4 passivation layer in contact with the solution produces large stray capacitances that might significantly degrade the instrument performance.

3. Materials and Methods

The monitoring platform is composed of two main elements: a cell culture device integrated with a microelectrode chip and a multichannel bipotentiostat [3]. The cell culture device consists of a micromilled poly(methyl methacrylate) (PMMA) holder characterized by an upper vial, containing up to 600 μL of volume, that allows cell seeding and culture medium exchange. The gold microelectrode chips have been fabricated using a standard lift-off process on a Si/SiO_2 substrate. On each chip, 12 electrode arrays are defined, each consisting of an interdigitated electrode (IDE) (functioning as the working electrode (WE)), a counter electrode (CE) and a reference electrode (RE). The WEs consist of 12 500 μm long fingers having 10 μm width and spacing.

HeLa cells and 3T3 Fibroblasts were grown in 75-cm^2 cell culture flasks at 37 °C with 5% CO_2 in a humidified atmosphere. Dulbecco's Modified Eagle's medium supplemented with 10% fetal bovine serum and 1% penicillin/streptomycin was used as culture medium for both cell lines. After reaching 80-90% confluency, the cells were washed with Dulbecco's phosphate buffered saline (PBS) and harvested through trypsinization.

Electrochemical impedance spectroscopic (EIS) tracking was used as a non-invasive biophysical technique to continuously monitor and compare cell adhesion and proliferation of HeLa cells and 3T3 Fibroblasts. Prior to using, in order to clean the gold surface of the microelectrode chip from the ambient contaminants, the chips were treated for 10 minutes with a solution containing 25% H_2O_2/ 50 mM KOH [5]. The culture chamber was sterilized using 500 mM NaOH. To facilitate cell adhesion, the electrode chip was coated with poly-L-lysine (10 µg/mL). Two different cell densities (about $2.5{\cdot}10^5$ and $3.5{\cdot}10^5$ cells) were initially seeded into the chamber and the measurements were performed while the monitoring platform was kept in the incubator. During the first 4 hours after cell seeding, impedance spectra were acquired from each sensor element every 20 minutes, then every hour. In order to limit the current flowing through the cells, a 200 µV sinusoidal signal was applied and 30 points were acquired in the frequency range between 100 Hz and 100k Hz, each with an averaging time of 2s. After 24 hours, cells were fixed (for 1 hour with 2% glutaraldehyde solution in PBS, followed by rinsing with PBS and cell culture tested water) and dehydrated with ethanol and imaged using SEM .

All calculations and graphs were obtained using MATLAB software. For each experiment, data acquired from the different sensor elements were averaged and processed to derive a parameter termed Cell Index (CI) [6] that represents the relative impedance change in the system. For each frequency and for each time point, the CI is calculated as $CI(t,f) = (|Z(t,f)| / |Z(0,f)|) - 1$, where $|Z(0,f)|$ is the magnitude of impedance acquired at the beginning of the experiment, when cells were not yet seeded into the well and $|Z(t,f)|$ is the magnitude of impedance acquired after cell seeding at different time points. Data are represented as mean between experiments ± standard error of mean.

4. Experimental Results

In order to evaluate the behavior of the detection platform and compare the two alternative detection configurations, electrochemical impedance spectroscopic monitoring of two different cell lines was performed. Fig. 2a shows the increase in CI for HeLa cell adhesion and spreading during 12 hours. The spectra were acquired using both the interdigitated coplanar and vertical configuration mode, and the CI was calculated and plotted at the characteristic peak frequency (about 24kHz and 100kHz for the vertical and the coplanar configuration, respectively). For both sensing configurations, the maximum value of CI was reached 10 hours after cell seeding. As can be observed, the coplanar mode provides

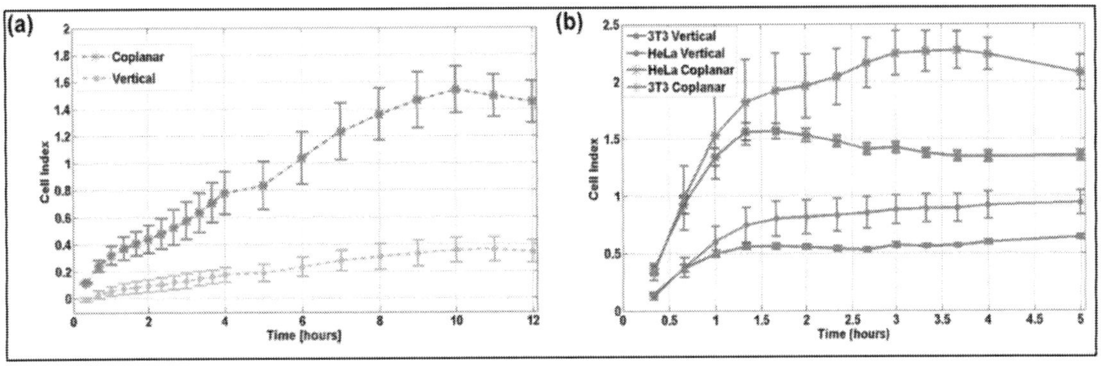

Figure 2. (a) Monitoring of HeLa cell adhesion and spreading: CI increase over time at 24kHz (vertical) and 100kHz (coplanar) (initially ~$2.5{\cdot}10^5$ cells were seeded). (b) Comparison between HeLa cell and fibroblast adhesion monitoring using vertical and coplanar configuration (initially ~$3.5{\cdot}10^5$ cells were seeded). 9 and 5 arrays were used for the statistical evaluation of HeLa and fibroblasts, respectively, the error bars represent standard error of mean.

Figure 3. SEM images of HeLa cells on the microelectrode chip 12 hours after cell seeding: (a) a population of HeLa cells and (b) a magnified view of the cell population.

the highest sensitivity reaching the CI of about 1.6. For the vertical configuration, the corresponding value was about 0.45. Fig. 3 shows SEM images of HeLa cells after 12 hours of measurements.

A comparison between the two cell lines, characterized by different adhesion properties, is shown in Fig. 2b. The CI trend over 5 hours of both HeLa cells and fibroblasts confirms that the interdigitated coplanar configuration provides the highest sensitivity for monitoring (CI values 5 hours after cell seeding for HeLa cells and fibroblasts were 0.65 ± 0.02 and 0.9 ± 0.1 using the vertical configuration, as well as 1.4 ± 0.05 and 2 ± 0.15 using coplanar configuration). Moreover, under the same culture conditions and the same cell density, fibroblasts show higher CI values, confirming their stronger adhesion in comparison with HeLa cells. A slightly higher standard error of mean is also observed in the case of fibroblasts, probably due to a larger variability of cell morphology.

5. Conclusions

The presented electrochemical impedance monitoring system allows parallel measurements to study biological phenomena. In order to optimize cell proliferation assays, two alternative configuration modes have been experimentally compared in parallel using different cell lines. It has been clearly demonstrated that for monitoring cell adhesion, the coplanar configuration provides the highest sensitivity and that, as expected, cell lines with higher adhesion to the electrodes provide a higher impedance variation. These experimental results validate the use of the platform and the adopted simple and general equivalent impedance model, representing a quantitative basis for future investigations of more subtle effects such as differentiation of stem cells or discrimination between necrosis and apoptosis.

6. References

[1] I. Giaever, C.R. Keese, "Monitoring fibroblast behavior in tissue culture with an applied electric field," PNAS, vol. 81, pp. 3761-3764, 1984.

[2] C. Spegel, A. Heiskanen, L.H. Dæhli Skjolding, J. Emnéus, "Chip Based Electroanalytical Systems for Cell Analysis – a ***invited*** review", Electroanalysis 20, 680-702, 2008.

[3] M. Vergani *et al.*, "Multichannel bipotentiostat integrated with a microfluidic platform for electrochemical real-time monitoring of cell cultures," IEEE Trans. Biomed. Circuits Syst., *DOI: 10.1109/TBCAS.2012.2187783.*

[4] M. Carminati, M. Vergani, G. Ferrari, L. Caranzi, M. Caironi, M. Sampietro, "Accuracy and resolution limits in quartz and silicon substrates with microelectrodes for electrochemical biosensors" Sens. Actuat. B, *in press.*

[5] L. M. Fischer, M. Tenje, A. R. Heiskanen, N. Masuda, J. Castillo, A. Bentien, J. Emnéus, M. H. Jakobsen, A. Boisen, "Gold cleaning methods for electrochemical detection applications, Microelectron. Eng. 2009, 86, 1282-1285.

[6] K. Solly, X. Wang, X. Xu, B. Strulovici, W. Zheng, "Application of real-time cell electronic sensing (RT-CES) technology to cell-based assays," Assay Drug Dev. Technol. Vol. 2, pp. 363-373, 2004.

Mirrored Modified Howland Circuit for Bioimpedance Applications: Analytical Analysis

P Bertemes-Filho[1], L H Negri[2], A. Felipe[2] and V C Vincence[2]
[1, 2]State University of Santa Catarina, Dept. Electrical Engineering,
Campus Universitario Prof. Avelino Marcante, Joinville-SC, 89219710, Brazil

E-mail: bertemes@joinville.udesc.br

Abstract. Multifrequency Electrical Bioimpedance (MEB) has been widely used as a non-invasive technique for characterizing tissues. Most MEB systems use wideband current sources for injecting current to load and instrumentation amplifiers for measuring the resultant potential. Current sources should present intrinsically high output impedance in a very wide frequency range. The objective of this work is to investigate the performance of the Mirrored Modified Howland Current Source (MMHCS) by comparing the analytical solution with the SPICE simulation. It was implemented four MMHCS circuits by using four different operational amplifiers each one. The output current was set to 1 mAp (peak) in the frequency from 1 Hz to 100 MHz. Both techniques presented similar results at lower frequencies. It can be concluded that the output impedance of the circuit is highly dependent of the open-loop gain of the operation amplifier. The analytical solution showed that it is possible to project a current source by using only theirs output current and impedance equations.

1. Introduction

Multifrequency Electrical Bioimpedance (MEB), also called Spectroscopy, has been widely used as a non-invasive technique for detecting cancerous tissues [1,2,3], tumors [4], meningitis [5] and brain cellular oedema [6], for analyzing body composition [7] and bovine milk quality [8]. The EIS is considered a fast, inexpensive, practical and efficient technique [9], which also can be found available for emerging wearable applications [10] and for on-chip systems such as the AD5933 [11].

Most EIS systems consist of applying a multi-frequency sinusoidal current of constant amplitude in the tissue sample, measuring the resulting potential and then calculating the transfer impedance (Zt) [9]. In order to get an accurate calculated transfer impedance, it is necessary to assure that the injecting current has a constant amplitude over a wide frequency range, which may be obtained by using a current source with high output impedance [12]. However, stray capacitances reduce the current amplitude at higher frequencies [13]. High performance current source with a wide bandwidth by using operational amplifiers has been widely developed [14,15]. Current sources in EIS may be based on the operational transconductance amplifier (OTA) approach [16,17], but they suffer from linearity and common mode voltage at the load. Most current sources in EIS use the modified Howland structure [18].

Our targeted measuring applications require an operating frequency band of 100 Hz to 1 MHz and current amplitudes of up to 1 mAp (peak). The objective of this work is to investigate the

[1] To whom any correspondence should be addressed.

analytical solution of the output current and impedance from a modified Howland current source by considering the non-idealities of the operational amplifier. It also compares SPICE simulations to the analytical solution.

2. Materials and Method

2.1. Modeling of the modified Howland circuit

The modified Howland current source is shown in *figure* 1. The gain is set by R_5, modified by the ratio of R_1/R_2 (which is typically 1/1). Consequently it can be used low values for R_5, and keep all the other resistors high in value, such as 100k or 1 MΩ. In this type of circuit, it can be noted that it is not just the ratio of R_3/R_4 that must match R_1/R_2, but it is also the ratio $R_3/(R_4+R_5)$ that must be equal to R_1/R_2. By doing an intuitive analysis as mentioned above, it can be found that if $R_1=R_2$, then R_4 will normally be (R_3-R_5). Conversely, R_3 can be made a little higher, to get the gain to balance out. Usually it is put a potentiometer in series with R_3.

In order to get good Common Mode Rejection Ratio (CMRR) and high output impedance it is necessary to trim the resistors and choose FET inputs amplifiers. On the other hand, the non-ideal characteristics of the operational amplifiers (i.e., input impedance Z_{in}, output impedance Z_o and open-loop gain A) reduce frequency bandwidth of both output current and impedance.

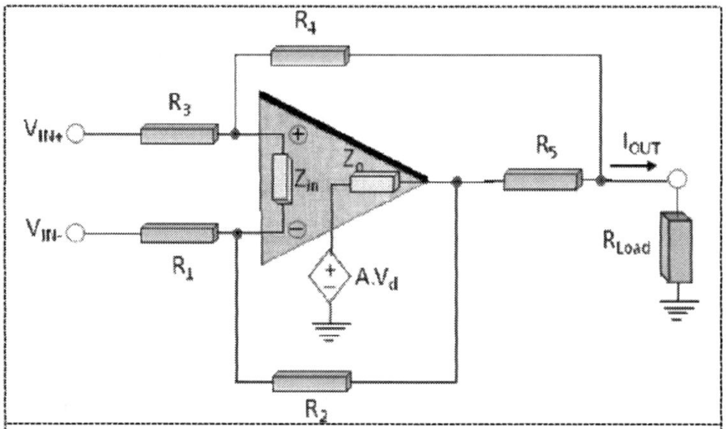

Figure 1. Schematic diagram of the modified Howland current source with grounded load.

The output current I_{out}, which is also called Norton current, is calculated by short-circuiting the output voltage V_L of the circuit and grounding the input signal V_{IN+}. The Norton current I_{out} is shown in (1), where $R_1=R_2=R_4=R$, $R_5=r$ and $R_3=R+r$.

$$\frac{I_{out}}{V_{IN}} = -\frac{(2Z_{in}A-r-2Z_o)R^2+(Z_{in}rA-r^2-3Z_or-2Z_{in}Z_o)R-Z_or-Z_{in}Z_or}{4(r+Z_o)R^3+(Z_{in}(2rA+4Z_o)+3r^2+(6Z_o+4Z_{in})r)R^2+(Z_{in}r(rA+4Z_o)+2(Z_o+Z_{in})r^2)R+Z_{in}Z_or^2} \qquad (1)$$

On the other hand, the output impedance Z_{out} can be calculated by short-circuiting the input V_{IN-} and adding a voltage source at the output of the circuit. The complete equation for Z_{out} is shown in (2), where $R_1=R_2=R_4=R$, $R_5=r$, $R_3=R+r$ and $V_{IN+}=0$.

$$\frac{1}{Z_{out}} = \frac{4R^3+(6r+4(Z_o+Z_{in}))R^2+\left(2r^2+4((Z_o+Z_{in})r+Z_oZ_{in})\right)R+Z_or(r+2Z_{in})}{4(r+Z_o)R^3+(Z_{in}(2rA+4Z_o)+3r^2+(6Z_o+4Z_{in})r)R^2+\left(2Z_or^2+Z_{in}\left((A+2)r^2+4Z_or\right)\right)R+Z_{in}Z_or^2} \qquad (2)$$

2.2. Mirrored Modified Howland Current Source (MMHCS)

A MMHCS circuit consists of two wide bandwidth operational amplifiers, as shown in Figure 2. The MMHCS is a voltage controlled current source with a floating load R_{Load} and it consists of two symmetrical modified Howland structures. Both output current and impedance were investigated by using the SPICE simulator. Four types of operational amplifiers were used in the simulations, as shown in table 1. For simulations purpose, the circuit is set to supply an output current $\pm I_{OUT}$ of 1 mAp (peak) over the frequency range 1 Hz to 100 MHz, which is ideally controlled by the ratio between the input voltage V_{in} (V_1=V_2= 1 Vp) and the resistor *r* (= 1 kΩ), where R_1=R_2=R_4=R=100 kΩ, R5=r and R_3=R+r.

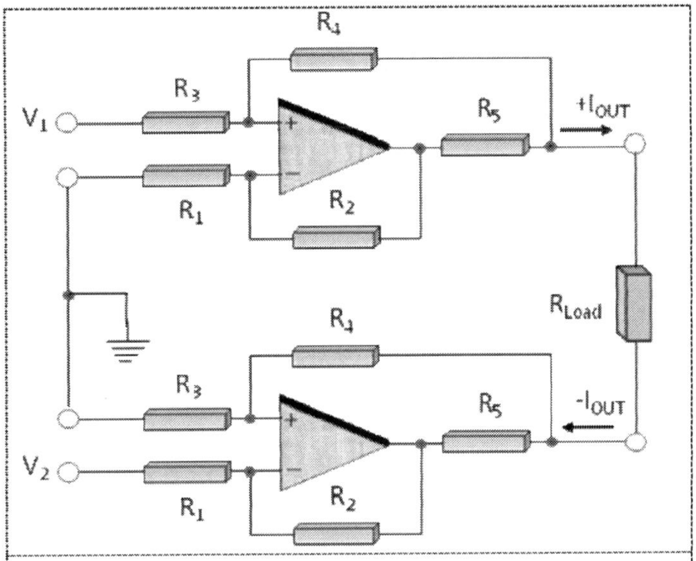

Figure 2. Schematic diagram of the simulated mirrored modified Howland current source.

The output impedance of this circuit is frequency dependent, as shown in (2). It was taken into account the non-ideal characteristics of the operational amplifier, such as: input impedance Z_{in}, output impedance Z_0, and the open-loop gain A (see Table 1). The values from Table 1 are used to calculate the analytical solutions for both output current and impedance of the MMHCS circuit.

Table 1. Values used for calculating the magnitude of both I_{out} and Z_{out}.

	$Z_{in}(s)=R_{in}/(1+sR_{in}C_{in})$			$A(s)=A_0/(1+s/\omega_{c0})$	
	R_{in} (Ω)	C_{in} (pF)	Z_0 (Ω)	A_0 (dB)	ω_{c0} (Hz)
OPA655	1G	1.2	0.04	58	400π k
TL081	1G	0	10	105	50π
OP07	50M	0	60	113	8π
uA741	2M	1.4	75	97	30π

Note. A_0 of the opamp has been extracted from the plots of the open loop gain versus frequency included in the datasheet.

3. Results

Both analytical and simulated results of the output current are showed in Figures 3(a) and 3(b), respectively. It can be seen that the output current keeps constant and equal to 1 mAp up to approximately 100 kHz for both results. However, SPICE simulations have showed a wider frequency response for the operational amplifier OPA655. It can also be observed that the operational amplifier μA741 showed analytically the worst frequency response whereas OP07 in the SPICE simulations.

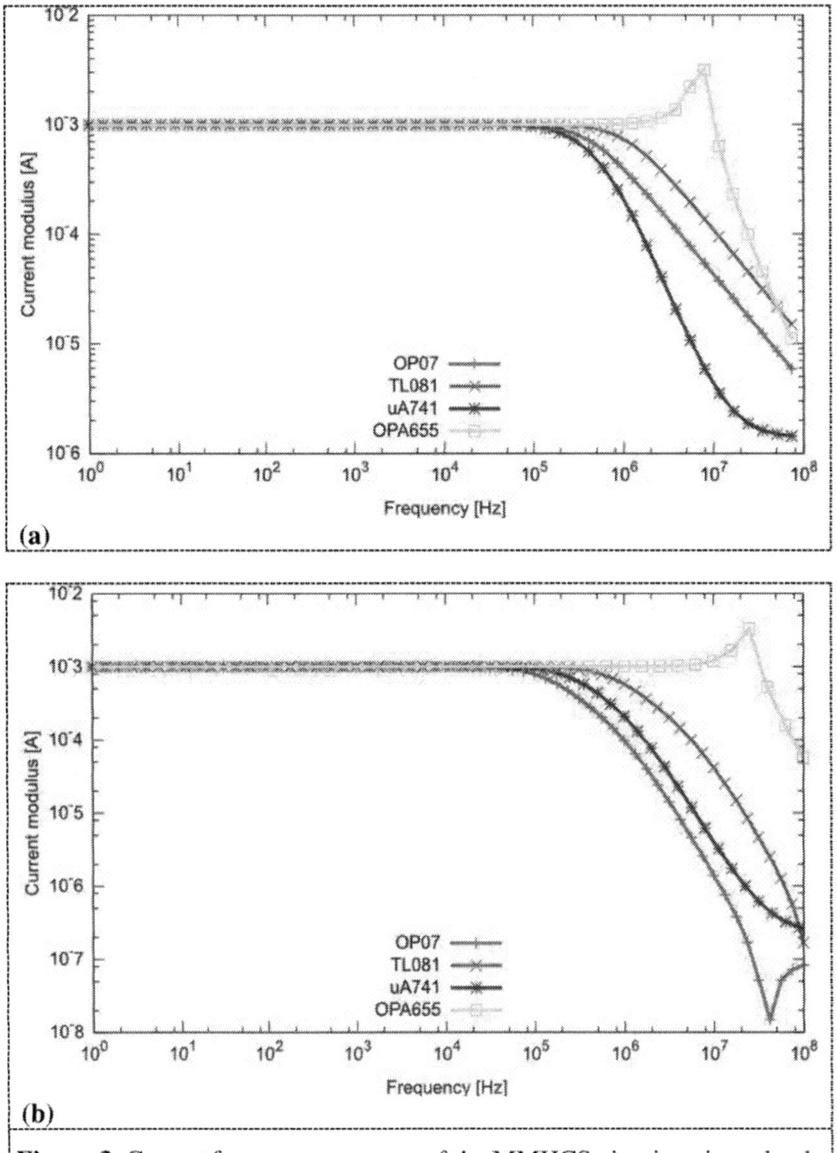

Figure 3. Current frequency response of the MMHCS circuit, using a load of 1 kΩ. **(a)** Analytical solution. **(b)** SPICE simulation.

The modulus of the output magnitude for both analytical and SPICE simulation are showed in Figures 4(a) and 4(b), respectively. It can be seen that the output impedance at very low frequency is approximately 200 MΩ when implementing the MMHCS with OP07. However, at 1 MHz, the operational amplifier OPA655 presented a better performance and showed an output impedance of approximately 100 kΩ.

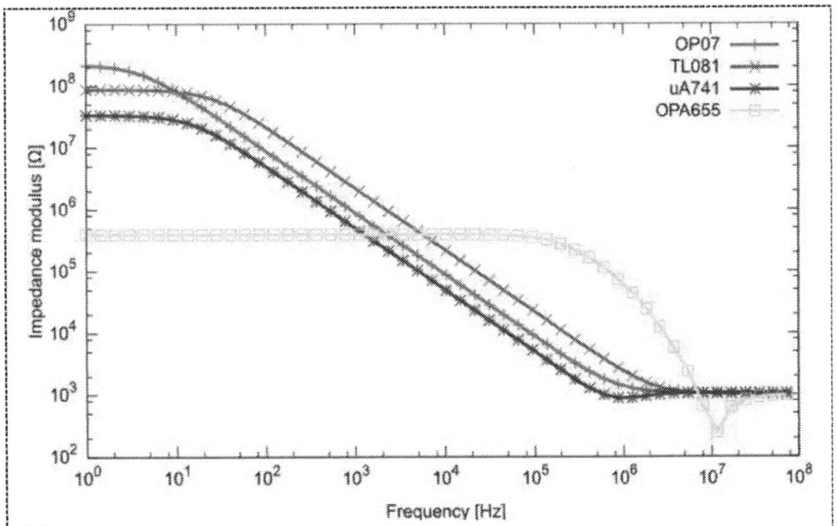

(a)

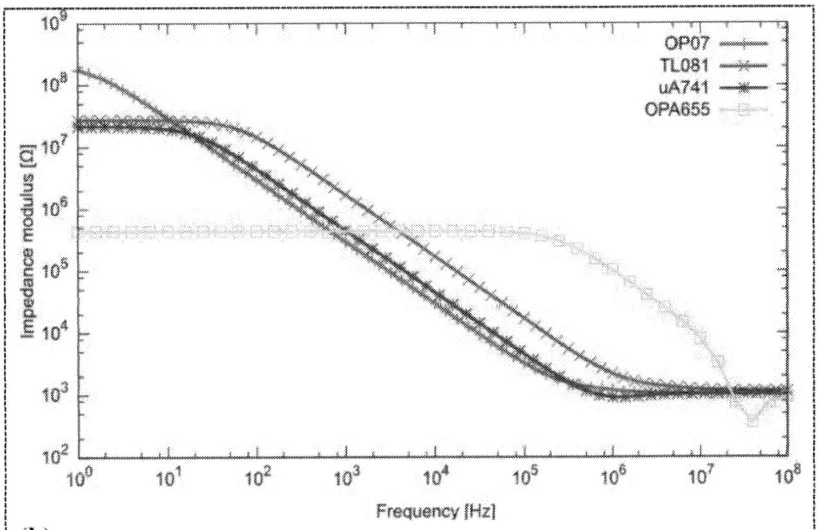

(b)

Figure 4. Output impedance frequency response of the MMHCS circuit. **(a)** Analytical solution. **(b)** SPICE simulation.

In order to investigate the variation in the output of the MMHCS circuit due to loading effects, Figure 5 shows the linearity of the output current by varying the load values from zero (short-circuit) to 5 kΩ at 100 kHz. It can be observed that both OPA655 and TL081 present good linearity response. The maximum linearity error occurs for a 5 kΩ load and they are approximately 17% for the OP07 whereas 13% for the μA741.

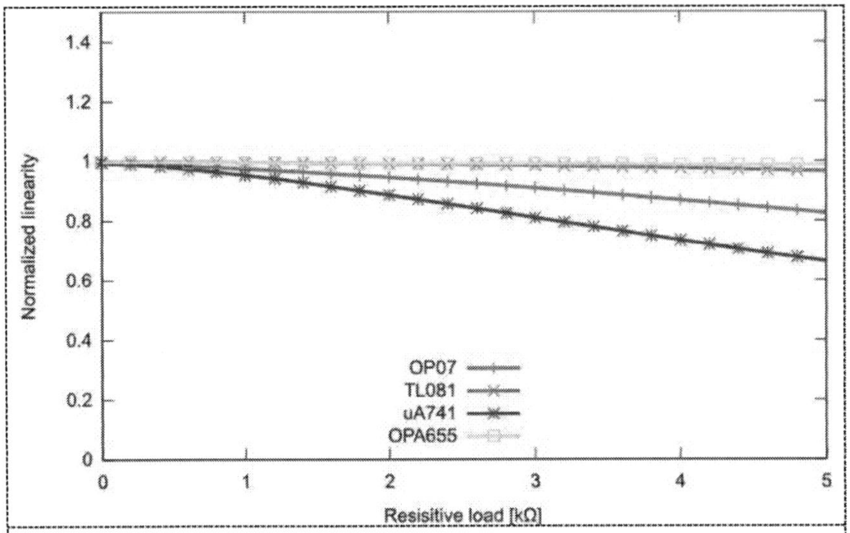

Figure 5. Linearity response of the MMHCS circuit as function of load at 100 kHz, using an sinusoidal input of 1 Vp.

4. Discussions

It was showed that the results from the analytical equations for both output current and impedance spectrum are very close to the SPICE simulations, especially at lower frequencies. Even the non-linear behavior of the operational amplifier OPA655 at high frequency presented in the simulations could be also observed in the analytical solutions.

The operational amplifier OP07 showed a difference between analytical equation and SPICE simulations for output current spectra, especially at higher frequencies. This might be explained by the fact that the SPICE simulator does take into account others non-idealities effects which are not considered in the analytical equations. The analytical equations were not taken into account the slew rate and the input common mode impedance of the operational amplifiers, for example. Furthermore, the simulator models a more realistic open loop gain of the amplifier, for example by considering a second-order relationship with frequency. Here the open-loop gain was considered to be a first-order function of frequency.

It has to be emphasized that the output current of the MMHCS circuit cross the load does depends on the output impedance of the circuit, as shown in Equation 3. The higher the output impedance of the circuit, more closely the load current would be to the output current projected in the circuit. Furthermore, it was not considered parasitic capacitive effects (i.e. stray capacitances from cables, electrodes, circuit layout, etc). Cautions should be considered when comparing SPICE simulation to real life circuit, as the presence of parasite capacitances can reduce the output impedance of the circuit, especially at higher frequencies.

$$I_{Load} = \frac{I_{out} Z_{out}}{Z_{out} + R_{Load}} \tag{3}$$

If we consider the input impedance of the operational equals to infinity and its output impedance equals to zero, it can be calculated from (2) that the output impedance of the MMHSC is directly proportional to the open-loop gain the opamp.

5. Conclusion

A comparison between the analytical solution for the output current and impedance of a MMHCS circuit and the SPICE simulation were performed. It can be concluded that it is possible to reproduce these results from simulations. These results may encourage the researches to simulate and project a MMHCS circuit by using only the analytical equations 1 and 2. Gain from these results might be important when projecting wide frequency bandwidth and high output impedance Howland modified current sources.

6. References

[1] Keshtkar A, Salehnia Z and Shokouhi, B 2012 *Pathology Research International* **2012** 1-5
[2] Aberg P, Nicander I nd Ollmar S 2003 *IEEE Proc. of the EMBS Annual Int. Conf.* (Cancun) vol 1 3211-14
[3] Aberg P, Nicander I, Hansson J, Geladi P, Holmgren U and Ollmar S 2004 *IEEE Transactions on Biomedical Engineering* **51** 2097-102
[4] Skourou C, Rohr A, Hoopes P J and Paulsen K D 2007 *Phys. Med. Biol.* **52** 347–63
[5] Van Kreel B K 2001 *Med. Biol. Eng. Comput* 39 551–7
[6] Lingwood B E, Dunster K R, Colditz P B and Ward L C 2002 *Brain Res.* **945** 97–105
[7] Moissl U M, Wabel M P, Chamney P W et al 2006 *Physiological Measurement* **27** 921-33
[8] Bertemes-Filho P, Negri L, Paterno A S 2011 *5th European Conference of the International Federation for Medical and Biological Engineering* (Budapest) vol 1 p 1275–8
[9] Grimnes S and Martinsen O G 2000 *Bioimpedance and Bioelectricity Basics* (Norway: Academic Press) p 360
[10] Yúfera A and Rueda A 2009 *16th IEEE International Conference on Electronics, Circuits and Systems* (Tunisia) vol 1 948-51
[11] Ferreira J, Seoane F, Ansede A and Bragos R 2012 *Journal of Physics: Conference Series* **224** 1-4
[12] Bertemes-Filho P 2002 *Tissue Characterisation using an Impedance Spectroscopy Probe* (Sheffield: The University of Sheffield Press) p. 96
[13] Seoane F, Macías R, Bragós R nd, Lindecrantz K 2011 *Measurement Science and Technology* **22** 1-11
[14] Bertemes Filho P, Lima R G, Amato M B P et al. 2006 *XX Brazilian Congress Biomed. Eng.* (São Paulo) vol 1 p 1167-70
[15] Seoane F, Bragós R and Lindecranz K 2006 *IEEE Proc. of the EMBS Annual Int. Conf.* (New York) vol 1 p 5121-5
[16] Jivet I and Dragoi B 2008 *Physiol. Meas.* **29** S193–201
[17] Casas O, Rosell J et al. 1996 *Physiol. Meas.* **17** A1–6
[18] Bertemes-Filho P, Brown B H and Wilson A J 2000 *Physiol. Meas.* **20** 1-6

Acknowledgment
We thank the State University of Santa Catarina (UDESC) for the financial support.

Rheology and BIA

M Piquemal[1] and JC Zanotti[2]
[1]Independent investigador, Assomption, Paraguay
[2]Emeritus professor, Chemistry department, National University of Assomption, Paraguay

E-mail:piquemal@gmail.com

Abstract. Is it possible to link erythrosedimentation to a global monofrequency of 50KHz bio impedance analysis parameters? Two populations, Control (N=20, average age of 38 ± 18 years, 55% female predominance) versus neopathology undergoing chemotherapy treatment (N=10, average age of 51 ± 10 years, 70% female predominance) were exposed to two simple tests: Blood composition and erythrosedimentation and global Bio Impedance mono frequency (50 KHz) assessment analysis. The Results of simple statistical tests (average, anova) show that the two populations were significantly different (1%), using both biological bood tests or BIA. Only the neopathological group showed a significant correlation between the erythrosedimentation the bio impedance parameters such as Phase angle or Xc (ratio - 0.93 and – 0.93)
Keywords: Erythrosedimentation, phase angle, Xc, blood composition

1. Introduction

The circulation of erythrocytes in the blood (GR) is made possible thanks to a control of blood viscosity. From a mechanical point of view, there is, an interaction between the blood flow, the axial force of the sanguinous current and the rate of deformation of the red cells. While the erythrocytes travel through the vascular system, they have a tendency to restructure themselves in cylindrical forms. This conformity prevents phenomenon such as internal abrasion, turbulence within the vascular system which, without said conformity, could destroy these carriers of hemoglobin.

This cellular behavior is due to the effects of long sanguineous proteins', such as fibrinogen, which maintains the cells in this cylindrical structure. The anticoagulated blood, later pipetted in a thin Westserngreen graduated pipet and placed vertically, shows an alignment of the erythrocytes, in the form of a long cylinder: the rolls. The time during which the erythrocytes deposit at the bottom of the vertical pipet is referred to as erythrosedimentation speed.

The results of this test are measured in millimeters of sedimentation of red blood cells (mm) per minute. The mechanism behind this cylindrical conformity of the erythrocytes is related to the presence of fibrinogen or any other sanguineous molecule of an anisometric structure with a high axial quotient [1].The distance between stacked erythrocytes also affects this process, being thinner than between red blood cell flowing in the bloodstream [2]. There is, therefore, reduction of the repulsive electrical force between the red blood cells. The erythrosedimentation represents an indicator of the inflammation without being an absolute indicator of the inflammation, since this process can also depend on the geometry of the red cells. Given the previously stated information, we would like to know if the measurement of the global cutaneous bioimpedance of a body can relate, to the state of the sanguineous composition and specifically, to erythrosedimentation. Therefore, two populations are subjected to this concept. The first group consists of subjects considered biologically and clinically healthy (control group). The second is made up of subjects undergoing treatment for neo pathologies (test group). Would it be possible to show that the variation of the angle alpha (marker of capacitive aspect of the cellular membrane), is equivalent, in blood, to an increase of the erythrosedimentation and a possible thrombosis?

2. Methods and material

2.1. Material

A blood sample is taken in fasting, after a 20 minute rest in a reclined position. This sample is analysed to determine: blood count and erythrocyte sedimentation rate. The reading of the blood count is performed manually, with microscopic observation of the lamina.
This entire procedure is performed following a bioimpedance measurement of coetaneous skin, with a single frequency of 50 kHz, in a tetrapole setup (right hand and foot) using the BIA 2000s device from the InputData Company, Germany) [3]

2.2. Method

The experiments population consists of 30 people from different socio-cultural origins, with an average age of 43 ± 17 years, predominantly female (60%). This main population is divided into two groups. The first group consists of 20 individuals without clinical or biological manifestation of disease (N = 20, average age of 38 ± 18 years, 55% female predominance). The second group is made up of 10 people with neoplasia: 50% breast cancer and 50% intestinal cancer (N = 10, average age of 51 ± 10 years, 70% female predominance), undergoing chemotherapy. Before the experiment all subjects were adequately informed and gave their written consent to participate in the measurement of the bio-impedance signals. All measurements were performed in compliance with the ethical principles set forth in the Declaration of Helsinki [4]. BIA electrodes are applied as follows:
-Hand (first electrode: placed in the middle of the union between the two styloids, at the back of the hand; second: placed in the middle of the space between the second and third metacarpal, 7 cm between the two)
-Feet: (first electrode: anterior side of the tibia-fibula; second: between second and third metatarsal, 7 cm between the two)
The bio-impedance measurement is performed three times, patient resting on supine position, each consecutive measurement taken 2 minutes apart. The presented data represents the average of three indidual measurements.
Analysis of the blood is performed manually, as a recount, in a biochemical analysis laboratory.

3. Results

Two simple statistical studies were performed, based on the average, standard deviation and variance analysis of the two data sets from these two populations, namely data originating from electrical properties (with the following five parameters: impedance, reactance, resistance at the hand and foot, Alpha angle) and biological data from the blood samples. The results of these test statistics are as follows:

3.1. Data originating from electrical studies

3.1.1. Simple statistical analysis

Bio impedance data are gathered in table 1.

Table 1. Basic statistical results regarding a monofrecuencial analysis of global coetaneous impedance. a) the clinically and biologically healthy population b) neopathological population. (Z, R, Xc in ohms, and alpha in °).

a)

Healthy subjects	Z 50kHz	Xc	R-Hand	R-Foot	Alpha
Average	518.30	61.15	148.70	128.75	6.80
Standard deviation	66.14	9.40	39.35	24.12	1.17

b)

Neopathological subjects	Z 50kHz	Xc	R-Hand	R-Foot	Alpha
Average	424.20	32.60	207.10	239.20	4.36
Standard deviation	35.22	6.85	56.59	25.21	0.69

A significant variation is noticed between BIA parameters throughout the coetaneous tissue of both groups. Would this data be discriminative?

3.1.2Analysis of variance between the two populations, from an electrical perspective:

Table 2. Analysis of variance.

Indicators	Value of p
Z 50 kHz	0.00025377
Xc	2.9834E-09
R-Hand	4.0607E-12
R-Feet	2.9792E-12
Alpha	1.0787E-15

All 5 parameters are discriminative data regarding the overall bioimpedance between the two populations. [5, 6, 7]

3.2. Biological blood data

3.2.1. Simple statistical analysis (Table 3).

Table 3. Simple statistical results of the blood data a) for the group of clinically and biologically healthy subjects, b) subjects with neopathologies.

a)

Healthy Subject	Hemoglobine	Red Cells	White Cells	Segmented neutrophils	Lymphocytes	Monocites	Eosinophils	Erythrosed. 1st hour
Average	12.82	4240500	5439	54%	43%	0 %	3%	21
Standard deviation	1.08	357734	838	12%	11%	0%	5%	15.5

b)

Subjects with neopathologies	Hemoglobine	Red Cells	White Cells	Segmented neutrophils	Lymphocytes	Monocites	Eosinophils	Erythrosed. 1st hour
Average	11.20	3724000	4249	67%	27%	3%	3%	39
Standard deviation	1.22	420270	1657	9%	11%	1%	4%	21. 5

3.2.2. Analysis of variance between the two populations, blood perspective. The variance analysis shows us that on the following blood parameters the two types of populations can be differentiated, see table I4.

Table 4. Value of the probability of the test between the two populations, regarding biological blood values.

Indicators	P-value
Hemoglobin	0.00093119
Red Blood Cells	0.0015001
White Blood Cells	0.0135427
Segmented Neutrophils	0.00513489
Lymphocytes	0.00106837
Monocytes	3.2515E-12
Eosinophils	0.74358988
Eritrosed 1st hour	0.01281566

Therefore, out of 8 biological indicators, only the rate of eosinophils would appear not to discriminate between these two populations. The reason probably comes from the data paths in a country of parasitological endemic (giardiasis).

The consistency of the results obtained by these two different methods (bio impedance and biology), leads us to think of a possibility of cross-correlation between these two sources of information (blood biology and overall bio impedance). Would each of these two populations intrinsically retain a similar pattern among biological data and those of bio impedance?

To answer this question, we performed a cross correlative study in each of the two subject groups, including biological datasets (8 parameters) and the impedance (5 parameters).

3.2.*Correlative study in each of this population: bio-impedance data, versus biological blood data.*

Table 5. shows the correlation coefficients between these two types of data. A) Healthy population, b) Population with neoplasia. (Abbreviations: RBC: red blood cells, WBC: white blood cells, Neutseg. polynuclear Neutrophil, erythrosedimentation 1st hour. Rate of erythrosedimentation in the first hour, erythrosedimentation. 2nd hour. Rate of erythrosedimentation at the second hour).

a) Healthy population

	Hemo -globin	Hema-tocrit	GR	GB	Neut. Seg.	Lymph -ocites	Eosin -ophils	Eritrosed. 1st Hour	Eritro -sed. 2nd Hour	MCV	MCH	MCHC
R 50kHz	-0.52	-0.52	-0.52	0.23	0.17	-0.05	-0.24	0.07	0.12	0.31	-0.32	0.14
XC	-0.17	-0.16	-0.16	0.21	0.09	0.09	-0.38	-0.03	-0.14	0.21	-0.28	-0.02
R-Hand	-0.34	-0.34	-0.34	-0.07	0.12	-0.04	-0.13	-0.05	-0.09	-0.20	0.34	-0.02
R-Foot	-0.15	-0.15	-0.15	-0.03	0.21	-0.06	-0.27	0.21	0.11	0.04	0.12	0.13
Alpha	0.35	0.35	0.35	-0.03	-0.08	0.17	-0.17	-0.13	-0.29	-0.07	-0.02	-0.14

b) Population with cancer

	Hemo -globin	Hema -tocrit	GR	GB	Neut. Seg.	Lymph -ocites	Eosin -ophils	Eritrosed. 1st Hour	Eritro -sed. 2nd Hour	MCV	MCH	MCHC
R 50kHz	0.82	0.81	0.35	-0.35	0.30	0.82	0.18	-0.80	-0.18	-0.12	0.03	0.18
XC	0.90	0.91	0.32	-0.63	0.65	0.90	-0.13	-0.93	-0.13	-0.43	-0.33	-0.13
R-Hand	0.00	-0.06	0.38	0.43	-0.38	0.00	0.20	-0.14	0.37	0.56	0.39	0.20
R-Foot	0.57	0.52	0.51	0.18	-0.16	0.57	0.22	-0.61	0.21	0.37	0.35	0.22
Alpha	0.88	0.89	0.32	-0.73	0.76	0.88	-0.25	-0.93	-0.13	-0.56	-0.49	-0.25

A weak correlative relation of data from biological and electrical origin can be identified within the clinically and biologically healthy population. The correlation coefficient is negative and only only points to the red cell blood count, as reverse dependency factor of the overall body impedance. However, there is no evidence of dependency between the erythrocyte sedimentation rate and biophysical parameters obtained.

This pattern is modified in the case of neoplasms. Here are strong correlative relationships between the sedimentation rate, red cell count, the white cell blood count, with four of the electrodes applied to the tissue: capacitive effect of the cell membrane and ionic modification of the surrounding tissues.

4. Interpretation

The rate of sedimentation depends on several factors. In vitro, once pipetted, the density change of blood cells versus plasma is one of the major factors of a high erythrosedimentation rate. The increase

of high axial ratio proteins (e.g. fibrinogen), usually present during inflammatory processes, accelerates this process. Another factor such as the change of structure of the red cell membrane and the reduction of its electrical charge allow the erythrocytes to form ever longer cylindrical structures. Hence, these floating red blood cell structures in the plasma, will precipitate more rapidly, due to increased density, in relation to liquid medium in which they are initially present.
The use of 5 fluorouracile, (responsible for the destruction of the cellular cytoskeleton) and the paclidal (which acts upon DNA replication), active chemotherapy principles, present in these patients, weaken the cellular membrane. The loss of cellular membrane capacitance, identified from an electronic perspective, through the alpha angle at an impedance of 50 KHz. On table 5b, a change of capacitive property is evident via a high correlation ratio affecting red blood cells and white cells associated with the overall reduction of the alpha. This correlative statistical relationship is direct: decreased value of the angle alpha in patients with neoplasia, increased repulsive charge between the red blood cell membranes, thereby increasing the acceleration of erythrocyte sedimentation. However, such a phenomenon is not observed in the control group (weak correlation = 0.51 and inverse), always referring to the red blood cells, where an increase of the capacitive effect of the red cell membranes can be identified which in turn leads to an increase of the repulsive charges between membranes and therefore to a reduction of erythrosedimentation rate (reduction of length of cellular structures).

5. Conclusion

Two groups of patients were subjected to the same protocol, in order to correlate the two types of measurements. The first set of data comes from the biology of the blood. The blood count and erythrocyte sedimentation are contemplated. The second one explores the bodily electrical properties under a single frequency of 50 kHz. This is the case of bio-impedance. The relation between the erythrocyte sedimentation process, by which red blood cells are pooled, in vitro into stacks or rolls of erythrocytes thus precipitating the test tube that contains it, and the properties of the cell membrane body captured by external electrical measurement is sought. Previous studies have shown that the electrical indicator of the bio-impedance, angle alpha, is related to the change of capacitive properties of the cell membranes.
A statistical analysis of a population of young people (N = 30, average age 43 ± 17 years old, predominantly female: 60%) divided into two sub groups, control group of subjects with normal biological results (N = 20, average age 38 ± 18 years, 55% female predominance) versus the neo pathological group undergoing chemotherapy (N = 10, average age 51 ± 10 years, 70% female predominance), has produced the following results. Between these two assessment techniques (biological and corporal bioimpedance), there is a correlation which could become a possibility to differentiate the two groups significantly to within 1% ($P < 0.01$). Regarding the healthy population, the error found in the chemotherapy-treated population is marked by a reduction from 22% of the value of the impedance at 50kHz (R50KHz), 88% of the reactance Xc, of 56% of the alpha angle . The skin resistance in the hand and foot has increased respectively by 28% and 26% compared to the control group.
The correlative study between the two sets of parameters of the two measurement techniques: biological (12 parameters) and electrical (5 parameters) shows that the correlative pattern between these two sources, changes according to the state of health of the studied group. When it comes to healthy subjects, there is only a weak inverse relationship (-0.52) linking biological data to the bio-

impedance. In the case of subjects with neopathology, the following electrical parameters are very strongly correlated with the rate of correlation: R 50 KHz (ratio: -0.8,); Xc (ratio: -0.98); Alpha (ratio : -0.93). These ratios indicate a great dependency between these electrical parameters (weakening effect of the cellular membrane capacitance and ESR).

This correlation could be useful as it would enable the oncologist to evaluate the risk of thrombosis during chemiotherapy, through a measurement by BIA. The measurement is non intrusive, it can be done fast, safely and can provide important preventative information regarding the patient´s immediate needs.

6. Reference

[1] Jandl JH and Castle WB 1956 Agglutination of sensitized red cell by large anisometric molecule *J. Lab Clin. Med* **47** 669

[2] Ponder E 1948 Hemolysis and related Phenomena. (New York: Grune & Stratorn)

[3] BIA 2000-s. Available from URL: http://www.data-input.de/_site/german/produkte/geraete/nutriguard-ms.php

[4] Lopez IFD 2001 Legal and Ethical Aspects of Informed Consent in medical care in Mexico *Rev. Mex Clin Pathol.* **48** 03-06

[5] Thibault R Cano 2011 Quantification of lean tissue Losses During cancer and HIV infection / AIDS *Curr. Opin. Clin. Nutr. Metab.* Care. **14** 261-7

[6] Shah S et al 2001 Severity of human immunodeficiency virus infection is Associated with Decreased phase angle, fat mass and body cell mass in adults with pulmonary tuberculosis infection in Uganda *J Nutr.* **11** 843-7

[7] Colin Ramirez E et al 2012 Bioelectrical impedance phase angle as a prognostic marker in chronic heart failure *Nutrition* **28** 901-5

AUTHOR INDEX

AUTHOR INDEX

Institute of Physics
Dirac House, Temple Back
Bristol BS1 6BE UK